PSYCHOLOGY LIBRARY EDITIONS:
COGNITIVE SCIENCE

Volume 6

NEURAL NETWORK MODELS OF CONDITIONING AND ACTION

NEURAL NETWORK MODELS OF CONDITIONING AND ACTION

Edited by
MICHAEL L. COMMONS,
STEPHEN GROSSBERG AND
JOHN E. R. STADDON

Routledge
Taylor & Francis Group

LONDON AND NEW YORK

First published in 1991 by Lawrence Erlbaum Associates, Inc.

This edition first published in 2017
by Routledge
4 Park Square, Milton Park, Abingdon, Oxon OX14 4RN
605 Third Avenue, New York, NY 10017

Routledge is an imprint of the Taylor & Francis Group, an informa business

British Library Cataloguing in Publication Data
A catalogue record for this book is available from the British Library

ISBN: 978-1-138-19163-1 (Set)
ISBN: 978-1-315-54401-4 (Set) (ebk)
ISBN: 978-1-138-19204-1 (Volume 6) (hbk)
ISBN: 978-1-138-19212-6 (Volume 6) (pbk)
ISBN: 978-1-315-64009-9 (Volume 6) (ebk)

Publisher's Note
The publisher has gone to great lengths to ensure the quality of this reprint but
points out that some imperfections in the original copies may be apparent.

Disclaimer
The publisher has made every effort to trace copyright holders and would welcome
correspondence from those they have been unable to trace.

NEURAL NETWORK MODELS OF CONDITIONING AND ACTION

A Volume in the
Quantitative Analyses
of Behavior Series

Edited by

MICHAEL L. COMMONS
Harvard Medical School

STEPHEN GROSSBERG
Boston University

JOHN E. R. STADDON
Duke University

LEA   LAWRENCE ERLBAUM ASSOCIATES, PUBLISHERS
1991   Hillsdale, New Jersey   Hove and London

Lawrence Erlbaum Associates, Inc., Publishers
365 Broadway
Hillsdale, New Jersey 07642

Library of Congress Cataloging-in-Publication Data
LC card number: 86-640678
ISBN: 0-8058-0842-6
0-8058-0843-4(pbk.)

Printed in the United States of America
10 9 8 7 6 5 4 3 2 1

Contents

1) Alliston K. Reid	12) William S. Maki
2) Jennifer L. Raymond	13) Richard S. Sutton
3) Gail Carpenter	14) Russell M. Church
4) Thomas P. Vogl	15) Nestor A. Schmajuk
5) Stephen José Hanson	16) James A. Mazur
6) John E. R. Staddon	17) John W. Moore
7) Hilary A. Broadbent	18) Douglas A. Baxter
8) Stephen Grossberg	19) John H. Byrne
9) Michael L. Commons	20) Dean V. Buonomano
10) Daniel L. Alkon	21) William B. Levy
11) Daniel S. Levine	

About the Editors

Michael L. Commons is Lecturer and Research Associate in the Department of Psychiatry at Harvard Medical School, Massachusetts Mental Health Center, and Director of the Dare Institute. He did his undergraduate work at the University of California at Berkeley, and then at Los Angeles, where in 1965 he obtained a B.A. in mathematics and in psychology. In 1967 he received his M.A., and in 1973 his Ph.D., in psychology from Columbia University. Before coming to Harvard University in 1977 as a postdoctoral fellow and then becoming research associate in psychology, he was an assistant professor at Northern Michigan University. He has co-edited *Quantitative Analyses of Behavior*, volumes 1–11 and *Beyond Formal Operations: Late Adolescent and Adult Cognitive Development*. His area of research interest is the quantitative analysis of the construction and understanding of reality as it develops across the life span, especially as these elements affect decision processes, life-span attachment and alliance formation, and ethical, social, cross-cultural, educational, legal, and private sectors.

Stephen Grossberg received his graduate training at Stanford University and Rockefeller University, and was a Professor at M.I.T. He is Wang Professor of Cognitive and Neural Systems at Boston University, where he is the founder and Director of the Center for Adaptive Systems, as well as the founder and Co-Director of the graduate program in Cognitive and Neural Systems. He also organized the Boston Consortium for Behavioral and Neural Studies, which includes investigators from six Boston-area institutions. He founded and was first President of the International Neural Network Society, and is editor-in-chief of the Society's journal, *Neural Networks*.

During the past few decades, he and his colleagues at the Center for Adaptive

Systems have pioneered and developed a number of the fundamental principles, mechanisms, and architectures that form the foundation for contemporary neural network research, including contributions to content-addressable memory; associative learning; biological vision and multidimensional image processing; cognitive information processing; adaptive pattern recognition; speech and language perception, learning, and production; adaptive robotics; conditioning and attention; development; biological rhythms; certain mental disorders; and their substrates in neurophysical and anatomical mechanisms.

John E. R. Staddon is James B. Duke Professor of Psychology, and Professor of Zoology and Neurobiology at Duke University, where he has taught since 1967. His research is on the evolution and mechanisms of learning in humans and animals. He is the author of numerous experimental and theoretical papers and two books, *Adaptive Behavior and Learning* (1983, Cambridge University Press) and *Learning: An Introduction to the Principles of Adaptive Behavior* (with R. Ettinger, 1989, Harcourt-Brace-Jovanovich).

About the Contributors

Adel M. Abunawass is a PhD candidate in Computer Science at North Dakota State University. He received his BS degree in Computer Science and Mathematics from Moorhead State University. His research interests are in artificial intelligence and he is investigating some basic properties of connectionist networks.

Daniel L. Alkon is internationally recognized for his discoveries of molecular and physical forms that associative memories take within brain structures, based on his pioneering work elucidating associative learning in *Hermissenda crassicornis* and, more recently, rabbit hippocampus. He is the author of numerous publications including the Cambridge University Press volume, "Memory Traces in the Brain" (1987).

Douglas A. Baxter received his PhD in 1981 from The Department of Zoology at The University of Texas at Austin. He is currently a Senior Research Scientist in The Department of Neurobiology and Anatomy at The University of Texas Medical School in Houston. His research interests are to analyze the cellular and network properties of neurons and neural circuits that are involved in adaptive information processing and to model these adaptive elements and circuits.

Eric W. Bing graduated in 1990 from Harvard University concentrating in Computer Science. He has been Research Assistant at the Cornell Artificial Intelligence Laboratory, the Dare Institute and Harvard Psychology Department. His research interests are computational approaches to intelligence, including artificial intelligence and neural networks, and neuropsychology.

Kim T. Blackwell received her veterinary degree and PhD in bioengineering from the University of Pennsylvania in 1988. She has performed research in color vision, neural networks, and medical imaging. Her present work includes the development of biologically motivated neural network algorithms and the evaluation of color display of single band medical and radar images.

Hilary A. Broadbent earned a Bachelor of Arts degree from Barnard College in 1986 with majors in biology and ancient Greek. She worked as a research assistant at the Project for Animal Intelligence at Columbia University under the direction of Herbert Terrace. She is presently a graduate student in Psychology at Brown University.

Dean V. Buonomano received his BS in Biology from The State University of Sao Paulo at Campinas in 1986. He is currently a PhD student in The Program in Neuroscience at The University of Texas Medical School in Houston. His research interests are neural bases of learning and memory and artificial neural networks.

John H. Byrne received his PhD in Bioengineering from The Polytechnic Institute of Brooklyn in 1973. He is currently Professor and Chairman of the Department of Neurobiology and Anatomy at The University of Texas Medical School in Houston. His research interests are the cellular bases of learning and memory, synaptic transmission, excitable membranes, neuronal modulation, and artificial intelligence.

Thomas J. Carew received his PhD in Physiological Psychology from the University of California, Riverside in 1970. He is currently Professor of Psychology and Biology at Yale University. His research interests are the cellular and molecular bases of learning and memory, development of the nervous system, development of learning and memory, and mechanisms of synaptic transmission.

Russell M. Church is a Professor of Psychology, Professor of Cognitive and Linguistic Sciences, and a member of the Graduate Program in Neural Science at Brown University. He is a Fellow of the Society of Experimental Psychologist and has been President of the Division of Experimental Psychology, APA. He was co-editor of Volume VII in this series (*Biological Determinants of Reinforcement*, Lawrence Erlbaum Associates, 1988).

Costa M. Colbert received the BES degree in Biomedical Engineering and the M.S.E. degree in Electrical Engineering and Computer Science in 1984 and 1986, respectively, from the Johns Hopkins University. He is currently in the Medical Scientist Training Program at the University of Virginia. His current research interests include the electrophysiology of the hippocampal formation and the development of neural network models.

David G. Cook received his BS in Psychology from the University of Utah in 1984. He is currently a PhD student in the Psychology department at Yale University. His research interests are the cellular and molecular mechanisms of neuronal plasticity.

Jim DiCarlo enrolled as a biomedical engineering student at Northwestern University in 1986. He has spent the past two years working on real-time models of hippocampal function in associative learning, particularly classical conditioning. Presently, he is studying the computational role of the hippocampus in configural learning and amnesia for an Honors Thesis in Biomedical Engineering with Dr. Nestor Schmajuk. DiCarlo received a BS in Biomedical Engineering with a concentration in physiological systems in June, 1990. He plans to pursue both an MD and PhD in preparation for a career in medical research.

Charla Cristal Griffy is a member of the class of 1991 at Harvard University, concentrating in East Asian Studies. Her research interests are computational approaches to intelligence, including artificial intelligence and neural networks. In 1986, she was a Westinghouse Science Talent Search Top Forty Winner. In 1987, she was Software Designer at the Kennedy Space Center for the National Aeronautics and Space Administration.

Stephen José Hanson is group leader of the Learning and Knowledge Acquisition research group at Siemens Research Center and a visiting research scientist in the Cognitive Science Laboratory at Princeton University. He received his PhD from Arizona State University at Tempe in Experimental and Quantitative Psychology. He spent two years at Indiana University at Bloomington as an assistant professor and then joined Bell Laboratories as a member of technical staff in 1982. At the breakup of the Bell system in 1984, he joined Bellcore in the AI and Information Science Research Group. In 1989 he joined Siemens Corporate Research in Princeton, N.J. He has done modeling and research in the areas of human-computer interaction, programming productivity and learning complex skills. He is specialized in learning theory and studied and published papers on learning in humans, animals, and machines. His interests include connectionist models, machine learning and general adaptive processes.

Frederick M. Kuenzi received his BA in Biology from Westminster College in 1983 and his MS in Entomology from the University of Wisconsin in 1985. He is currently a PhD student in the Biology department at Yale University. His research interests are the mechanisms of motor control in invertebrates and kinematics.

Daniel S. Levine is Associate Professor of Mathematics at the University of Texas at Arlington. He is the author of *Introduction to Neural and Cognitive Modeling* and the senior editor, with Samuel J. Leven as co-editor, of *Motivation, Emotion, and Goal Direction in Neural Networks,* both to be published late in 1990 by Lawrence Erlbaum Associates. He is a member of the editorial board of *Neural Networks.*

William B Levy is the director of research in the Department of Neurosurgery at the University of Virginia. He has studied and published on the topic of associative synaptic modification for many years using both electrophysiological and anatomical techniques. In 1979 he co-discovered, with O. Steward, a long-term depression that complements long-term potentiation and also confirmed the existence of associative synaptic modification. Currently his research includes the functional implications of synaptic modifications for neural networks.

William S. Maki received his PhD in Experimental Psychology in 1974 from the University of California at Berkeley. He has been on the faculty at North Dakota State University since 1973 and is currently Professor of Psychology and Adjunct Professor of Computer Science. Maki's research has been focused on cognitive processes in animals, especially short-term memory and attention.

John W. Moore is a professor of psychology and computer and information science at the University of Massachusetts at Amherst, where he has taught since 1962. He obtained his BA from Lawrence College in 1958 and his PhD from Indiana University in 1952. In 1969–70 he was a visiting professor of neurology at University of Miami (Coral Gables, Florida). From 1976–78, he was senior staff scientist at University College, London, where he was affiliated with the MRC Neural Mechanisms of Behaviour unit. His research interests include quantitative learning theory and computational neuroscience.

Paul S. Prueitt is Assistant Professor of Mathematics at Hampton University. He received his PhD in 1988 from the University of Texas at Arlington, with a dissertation entitled "Some Techniques in Mathematical Modeling of Complex Biological Systems Exhibiting Learning." He has published one article on modeling of the immune response (with J. Eisenfeld) and two on modeling of the frontal cortex (with D. S. Levine and S. J. Leven).

Jennifer L. Raymond received her B.A. in 1987 from The Department of Mathematics at Williams College. She is currently a PhD student in The Program in Neuroscience at The University of Texas Medical School in Houston. Her research interests are the cellular network processes underlying learning and memory.

Nestor Schmajuk graduated from the University of Buenos Aires in 1975 and became a professor of Biomedical Engineering in Argentina. In 1984, he obtained a Masters of Arts degree from the State University of New York at Binghamton. In 1986, he obtained a PhD in Psychology from the University of Massachusetts at Amherst. He was a post-doctoral fellow at the Center for Adaptive Systems at Boston University in 1987. In 1987, he became Assistant Professor of Psychology at Northwestern University. He applies experimental and theoretical tools to the study of hippocampal involvement in attention, learning, and memory including mathematical and neural models of the hippocampus, and the effect of hippocampal lesions on classical conditioning, and hippocampal impairment on schizophrenia.

David C. Tam holds BSc degrees in Computer Science, Physics, and Astrophysics, and a PhD degree in Physiology (1987), all from the University of Minnesota. Currently, he is establishing a computational neuroscience laboratory and conducting research in theoretical models of biological neural networks, mathematical analysis of neurophysiological data, and experimental studies in multi-unit recordings of neurons at Baylor College of Medicine.

Edward J. Trudeau is a member of the class of 1991 at Harvard University concentrating in Applied Mathematics. His research interests are computational approaches to intelligence, including artificial intelligence, neural networks, and recursive functions.

Thomas P. Vogl has performed and directed research in optical physics, the biological effects of light, non-linear optimization techniques, and computer modeling. His many publications address problems in infrared detection and imaging, nonlinear optimization, automatic design of optical and illuminating systems, phototherapy for newborn jaundice, and neural networks.

Ying Zhang graduated in electrical engineering from the University of Science and Technology of China in Hefei. She is currently a graduate student in Neurobiology at Duke University.

Preface

This book grew out of a symposium of the same name that was held at Harvard University on June 2–3, 1989. The meeting's participants included experimental and theoretical cognitive scientists and neuroscientists. The goal of the meeting was to present and discuss some of the exciting interdisciplinary developments that are clarifying how animals and people learn to behave adaptively in a rapidly changing environment.

This is too vast a subject to fully explore in one event or book. The meeting's subject matter focused on conditioning and action; in particular, on aspects of how recognition learning, reinforcement learning, and motor learning interact to generate adaptive goal oriented behaviors that can satisfy internal needs. This topic is as important as understanding brain function as it is for designing new types of freely moving autonomous robots.

All the meeting participants shared the view that a dynamical analysis of system interactions is needed to understand these challenging phenomena. Neural network models provide a natural framework for representing and analyzing such interactions, which is why all the articles either develop neural network models or provide biological constraints for guiding and testing their design.

The book is roughly broken into two parts: one part focuses on classical conditioning and the other on instrumental conditioning, but results in either part are often relevant to results in the other part due to the overlap of neural mechanisms that are engaged by the two learning paradigms. For example, the adaptive timing mechanisms described by Grossberg for classical conditioning and by Church for instrumental conditioning share similar properties that are worth exploring further. Also of interest are the similar organizational principles that authors have used to model different parts of the mammalian brain, such as hippocampus and frontal lobes, and the similarities of certain basic circuit designs that are being described in both invertebrate and vertebrate learning models.

These interdisciplinary connections illustrate that the field of conditioning and action has entered an exciting phase in its development, one in which neural network models can begin to speak with equal ease and value to the full spectrum of scientists, from biologists to engineers, who are interested in how we learn to act to achieve a rewarding goal.

The symposium was supported in part by the Center for Adaptive Systems at Boston University, the Society for the Quantitative Analyses of Behavior, the Department of Psychology at Harvard University, and by the Dare Association, Inc.

We wish to thank Rebecca M. Young, Patrice M. Miller, and Dean Gallant, with assistance from Basil Arabos, David Lane, Rachel Levine, Jin Li, Edward Miech, Julia H. Weaver, and Laura Whitton for making local symposium preparations at Harvard, and Sheila M. McDonald for helping to organize the book. We also thank Carol Yanakakis Jefferson, Diana Meyers, and Cynthia Suchta at the Boston University Center for Adaptive Systems for their support in planning the symposium program and preparing the book.

Michael L. Commons
Harvard Medical School
Stephen Grossberg
Boston University
John E. R. Staddon
Duke University

MODELS OF
CLASSICAL CONDITIONING

1 Memory Function in Neural and Artificial Networks

Daniel L. Alkon
National Institutes of Health,
Laboratory of Molecular and Cellular Neurobiology
Bethesda, Maryland

Thomas P. Vogl
Kim T. Blackwell
ERIM, Arlington, Virginia

David Tam
University of California, Irvine

ABSTRACT

The DYSTAL (Dynamically Stable Associative Learning) model is an artificial modifiable neural network based on observed features of biological neural systems in the mollusc *Hermissenda* and the rabbit hippocampus. In the DYSTAL network, synaptic weight modification depends on (a) convergence of modifiable "collateral" and unmodifiable "flow-through" inputs, (b) temporal pairing of these inputs, and (c) past activity of elements receiving the inputs. Modification is independent of element output. As a consequence, DYSTAL shows (a) linear scaling of computational effort with network size, (b) rapid learning without an external "teacher," and (c) ability to complete patterns, independently associate different ensembles of inputs, and to serve as a classifier of input patterns.

Our memories are of complex patterns of stimuli—not of individual stimuli in isolation of each other. An image of a face is one such pattern, a melodic refrain is another. It is as if the pieces of a pattern are linked to or associated with each other so that awareness of a critical number of pattern elements or pieces activates the links to restore or recall the entire pattern. Attention to a distinctive scar can cause us to recall the appearance of a face encountered in the past. Hearing a few notes in a particular sequence may trigger a memory of a symphonic movement.

Mammalian memory formation may ultimately be reducible to link formation between stimulus elements in a pattern. Two lines of evidence from our laboratory provide support for this hypothesis. Our physiologic observations demonstrate remarkable similarities between molecular and biophysical mechanisms for learning links or associations in the snail *Hermissenda* and in the rabbit hippocampus

1

(Alkon, 1989; Alkon, Quek, & Vogl, 1989). Association formation within a system of a relatively few Hermissenda neurons involves cellular substrates identical to those of hippocampal systems with vast arrays of neurons. Storage in such hippocampal arrays could arise, therefore, from multiplication of storage events within minimal neuronal nets such as the visual-vestibular network of *Hermissenda*. Induction of theoretical networks from these biological networks provides additional support for such an inference. My colleagues, Tom Vogl and Kim Blackwell, and I have started to construct artificial networks whose elements interact with each other and modify their weight according to principles abstracted from the *Hermissenda* and rabbit neural systems. A number of promising features of our artificial network, such as linear scaling, rapid operation, and lack of an external teacher, suggest that by simply increasing the number of elementary associatively modified links between neurons, nature may have enormously eased the computational burden on our brains during cognitive functions.

Our physiologic studies have focused on learned associations as a result of Pavlovian conditioning. Pavlovian conditioning results from a rudimentary temporal pattern: one stimulus followed after a constant delay by a second stimulus. We learn that a flash of lightening is followed by thunder. A snail learns that a flash of light is followed by rotation induced turbulence. For the snail, the first step in establishing a temporal link between a light stimulus and a rotation stimulus occurs when electrical signals in the snail's visual pathway arrive at particular network locations together in time with electrical signals from the snail's vestibular pathway. At these common visual-vestibular loci (one being a neuron called the type B cell) the electrical response, depolarization, is unique to the temporal relationship of the light and rotation stimuli. Chemical signals unique to the temporal pairing of light and rotation continue the linkage formation. Calcium flows through channels within the B cell outer membrane and becomes elevated within the cytoplasmic compartment. Our observations suggest that the linkage process extends its duration when calcium and diaclyglycerol together cause movement of protein kinase C (PKC) from the cytoplasm into the membrane. Observations implicating PKC in *Hermissenda* conditioning include:

1. Learning-specific differences in phosphorylation of PKC substrates and the duplication of such differences by exposure to phorbal ester.

2. Blocking of learning-induced biophysical changes by blocking PKC translocation with sphingosine and PKC-mediated phosphorylation with H7.

3. Production of learning-induced biophysical changes by activating PKC with phorbol ester, etc.

4. Learning-specific changes in the amounts of PKC protein substrates and the m-RNA which encodes for them.

The light rotation temporal link in *Hermissenda* is recalled when depolarization of the Type B cell elicits a markedly reduced K^+ flow through specific channels. At the resting potential of the cell membrane, after the acquisition phase, there is no difference in K^+ flow. Thus, recall depends on the voltage dependence of the K^+ channel transformation. Pharmacologic blocking experiments indicate that this voltage dependence involves calcium flowing into the cell to activate a much more sensitive membrane-associated PKC. This PKC-regulated K^+ flow remains reduced weeks after the training. This then is an entirely new time domain for biophysics, beautifully designed for associative memory and never encountered before in fully differentiated neurons. In *Hermissenda*, it was this biophysical difference that was causally implicated for storage of the associative memory. Furthermore, we found evidence of this same new biophysics for biological signaling in the rabbit hippocampus where causal implication was not possible but correlation was.

More permanent memory storage almost certainly involves changes of protein synthesis, most likely as an extension of second messenger function into more prolonged temporal domains. In *Hermissenda,* regulation of protein synthesis profoundly affects calcium-stimulated reduction of K^+ ion flow, which occurs as a consequence of PKC translocation. That associative memory storage involves changes in the synthesis of specific proteins was more directly demonstrated by relating the learning behavior of living *Hermissenda* to protein and RNA metabolism of neurons functionally implicated in memory storage. Up to four days after the cessation of all training, the efficacy of memory storage was closely correlated with increases in m-RNA synthesis within the *Hermissenda* eye (whose only neuronal elements are the five photoreceptors that include 3 type B cells). There was also close correlation of memory storage with the quantity of specific proteins, one of which, the 20,000 kDa protein, a PKC and Ca^{2+}/CAM-II kinase substrate, showed differences in phosphorylation shortly after memory acquisition. Furthermore, synthesis of m-RNA with distinct molecular weights was also closely related to memory storage. Again, the PKC substrate with 20,000 kDa m.w., which we found to have GTP-ase activity and binds GTP, appeared to correspond to an m-RNA species (of similar m.w.) that was clearly related to the efficacy of memory storage. (In addition, two other proteins, one a GTP-ase, the other most likely a structural constituent, were correlated with memory storage). Injection of the 20,000 m.w. protein, now known to be a GTP-binding protein, exactly simulated the effects of conditioning on the Type B cell. The number and quantity of species of m-RNA and proteins altered with memory storage suggest profound alterations of cellular metabolism. Such alteration might occur during substantial structural changes as can occur during growth and development.

Marked structural alterations were in fact bound to be specific to associative memory storage in *Hermissenda*. Blinded comparisons of type B cell among the three groups demonstrated a conditioning-induced focusing of terminal branches where synaptic interactions occur. The magnitude of branching volume was also clearly related to the magnitude of K^+ flow reduction. Such learning-induced focusing may share properties with "synapse elimination" studied in developmental contexts.

In the snail *Hermissenda,* extension of memory storage from one major cellular compartment, the cell body, to another, the terminal branches, accompanies extension of the time domain from short-term to more permanent periods. An equally dramatic example of the same parallelism of temporal and spatial domains was found in the rabbit hippocampus. One day after the rabbit was learned to link a tone with an air puff to its eye, we found changes in the electrical properties of CA1 pyramidal cells within the rabbit's hippocampus. In slices isolated from conditioned vs. control animals there was reduced flow through the same K^+ channels that we had previously found to be altered when the snail was conditioned. Just as with the snail neurons, drug-induced movements of PKC from the cytoplasm into the membrane of the CA1 cells produced the same reduction of K^+ ion flow. Furthermore, fractionation of the CA1 cell region into membrane and cytosolic fractions revealed an unequivocal movement of the PKC from the cytoplasm into the membrane compartment only in conditioned animals.

Autoradiographic labeling experiments confirmed this conditioning-specific movement of PKC in the region of the cell bodies as well as the sites of synaptic interactions called dendrites (Fig. 1.1). The spatial temporal parallelism was most dramatically demonstrated in the distribution of the PKC label between cell body and dendritic layers. One day after conditioning, the cell body region had more PKC label than did the dendritic region. But three days after conditioning, this relationship was completely reversed. As the memory extended its time domain, the enzyme label moved its spatial domain from the cell body layer to the dendritic layer and also began focusing on fewer cells.

Why is it that we can see such dramatic changes with association formation? We might speculate that even input signals restricted to small compartments of the CA1 dendritic tree activate the cell bodies to change their K^+ currents, PKC distribution, protein synthesis, etc. Activated CA1 cell bodies then would increase transport of crucial molecules into all of the main branches of the dendritic tree, but these molecules would either localize or have localized effects in only those small compartments that initially received input signals (Fig. 1.2).

Ultimately, molecular probes, such as the radioactive phorbol ester just described, should generate a three-dimensional image of label distribution within neural systems. Precise mathematical modelling of these images might then be used to assess theoretical models of how neural systems within the brain acquire and store information.

Artificial Network Design

The snail and rabbit observations suggested some critical features for the design of an artificial system:

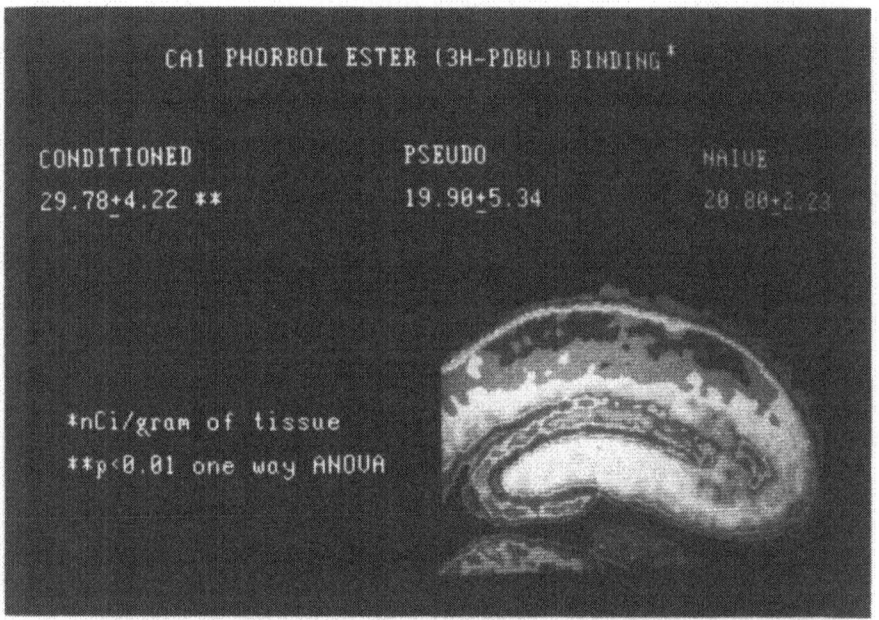

FIG. 1.1.   PKC is a ubiquitous regulatory enzyme that is highly enriched in the vertebrate brain. Here we employ a molecular probe to image learning-specific changes in the distribution of this enzyme within the hippocampus of rabbit after classical conditioning of the nictitating-membrane reflex. The molecular probe (3H-phorbol–12, 13-dbutyrate) binds with high affinity to PKC and can be detected across the brain structure by means of quantitative autoradiography. Computer digitizing techniques were employed to produce the image, which shows the heterogeneous distribution of this enzyme, that is most evident in trained animals, within the cell body layer and dendritic fields of the pyramidal neurons of the CA1 cell field (red area in image). The molecular probe demonstrates a statistically significant increase in the distribution of this important regulatory enzyme from animals that have undergone this form of associative learning (conditioned), compared with control (pseudo, naive) groups.

1. High-weight pathways. Some neural pathways and the weight of synaptic interaction between the elements within such pathways allow rapid and reliable transmission signals to effect well defined stereotypic or reflexive behavior. The unconditioned stimulus, rotation, for example, reliably elicits foot contraction of *Hermissenda*. The unconditioned stimulus, pressure (provided by an air puff) on the rabbit's corneal surface, causes eye blink and contraction of the nictitating membrane. Pathways (called "flow-through") were therefore distributed throughout the layers of our artificial network (called "DYSTAL") with initially high, fixed weights of interaction between elements (Alkon, Blackwell, Barbour, Rigler, & Vogl, 1990; Vogl, Alkon & Blackwell, 1989).

INPUT: SYNAPTIC POTENTIALS

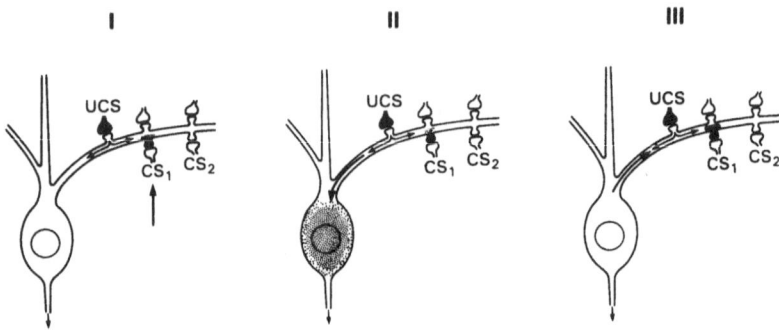

OUTPUT: IMPULSES

FIG. 1.2. Schematic of hypothetical model demonstrating sequence for activation of PKC in soma and dendrites. I. Initial activation of CS_1 contingent with UCS resulted in local activation of PKC at the post-synaptic spine. II. Subsequent to the initial activation of CS_1 and UCS (and dependent on the contingency of the two activating inputs), a "retro-signal" causes an increase in PKC synthesis within the cell body (1 day retention condition). III. Finally in the last stage of "consolidation," specifically targeted newly synthesized proteins (including PKC itself) make their way back to the initially activated post-synaptic dendritic region resulting in a long-term change in the biophysical characteristics of the spine in response to new inputs via the CS_1 input pathway.

2. Low-weight pathways. Similarly, the remaining DYSTAL pathways called "collaterals" were conceived as analogous to conditioned stimulus pathways in the snail and rabbit. Collateral pathway elements have interactions that are initially low and vary according to well defined associative learning rules (see #3).

3. Local interaction. The learning rules were based on system mechanisms of weight modification found in *Hermissenda* and the rabbit. In these species weight modification resulted from local interaction of distinct signals in response to precise temporal pairing of stimuli. Such local interaction occurs on post-synaptic membrane and can be induced in the absence of firing in the post-synaptic cell. This non-Hebbian mechanism of associative change dictates local weight modification in DYSTAL based on temporal pairing (here represented as two successive presentations) only of inputs. Such local weight modification in DYSTAL occurs independent of the output activity in elements and without back-propagation of signals.

4. Temporal history of element responses. In *Hermissenda* where it was possible to reconstruct a causal sequence during association formation, the Type B cell (a critical locus of visual-vestibular convergence) response to a single

pairing of visual with vestibular stimuli persisted until the next paired presentation during training. Although the Type B cell response relaxed considerably, the residual depolarization that remained added to and enhanced the next response to stimulus pairing. The DYSTAL operation incorporates this response persistence in the learning rule that allows for positive weight modification only if collateral and flow-through activity occurs on a common element for at least two successive pattern presentations (Fig. 1.3).

5. Other features. Additional design characteristics of DYSTAL were derived from what was known to be common to many biological neural systems such as receptive fields that extend beyond immediately adjacent elements, lateral inhibition, and convergence of signals from elements in one of many layers to elements in the next, and the absence of any external "teacher." The importance of some of these features as components of practical artificial neural networks have been receiving increasing recognition. Particularly noteworthy in this regard is the work of Grossberg et al. (Grossberg & Levine, 1987; Grossberg & Schmajuk, 1987) and Klopf (1988).

Function of Artificial Network (DYSTAL)

The objective of the initial design of DYSTAL, (which is amenable to different design configurations) was pattern restoration or image completion. A pattern or set of noisy patterns is presented to the input layer during training. Alternatively, a specific noisy or reduced test pattern is presented in association with a non-

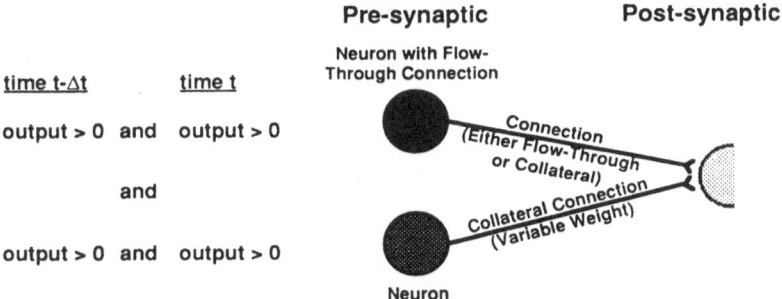

DYSTAL Learning Rule Case 1:
The only condition permitting weight increase

FIG. 1.3. The DYSTAL learning rule for the only condition permitting weight increase. The weight of a collateral connection is increased if and only if the collateral connection, as well as the connection from a neuron with a flow-through connection have both carried a signal two consecutive times. The weight of a collateral connection is not changed if the collateral connection has carried a signal twice successively, but the connections from neurons with flow through connections have not carried a signal twice successively. In all other cases, the weight of a collateral is decreased.

noisy exemplar. Subsequently, the network is "tested" with the presentation of reduced or noisy test patterns. DYSTAL then recreates a more complete version of the pattern, that is, "restores" it, and in so doing, shows some capacity for recognition of that image (Fig. 1.4). Quantitative assessment of such recognition can be made by adding a classifying rule based on some distance of the output from an image ideal such as an Euclidean distance measured from a mean output derived from presentations of the training pattern set. By such a measure, even a highly reduced network (3 × 3 elements in input and output layers, 7 × 7 on each of the two hidden layers) shows significant classification capability and improvement of classification capability with practice. The ability of DYSTAL,

FIG. 1.4. An illustration of DYSTAL's pattern completion ability. The patterns shown were not part of the training set. The darkness of the pattern is proportional to signal strength. Upper patterns are input patterns' lower patterns are output patterns. The left three pattern sets show pattern completion ability of a network with 3 x 3 input and output fields, trained to recognize "C"s and left shifted "C"s, in addition to "T"s. The right three pattern sets show pattern completion ability of a network with 19 × 19 input and output fields, trained to recognize "a"s and "O"s, in addition to "E"s. The number of pattern presentations required for training was the same for both networks. For additional illustrations of the pattern completion and classification performance of DYSTAL, see Alkon et al.(1990).

after training, to complete patterns in response to noisy incomplete test patterns was indicated by the significantly reduced Euclidean distance as well as by visual comparisons of test and output patterns.

Similar classification and completion capabilities characterize the function of networks trained by the steepest descent (back propagation) algorithm (Rumelhart, Hinton, & Williams, 1986; Vogl, Mangis, Zink, Rigler, & Alkon, 1988). However, it is inherent in the nature of steepest descent that the number of presentations required to achieve convergence is problem dependent and superlinearly related to the number of connections and elements in the network; this is not the case for DYSTAL. The number of pattern presentations required by DYSTAL depends solely on how noisy the patterns are; for most signal to noise ratios, a weight change step size that allows the system to reach equilibrium in 120 presentations or less of exemplars of each kind of pattern is sufficiently small. An accelerated back propagation network that we used for purposes of comparison contained the same number of layers (3) but a slightly smaller total number of elements as DYSTAL, and required about 6,000 presentations for convergence. With fewer connections and/or layers, the back-propagation network required significantly fewer presentations; with more connections or elements, as are essential for the solution of larger problems, the number of presentations required for convergence rapidly increases.

Potential Advantages of DYSTAL:

1. *Linear Scaling*. Because its design eliminates the output as a determinant of connection weight, DYSTAL computational requirements scale linearly as the number of elements increase. Thus, for much more useful input matrices such as 19×19, the same number of presentations as 3×3 or 64×64 networks are needed for DYSTAL to reach equilibrium. As the connections and/or elements in a back propagation network increase, it requires vastly increased numbers of iterations for convergence and thus dramatically increased computational effort.

2. *No external teacher*. The lack of requirements for an external teacher allows DYSTAL to classify and restore patterns without predetermined input standards. Thus, unlike networks that depend on back propagation and/or error minimization, DYSTAL shows the potential for "true learning" in the sense that patterns learned need not be anticipated and preprogrammed into a comparator function.

3. *Closed form solutions*. For a given input pattern set it is possible to arrive at a closed form solution for DYSTAL's representation of that pattern. The solution describes an equilibrium formulation that is quantitatively identical to the formulation derived by repeated presentation of the training set to the network until a stable equilibrium is reached. Descriptions with closed form solutions allow inspection of the actual working of the network layers as it learns from a test pattern set.

4. *Robustness to parameter variation.* There are only a few global parameters that control the performance of DYSTAL and we have shown that the performance of the network is insensitive to changes by factors of two or more in all but one of these parameters; the sensitive parameter is the ratio of weight increment to weight decrement step size. Thus a few trials of that single parameter for a given class of problems will quickly establish optimum system parameters.

5. *Robustness of architectural features.* It is possible to radically alter the architectural design of the DYSTAL network by adding lateral, feedback, or interlayer feed-forward connections to suit different functional objectives without significantly altering the network's learning rules or compromising its desirable properties such as linear scaling of computational effort with network size.

6. *Hardware Implementation.* The extensive feedback required by neural network models with external teachers are difficult to implement in large scale semiconductor (VLSI) chips because the multitude of feedback connections make such chips difficult to design, with a resulting low packing density. Because DYSTAL does not require any feedback connections, it can efficiently be implemented on large scale VLSI chips.

These apparent functional advantages of the DYSTAL design offer considerable promise for reducing the computational burden for artificial networks during pattern storage, reconstruction and classification. This promise suggests that biological architectures—the designs of neural systems that learn discrete links or associations, such as the visual-vestibular network of *Hermissenda*—can underlie much more complex pattern recognition such as occurs in mammalian brain structures, perhaps even our own.

REFERENCES

Alkon, D. L. (1989). Memory storage and neural systems. *Scientific American, 260,* 42–50.

Alkon, D. L., Quek, F., & Vogl, T. P. (1989). Computer modelling in associative learning in Hermissenda. In D. S. Touretzky (Ed.), *Advances in neural information processing systems I* (pp. 419–435). San Mateo, CA: Morgan Kaufman.

Alkon, D. L., Blackwell, K. T., Barbour, G., Rigler, A. K., & Vogl, T. P., (1990). Pattern recognition by an artificial network derived from biological neuronal systems, *Biological Cybernetics, 62,* 363–376.

Grossberg, S., & Levine, D. S. (1987). Attentional mechanisms in neural information processing: Examples from pavlovian conditioning. In M. Caudill & E. Butler (Eds.), *Proceedings of the 1st International Conference on Neural Networks, San Diego, CA* IEEE Cat. # 87th0191-7, pp. IV-49–IV-57.

Grossberg, S., & Schmajuk, N. A. (1987). Neural dynamics of attentionally modulated pavlovian conditioning: conditioned reinforcement, inhibition, and opponent processing. *Psychobiology, 15,* 195–240.

Klopf, A. H. (1988). A neuronal model of classical conditioning, *Psychobiology, 16,* 85–125.

Rumelhart, D. E., Hinton, G. E., & Williams, R. J. (1986). Learning internal representations by error propagation. In D. E. Rumelhart, J. L. McClelland, (Eds.), *Parallel distributed processing: Explorations in the microstructures of cognition,* (Vol. 1., 318–362). Cambridge, MA: MIT Press.

Vogl, T. P., Mangis, J. K., Zink, W. T., Rigler, A. K., Alkon, D. L. (1988). "Accelerating the convergence of the back propagation method." *Biological Cybernetics, 59,* 257–263.

Vogl, T. P., Alkon, D. L., & Blackwell, K. T. (1989). Dynamically stable associative learning (DYSTAL): A biologically motivated artificial neural network. *Proceedings of the IJCNN89 International Joint Conference on Neural Networks, Washington, D.C.* June 21, 1989, pp. II-101–II-103.

2 Empirically Derived Adaptive Elements and Networks Simulate Associative Learning

Douglas A. Baxter
Dean V. Buonomano
Jennifer L. Raymond
University of Texas Medical School, Houston

David G. Cook
Frederick M. Kuenzi
Thomas J. Carew
Yale University

John H. Byrne
University of Texas Medical School, Houston

ABSTRACT

A goal common to both neurobiologists and neural network theorists is to understand the events occurring within single elements and networks that contribute to learning and memory. Our approach is to analyze the properties of neurons and neural circuits that mediate simple forms of learning in the marine mollusck *Aplysia*, and to develop mathematical models of these neuronal elements and networks. A single neuron-like adaptive element, which reflects the biochemical and biophysical properties of sensory neurons, simulates many features of nonassociative learning and classical conditioning. Moreover, relatively simple neural networks that incorporate an empirically derived associative "learning rule" exhibit some higher order features of classical conditioning and some elementary features of operant conditioning. These results illustrate that biologically derived models of adaptive elements and networks are capable of simulating a variety of associative learning phenomena.

The computational capabilities of artificial neural networks emerge from at least three key factors: the properties of the individual adaptive elements, the patterns of connectivity within the networks, and the rules governing the interactions between the elements. Many artificial neural networks are composed of simple elements (often referred to as neurons). Each element sums its weighted inputs (often referred to as synaptic inputs), and if the sum of these synaptic inputs equals or exceeds a threshold value, the element is activated (the equivalent of a neuronal action potential). Generally, there is little structure within these networks, and the elements are either highly interconnected or arranged as two or three layers of elements that receive converging synaptic input from the preceding layer. The strengths, or weights, of the synaptic connections between the elements change according to rules or algorithms that are referred to as *learning rules*. For example, with a Hebbian learning rule synaptic efficacy changes as a function of simultaneous activities in the presynaptic and postsynaptic elements. As a result of training, the synaptic weights are altered *via* the learning rules and the arbitrarily connected network develops a functional structure that is appropriate for solving a particular problem. Theoretical work illustrates that networks with such apparently simple characteristics are capable of quite complex collective computations (e.g., Anderson & Rosenfeld, 1988; Bear, Cooper, & Ebner, 1987; Dobbins, Zucker & Cynader, 1987; Fukushima, Miyake & Ito, 1983; Hopfield, 1982, 1984; Hopfield & Tank, 1985, 1986; Koch & Segev, 1989; Lehky & Sejnowski, 1988; Linsker, 1986; Lippmann, 1989; see also, McClelland, Rumelhart and the PDP Research Group, 1986; Pearson, Finkel & Edelman, 1987; Rumelhart, McClelland and the PDP Research Group, 1986; Sejnowski, Koch, & Churchland, 1988; Sejnowski & Rosenberg, 1986; Zipser & Anderson, 1988). Other artificial neural networks (e.g., Grossberg, 1971; Grossberg & Levine, 1987; Schmajuk, this volume) include distinct substructures and interconnections that more closely approximate those in certain brain regions, but their units do not correspond to individual neurons. Therefore, an intriguing question, which we are pursuing, is what computational capabilities emerge as the properties of the adaptive elements, the patterns of connectivity and the learning rules within simulated neural networks are made more reflective of the details of neuronal biochemistry and physiology.

Substantial progress toward understanding the physiological basis of learning and memory has been made in a variety of vertebrate and invertebrate preparations (for recent reviews, see Alkon & Woody, 1986; Byrne, 1985, 1987, 1990; Carew & Sahley, 1986; Crow, 1988; Farley & Alkon, 1985; Hawkins, 1983; Kandel & Schwartz, 1982; Lynch, McGaugh, & Weinberger, 1984; Mpitsos & Lukowiak, 1986; Quinn, 1984; Thompson, Berger, & Madden, 1983; Thompson, 1986; Thompson, 1988; Weinberger, McGaugh, & Lynch, 1985; Woody, 1986). One preparation that has been particularly useful is the marine mollusck *Aplysia*. This animal has a relatively simple nervous system with large, identifiable neurons that are accessible for detailed biochemical, biophysical, and anatomical studies. Recent work indicates that the learning repertoire of *Aplysia* is quite extensive. There is evidence for both nonassociative learning (such as habituation, dishabituation, and sensitization) and associative learning (such as classical conditioning, some higher order features of classical conditioning, contextual conditioning, and operant conditioning) (Abrams & Kandel, 1988; Carew, Castellucci, & Kandel, 1971; Carew, Hawkins & Kandel, 1983; Carew, Walters, & Kandel, 1981; Colwill, Absher, & Roberts, 1988a,b; Cook & Carew, 1986, 1989a, 1989b, 1989c; Hawkins, Carew & Kandel, 1986; Hawkins, Clark & Kandel, 1985, 1987; Levy & Susswein, in press; Pinsker, Kupfermann, Castellucci & Kandel, 1970; Scholz & Byrne, 1987; Susswein & Schwarz, 1983; Susswein,

Schwarz, & Feldman, 1986; Walters, Byrne, Carew, & Kandel, 1983; Walters, Carew & Kandel, 1979; Walters, Carew, & Kandel, 1981). Thus the relatively simple nervous system of *Aplysia* provides an ideal model system in which to investigate the adaptive properties of biological elements, circuits and learning rules that underlie examples of different types of learning and memory.

Nonassociative Learning

The three major forms of nonassociative learning are habituation, dishabituation and sensitization. Habituation refers to a decline in magnitude of a response due to repeated elicitation of that response. Dishabituation refers to the enhancement of a habituated response by delivering a second stimulus (typically a strong or surprising one); and sensitization refers to the enhancement of a non-habituated response, also usually produced by delivering a strong stimulus to the animal. The neuronal mechanisms contributing to these forms of nonassociative learning have been analyzed extensively in two defensive behaviors of *Aplysia*; the siphon withdrawal and the tail withdrawal reflexes (for reviews, see Byrne, 1985, 1987; Carew, 1987; Hawkins, Clark & Kandel, 1986; Kandel & Schwartz, 1982). The synaptic connections between the sensory and motor neurons that mediate these reflexes exhibit a number of plastic properties, which in turn can be related to nonassociative learning phenomena. For example, habituation of the reflex is paralleled by the depression of the release of transmitter from the sensory neurons. Repeated stimulation of the sensory neurons causes cumulative inactivation of the influx of Ca^{2+} and depletion of transmitter (Bailey & Chen, 1983, 1988a; Klein, Shapiro, & Kandel, 1980). This synaptic depression leads to progressively less activation of the motor neurons, which in turn leads to a decrement in the behavioral response (Byrne, Castellucci, & Kandel, 1978; Byrne, 1982; Castellucci & Kandel, 1974; Castellucci, Pinsker, Kupfermann, & Kandel, 1970; Kupfermann, Castellucci, Pinsker, & Kandel, 1970).

The sensory neurons are also a cellular locus for dishabituation and sensitization. Dishabituation and sensitization are associated with an increase in the release of transmitter from the terminals of the sensory neurons (presynaptic facilitation) (Carew, Castellucci, & Kandel, 1971; Castellucci & Kandel, 1976; Pinsker, Kupfermann, & Kandel, 1970; Walters, Byrne, Carew, & Kandel, 1983). Novel or sensitizing stimuli are thought to produce presynaptic facilitation by activating facilitatory neurons. These facilitatory neurons release modulatory transmitters (at least one that is serotonin, 5-HT) that activate adenylate cyclase and increase the levels ·of the intracellular second messenger cyclic AMP (cAMP) in the sensory neurons (Bernier, Castellucci, Kandel, & Schwartz, 1982; Ocorr & Byrne, 1985; Ocorr, Tabata, & Byrne, 1986). The modulatory transmitter contributes to presynaptic facilitation in two ways. First, the modulatory transmitter, *via* an increase in the level of cAMP, mediates the closure of K^+ channels; this results in a broadening of the action potential and thus an increase in the influx of Ca^{2+} (Baxter & Byrne, 1987, 1989; Klein, Camardo, & Kandel, 1982; Pollock, Bernier, & Camardo, 1985; Shuster, Camardo, Siegelbaum, & Kandel, 1985; Siegelbaum, Camardo, & Kandel, 1982). Second, the modulatory transmitter, perhaps *via* cAMP, contributes to the regulation of the amount of transmitter that is available for release from the sensory neurons (Bailey & Chen, 1983, 1988b; Gingrich, Baxter & Byrne, 1988; Braha, Dale, Hochner, Klein, Abrams & Kandel, 1990; Gingrich & Byrne, 1987; Hochner, Klein, Schacher, & Kandel, 1986). Presynaptic facilitation leads to enhanced activation of motor neurons, which in turn leads to an enhanced behavioral response.

Associative Learning: Classical Conditioning

Associative learning is typically divided into two major categories, classical conditioning and operant conditioning (Mackintosh 1974). In classical conditioning, an animal learns to associate two stimuli, a conditioned stimulus (CS), which serves as a predictor or signal, and an unconditioned stimulus (US), which serves as reinforcement (Pavlov, 1927; Rescorla 1988). In operant conditioning, an animal learns to associate a response that it emits with the subsequent occurrence of reinforcement (Mackintosh 1974, Skinner, 1938). *Aplysia* are capable of exhibiting both types of associative learning in response systems that are well suited for cellular and computational analyses. In this part of the chapter, we discuss classical conditioning; operant conditioning will be taken up in a later section.

Behavioral Analyses of Classical Conditioning. Classical conditioning of the gill and siphon withdrawal reflex in *Aplysia* can be produced by pairing a weak tactile stimulus (CS) to a mantle organ, such as the siphon, with a strong shock to the tail (US). Following such a pairing procedure, the originally small response to the siphon stimulus is significantly augmented compared to a variety of control procedures, such as random CS-US presentations, or presentation of the US alone (Carew, Walters, & Kandel 1981). Another procedure for demonstrating classical conditioning in *Aplysia* is shown in Fig. 2.1, which illustrates differential classical conditioning (Carew, Hawkins, & Kandel 1983). With this procedure each animal receives two CSs, each delivered to a different mantle organ. One CS (CS+) is paired with a tail shock US, the other CS (CS−) is specifically unpaired (Fig. 2.1B). Conditioning is assessed by comparing the differential responses of the animal to each CS after training. Thus conditioning can be demonstrated within a single animal, since each animal serves as its own control. As is shown in Fig. 2.1C and 2.1D, *Aplysia* can acquire this task readily. Moreover, this form of conditioning in *Aplysia* shares several features in common with vertebrate conditioning, including a steep inter-stimulus interval function and sensitivity to the contingent relationship between the CS and US (Hawkins, Carew & Kandel, 1986). These features of classical conditioning in *Aplysia* not only permit comparison with conditioning in higher animals, they also provide important information

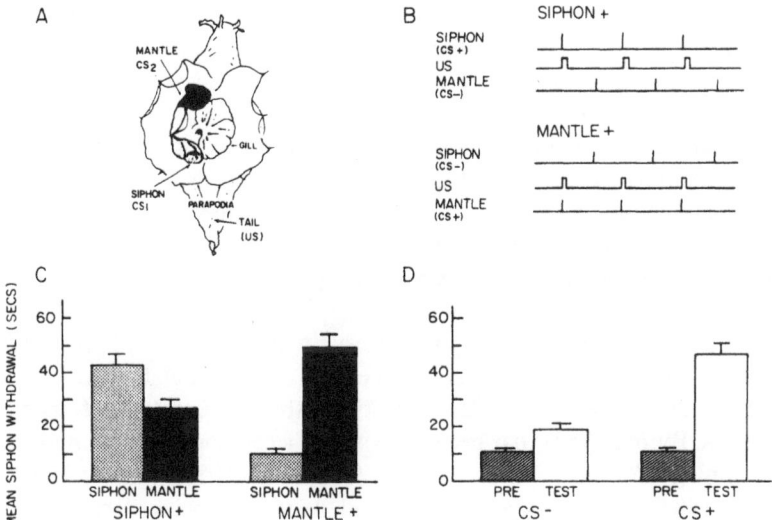

FIG. 2.1. *Differential Conditioning.* **A:** Dorsal view of *Aplysia* illustrating the two sites used to deliver conditioned stimuli: the siphon and the mantle shelf. The unconditioned stimulus (US) was an electric shock delivered to the tail. *B:* Paradigm for differential conditioning: one group (Siphon+) received the siphon CS (CS+) paired with the US and the mantle CS (CS−) specifically unpaired with the US; the other group (Mantel+) received the mantel stimulus as CS+ and the siphon stimulus as CS−. The intertrial interval was 5 min. **C:** Results of an experiment using the paradigm illustrated in B. Testing was carried out 30 min after 15 training trials. Responses in the Siphon+ group (N=12) were significantly more prolonged (p<0.05) to the siphon CS than to the mantle CS, whereas responses in the Mantle+ group (N=12) were significantly more prolonged (p<0.01) to the mantle CS than to the siphon CS. **D:** Pooled data from C. Test scores from the unpaired (CS−) and paired (CS+) pathways are compared with the respective pretest scores for these pathways. The CS+ test scores are significantly greater than CS− (p<0.005), demonstrating that differential conditioning has occurred. (Modified from Carew, Hawkins & Kandel, 1983)

(and critical constraints) for cellular and computational models of this form of associative learning.

Cellular Analysis of Classical Conditioning. The differential conditioning exhibited by *Aplysia* (Fig. 2.1) suggests that different CS pathways in the animal are independently capable of supporting associative learning. Cellular evidence supporting this hypothesis has been obtained in two different studies. Hawkins, Abrams, Carew, and Kandel (1983) found that synaptic transmission from individual siphon sensory neurons onto their follower cells could be significantly augmented by pairing activity induced in those sensory neurons with tail shock. This effect appeared to be an activity dependent enhancement of the same cellular process, presynaptic facilitation, that contributes to sensitization in the gill and siphon withdrawal reflex.

In a parallel and independent study, Walters and Byrne (1983) found that tail sensory neurons in the pleural ganglion of *Aplysia* also exhibited temporally specific augmentation of their synaptic output, when the activity of these neurons was paired with tail shock (see Fig. 2.4A). The associative effect was acquired rapidly and was expressed as a temporally specific amplification of presynaptic facilitation.

Thus a form of associative synaptic plasticity in two classes of sensory neurons of *Aplysia,* which is termed activity dependent neuromodulation (Buonomano & Byrne, in press; Hawkins, 1984; Hawkins, Abrams, Carew, & Kandel, 1983; Walters & Byrne, 1983), has been proposed as a cellular mechanism for classical conditioning (Abrams, 1985; Abrams & Kandel, 1988; Carew, Hawkins, & Kandel, 1983; Hawkins, Carew, & Kandel, 1986; Lukowiak, 1986). A proposed general cellular scheme of activity dependent neuromodulation and how it might be related to classical conditioning is illustrated in Fig. 2.2A. Two sensory neurons (SN 1 and SN 2), which constitute the pathways for the conditioned stimuli (CS), make weak subthreshold connections to a response system (e.g., a motor neuron). Delivering a reinforcing or unconditioned stimulus (US) alone has two effects. First, the US activates the response system and produces the unconditioned response (UR). Second, the US activates a diffuse modulatory system that nonspecifically enhances transmitter release from all the sensory neurons. This nonspecific enhancement contributes to sensitization (as mentioned earlier). Temporal specificity, characteristic of associative learning, occurs when there is pairing of spike activity in SN 1 with the US, which causes a selective amplification of the modulatory effects in that specific sensory neuron (Buono-mano & Byrne, in press; Hawkins, Abrams, Carew, & Kandel, 1983; Walters & Byrne, 1983). The cellular mechanism, at least in part, for this associative synaptic plasticity is an enhancement in the levels of cAMP beyond that produced by the modulator alone. It appears that the influx of Ca^{2+} associated with the CS (spike activity in SN 1) interacts with a Ca^{2+}-sensitive component of adenylate cyclase and amplifies the modulatory effects of the US (Abrams, Bernier, Hawkins, & Kandel, 1984; Abrams, Eliot, Dudai, & Kandel, 1985; Abrams & Kandel, 1988; see also Dudai, 1987; Ocorr, Walters, & Byrne, 1985). Activity in SN 2 does not amplify the effects of the US since the increased levels of Ca^{2+} do not occur in the appropriate temporal relationship to the US. The associative amplification of synaptic strength in the paired sensory neuron leads to an en-hancement of the ability of SN 1 to activate the response system and produce the conditioned response (Fig. 2.2B).

The empirical data discussed earlier support the hypothesis that correlates of

A. LEARNING   B. MEMORY

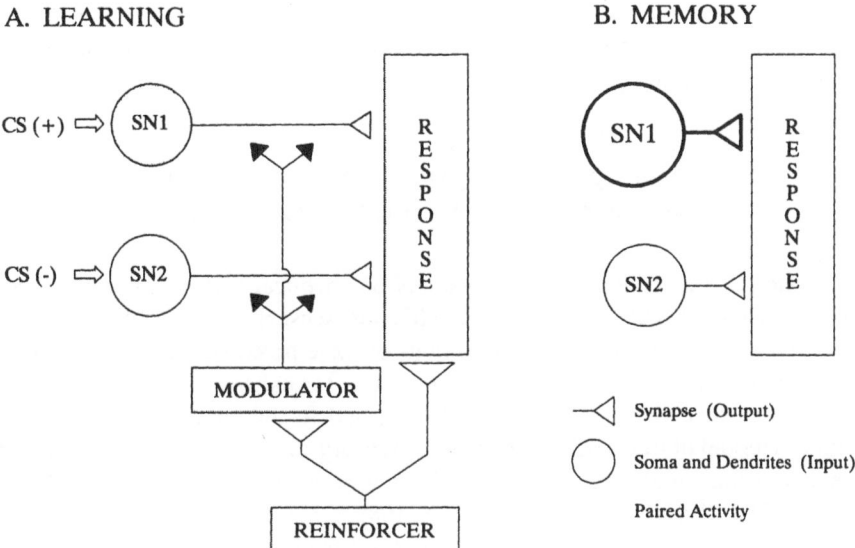

FIG. 2.2. *General Model of Activity Dependent Neuromodulation.* **A:** Learning. Stippling indicates temporally contiguous activity. A motivationally potent reinforcing stimulus (the unconditioned stimulus, US) activates a neural response system and a modulatory system that regulates the efficacy of diffuse afferents to the response system. Increased spike activity (the conditioned stimulus, CS) in the paired afferent (SN 1) immediately before (CS+) the modulatory signal amplifies the degree and duration of the modulatory effects, perhaps through the Ca^{2+} sensitivity of the modulatory evoked second messenger. The unpaired (CS−) afferent neuron (SN 2) does not show an amplification of the modulatory effect. **B:** Memory. The amplified modulatory effects cause increased in transmitter release and/or excitability of the paired neuron (SN 1), which in turn strengthens the functional connection between the paired neuron and the response system. (Modified from Walters & Byrne, 1983)

many forms of learning exist within single neurons. At least some elementary components of associative specificity are being explored at a molecular level (e.g., Abrams & Kandel, 1988; Crow, 1988; Gustafsson & Wigström, 1988; Ocorr, Walters & Byrne, 1985). Thus simple forms of learning such as classical conditioning need not require complex neural networks. Rather, the associative relationship that characterizes classical conditioning may result from intrinsic associative capabilities of certain cellular processes within individual neurons (see also, Kelso, Ganong, & Brown, 1986; Nicoll, Kauer, & Malenka, 1988). Our modelling strategy consists of developing neuron-like elements and small networks that incorporate as much detail as is available on the cellular processes that contribute to nonassociative and associative synaptic plasticity and testing the computational capabilities of these empirically derived models by attempting to simulate learning phenomena.

COMPUTATIONAL CAPABILITIES OF A SINGLE
EMPIRICALLY DERIVED ADAPTIVE ELEMENT

Our current understanding of the various biochemical and biophysical mechanisms that contribute to plasticity within the sensory neurons of *Aplysia* has allowed us to develop formal descriptions of these processes. The approach was to transform these processes into mathematical formalisms, assign values to the parameters that agree with published data, and fit the components together to create a model of transmitter release at the sensory-to-motor synapse. The specific details of this single cell model have been described previously (Byrne & Gingrich, 1989; Byrne, Gingrich, & Baxter, 1989; Gingrich, Baxter & Byrne, 1988; Gingrich & Byrne, 1985, 1987). The general features of this adaptive element are illustrated in Fig. 2.3 and discussed later.

Properties of Empirically Derived Adaptive Element

Release, Storage, and Mobilization of Transmitter.   The model contains equations describing two pools of transmitter, a readily releasable pool (P_R) and a storage pool (P_S). The releasable pool contains vesicles that are in close proximity to release sites. During an action potential, an influx of Ca^{2+} (I_{Ca}) causes the release of transmitter (T_R) from this pool, thus, the amount of transmitter that is released is a function of both the dynamics of the Ca^{2+} influx and the number of vesicles in the releasable pool. As a consequence of release, P_R is depleted. In order to offset depletion, transmitter is delivered (mobilization) from a storage pool (P_S) to the releasable pool. Vesicles move from one pool to the other via three fluxes, one driven by diffusion (F_D), another driven by Ca^{2+} (F_C) and the third driven by levels of cAMP (F_{cAMP}). The storage pool is replenished slowly by synthesis of new vesicles (Flux F_N). The amount of transmitter released is a direct measure of the ability of an adaptive element to activate the response system (see Fig. 2.2); thus, the amplitude of the excitatory postsynaptic potential (EPSP) may be thought of as a measure of the strength of the CS (see Fig. 2.6).

Regulation of Ca^{2+}.   The influx of Ca^{2+} (I_{Ca}) during simulated action potentials leads to the release of transmitter (T_R) and an accumulation of intracellular Ca^{2+}. The pool of Ca^{2+} is contained in two volumes (not shown): the submembrane compartment, which represents a thin layer of the cytosol lining the membrane immediately adjacent to the membrane, and the interior compartment, which represents a larger fraction of the cytosol further away from the membrane. The

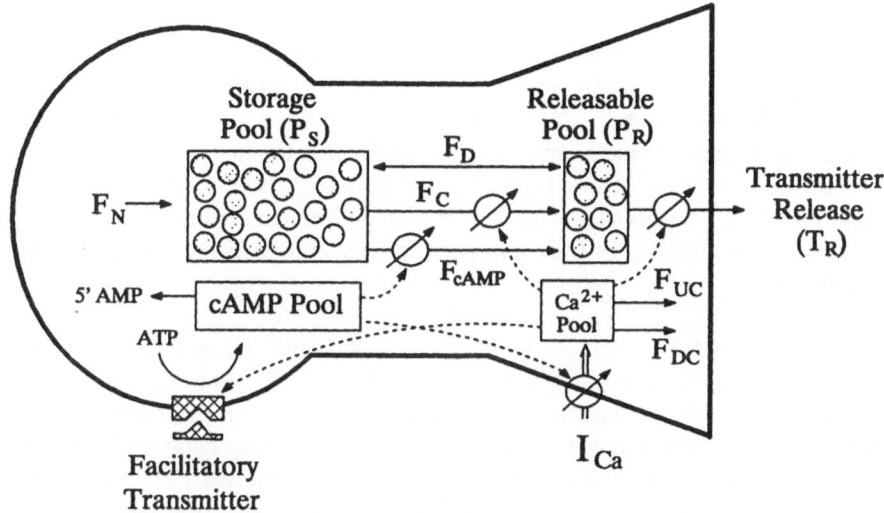

FIG. 2.3. *Empirically Derived Associative Element.* Components of the model of a sensory neuron that is described by Gingrich and Byrne (1985, 1987) (also see Gingrich, Baxter, & Byrne, 1988; Byrne & Gingrich, 1989; Byrne, Gingrich & Baxter, 1989). Action potentials lead to the opening of voltage dependent Ca^{2+}-channels and subsequent increase in the Ca^{2+}-current (I_{Ca}). The calcium influx leads to an increase in the level of intracellular Ca^{2+} (Ca^{2+} Pool), which in turn triggers the release of transmitter (T_R) from the releasable pool (P_r), mobilization of transmitter (F_C) from a storage pool (P_S) to P_R, and potentiates the synthesis of cAMP by the facilitatory transmitter. F_N represents the delivery of vesicles to the storage pool from unmodeled sources and results from transport from other storage pools or synthesis triggered by depletion of the storage pool. Ca^{2+} is removed from the Ca^{2+} pool by F_{UC}, active uptake, and by F_{DC}, passive diffusion. F_D represents diffusion of vesicles between the storage and releasable pools. The US results in the release of a facilitatory transmitter that activates adenylate cyclase and results in the synthesis of cAMP. The cAMP has two actions; first it enhances mobilization of transmitter (F_{cAMP}), and second, it indirectly increased Ca^{2+} influx (I_{Ca}) *via* cAMP-dependent increases in the duration of the action potential. Paired presentation of the CS and US result in increased levels of cAMP due to the priming of cAMP synthesis by Ca^{2+}. The pairing-specific elevated levels of cAMP produce increased mobilization of transmitter and spike broadening beyond that produced by the facilitatory transmitter alone. Consequently, when subsequent action potentials are elicited, there is enhanced Ca^{2+} influx and release to transmitter. The circles with arrows through their center represent elements of the model that are modulated positively by other variables. Unless otherwise noted all equation and parameters are as described in Gingrich and Byrne (1985, 1987). (Modified from Buonomano, Baxter, & Byrne, in press.)

Ca^{2+} within the submembrane compartment regulates the release of transmitter, whereas the Ca^{2+} within the interior compartment contributes to the regulation of the mobilization of transmitter and the synthesis of cAMP (discussed later). During an action potential, Ca^{2+} enters the submembrane compartment through voltage dependent channels, and diffuses rapidly into the interior compartment. Two fluxes remove Ca^{2+} from the interior compartment; one that represents active buffering of Ca^{2+} by organelles such as the endoplasmic reticulum (F_{UC}) and one that represents diffusion of Ca^{2+} into an innermost compartment that serves as a sink (F_{DC}).

Regulation of cAMP.   This model of synaptic plasticity also includes equations describing the concentration of cAMP and its effects on the release of transmitter. The cAMP cascade is a sequence of multiple events, but some evidence indicates that the rate-limiting step is the decay of activity of adenylate cyclase (Schwartz, et al., 1983). Therefore, as a first approximation, we assume that the concentration of cAMP is described by a single lumped-parameter dynamic equation. Presentation of a US rapidly activates adenylate cyclase, which leads to increased synthesis of cAMP, which is broken down eventually into nonactive 5' AMP. The effect of the CS (Ca^{2+} entry during spikes in the sensory neuron) on levels of cAMP is simulated by making the synthesis of cAMP a function of the intracellular concentration of Ca^{2+}. Thus the elevation of Ca^{2+} produced by a CS, which precedes an US, primes the cyclase and amplifies the subsequent US-mediated cAMP synthesis. The higher levels of cAMP result in greater presynaptic facilitation; that is, activity dependent neuromodulation.

Simulations of Learning Phenomena by a Single Adaptive Element

Simulations of Nonassociative Learning.   While this chapter is concerned primarily with simulations of associative learning phenomena, the importance of nonassociative learning should not be overlooked. In *Aplysia,* the cellular mechanisms underlying associative learning seem to be an elaboration of the mechanisms involved in nonassociative learning. Activity dependent neuromodulation, a neuronal analog of classical conditioning, involves a Ca^{2+}-dependent enhancement of the mechanisms contributing to presynaptic facilitation, which in turn contributes to sensitization and dishabituation (Abrams, 1985; Abrams, Bernier, Hawkins, & Kandel, 1984; Abrams & Kandel, 1988; Kandel, Abrams, Bernier, Carew, Hawkins, & Schwartz, 1983; Ocorr, Walters, & Byrne, 1985).

Our previous simulations illustrated that this neuron like adaptive element accurately simulates the neural analogs of many forms of nonassociative learning (for a detailed presentation of these results see Byrne & Gingrich, 1989; Byrne, Gingrich, & Baxter, 1989; Gingrich, Baxter, & Byrne, 1988; Gingrich & Byrne 1985). For example, repetitive action potentials in this single cell model result in the cumulative inactivation of I_{Ca}, depletion of transmitter in P_R, and thus, depression of synaptic transmission. This synaptic depression is the neural analog of habituation. In addition, the single cell model accurately simulates presynpatic facilitation, which is the neural analog of dishabituation and sensitization. Previously, dishabituation and sensitization were believed to involve a single, common mechanism. However, simulations with the single cell model indicate that these two similar forms of nonassociative learning result from different processes. Simulations indicate that sensitization is mediated primarily by spike broadening, whereas, dishabituation requires the mobilization of transmitter. Several lines of recent experimental evidence support the basic assumptions and conclusions of our model (Hochner, Klein, Schacher, & Kandel, 1986; Rankin & Carew, 1988).

Simulations of Differential Classical Conditioning. Figure 2.4 illustrates the ability of this single adaptive element to simulate the empirical data on a neural analog of differential classical conditioning. Part A shows the results of a previous experimental analysis (Walters & Byrne, 1983), and Part B shows the results of the computer simulations. As in the experimental analysis, the simulation study had a three phase test-train-test procedure. The first test-phase consisted of simulating single action potentials in the single cell model at 5 min intervals and establishing the baseline EPSP (labeled B in Fig. 2.4). Two training conditions

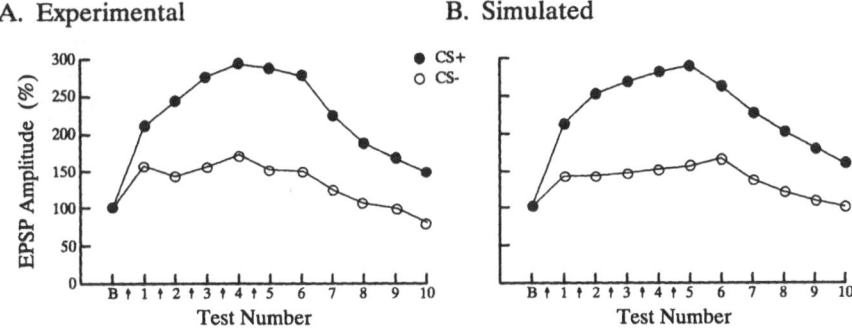

FIG. 2.4.   *Simulation of Cellular Analog of Differential Classical Conditioning* **A**: Experimental data. Three phase test-train-test procedure for paired (CS+, CS precedes US by 600 msec) and explicitly unpaired training (CS−, CS follows US by 2 minutes). The B on the *abscissa* is the normalized baseline excitatory postsynaptic potential (EPSP) evoked in a motor neuron in response to a test stimulus (single action potential in a sensory neuron). Training trials occur between test numbers B to 5 at the arrows. EPSPs produced by the training trials are not shown. Test numbers 5 to 10 represent posttraining period. Paired training (CS+) results in significant enhancement of the test EPSP compared to unpaired (CS−). **B**: Simulation of experimental data. Using the same test-train-test procedure as in the experimental data, the simulation captures salient qualities of empirical data of Part A. (Modified from Gingrich & Byrne, 1987).

were simulated; paired CS and US, with the CS initiated 600 msec before the US (CS+) and explicitly unpaired CS and US, with the CS initiated 2 min after the US (CS−). The CS was simulated with a 400 msec train (25 Hz) of spikes in the single cell model and the US was simulated by activating the cyclase for 200 msec. There were five training trials, each separated by 5 min (arrows in Fig. 2.4). Both of the conditioning procedures (CS+ and CS−) produced enhancement of transmitter release. The CS− group underwent nonassociative enhancement of synaptic strengths that is analogous to sensitization or dishabituation. The CS+ group, however, underwent an associative enhancement of synaptic strength that was beyond that of the CS− group. A comparison of differential conditioning of the CS+ and CS− groups in Figs. 2.4A and 2.4B indicate that there is a good fit between the experimental and simulated data.

A characteristic feature of associative learning is the requirement for a close temporal association between the CS and US for effective conditioning. For conditioning of many simple reflex responses, the optimal interstimulus interval (ISI) is generally about 500 msec. Longer ISIs are less effective and backward conditioning is generally ineffective. We were therefore interested in determining whether the single cell model could demonstrate a dependence on ISI similar to that observed in conditioning studies in *Aplysia* and other animals (Frey & Ross, 1968; Hawkins, Carew, & Kandel, 1986; Hull, 1943, 1952; Smith, Coleman, & Gormezano, 1969). Figure 2.5 illustrates the ISI function for the empirically derived adaptive element. An ISI of about 200 msec is optimal whereas longer or shorter ISIs are less effective. The simulated ISI function in the single cell model is a direct consequence of the kinetics of the buffering of intracellular

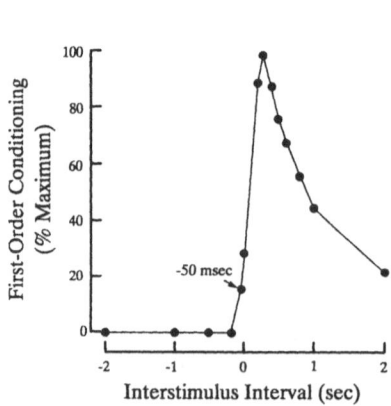

FIG. 2.5. *Interstimulus Interval (ISI) Function for the Associative Element.* The effective conditioning is the normalized difference between responses to the second test pulse after 10 training trials in paired (CS+) and explicitly unpaired (CS−) simulations as a function of the time period (ISI) between paired CS and US applications. The relationship between the effectiveness of conditioning and the CS-US interval is due to the time-course of the accumulation and buffering of Ca^{2+} in the cytoplasm. The ISI function of the model is similar to those obtained from behavioral experiments. (Modified from Gingrich & Byrne, 1987).

Ca^{2+}. When the ISI is short, high levels of Ca^{2+} are present at the time of the US and therefore the CS-mediated amplification of the effects of the US are greatest. As the interval between the CS and US increases, Ca^{2+} levels are buffered, consequently with longer ISIs, there is less amplification of the effects of the US. Thus the elevation of intracellular Ca^{2+} levels produced by the CS serves as a "trace" that becomes associated with a closely paired US. This ISI function (Fig. 2.5) is similar to those observed in many types of animal learning (see Mackintosh, 1974; Rescorla, 1988).

In summary, this single cell model accounts for many features of both nonassociative and associative plasticity at the sensory neuron synapses. The formalisms that we have used to describe complex processes such as the regulation of Ca^{2+} and cAMP levels, spike broadening and mobilization of transmitter are necessarily simplifications of the actual mechanisms. Nevertheless, it is our belief that these descriptions provide for a substantial biological underpinning for this model of learning.

Although our mathematical model is capable of predicting some features of associative learning, it cannot predict aspects of associative learning that depend on an interplay of multiple stimuli at different sites or on more than one stimulus modality. This limitation, however, is due to the fact that the model is based only on a single adaptive element. By incorporating the present single cell model into a circuit that includes multiple adaptive elements as well as facilitatory neurons, it will be possible to test its ability to predict more complex features of associative learning.

EMERGENT COMPUTATIONAL CAPABILITIES OF SMALL EMPIRICALLY DERIVED NEURAL NETWORKS

Although little is known about the neural mechanisms responsible for higher order forms of conditioning, several quantitative models of small networks have shown that in theory the same learning rules that are capable of simulating classical conditioning can also simulate some higher order features of associative learning (Aparicio, 1988; Buonomano, Baxter, & Byrne, in press; Byrne, Buonomano, Corcos, Patel, & Baxter, 1988; Gluck & Thompson, 1987; Grossberg & Levine, 1987; Hawkins, 1989a, 1989b; Klopf, 1988, 1989; Klopf & Morgan, in press; Rescorla & Wagner, 1972; Sutton & Barto, 1981; Sutton & Barto, in press). In this chapter, we present some of our results of studies that examine the computational capabilities of small networks where the properties of the adaptive elements, the patterns of synaptic connectivity, and the associative learning rule are derived from empirical observations in the nervous system of *Aplysia*.

Figure 2.6 illustrates a three cell artificial neural network that is derived from the characteristics of the neural circuit mediating nonassociative and associative learning in the withdrawal reflexes of *Aplysia*. Two sensory neurons (SN 1 and

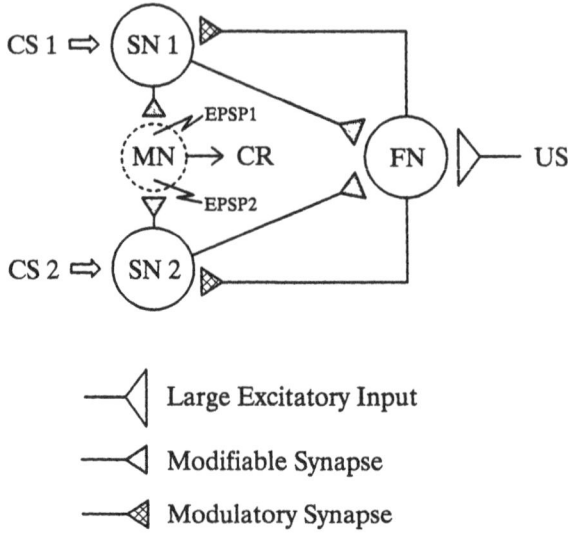

FIG. 2.6. *Network Architecture for Higher Order Features of Classical Conditioning.* The model of a sensory neuron that is shown in FIG. 2.3 was incorporated into a circuit consisting of two identical sensory neurons (SN 1 and SN 2) and a facilitatory neuron (FN). The sensory neurons can be activated independently by separate conditioned stimuli (CSs). The facilitatory neuron is activated by the US and can also be activated by a sensory neuron if the level of the input exceeds threshold. The facilitatory neuron is a nonplastic element that releases a transmitter that in turn induces the synthesis of cAMP in the sensory neurons. Sensory neuron input to the motor neuron (MN) and facilitatory neuron represent the conditioned response (CR). Because the input to the facilitatory neuron and MN are identical, we only simulated the facilitatory neuron in order to simplify the network. (Modified from Buonomano, Baxter, & Byrne, in press.)

SN 2) with identical properties (see Fig. 2.3) make synaptic contact with a motor neuron (MN). Activity in the individual sensory neurons represents separate pathways for conditioned stimuli (CS 1 and CS 2). The amplitude of the EPSPs at the sensory-to-motor synapses represent the ability of each CS to activate the response system and produce a CR. A key aspect of the architecture of this network is that the sensory neurons also make excitatory synaptic connections with a facilitatory neuron (FN). This synapse has two consequences, one practical and the second theoretical. First, from a practical point of view, the circuit can be simplified by removing the motor neuron and using the EPSP at the sensory-to-facilitatory neuron synapse as measure of changes in the synaptic strength of the CS. Second, from a theoretical point of view, a sensory neuron can 'take control' of the facilitatory neuron as the strength of its sensory-to-facilitatory neuron synapse increases. This possibility has fundamental implications with respect to neural models of higher order features of classical conditioning (discussed later; Grossberg, 1971; Hawkins & Kandel, 1984).

Because there is little experimental data on the properties of the facilitatory neuron, it is modelled as an element that simply sums its synaptic inputs, and if the inputs equal or exceed a threshold, a simulated action potential is initiated (for specific details, see Buonomano, Baxter, & Byrne, in press; Byrne, Buonomano, Corcos, Patel, & Baxter, 1988). As suggested by Hawkins and Kandel (1984), an important assumed property of the facilitatory neuron is that its output diminishes or accommodates rapidly, and that the recovery from accommodation is relatively slow (see FN Output in Figs. 2.8B and 2.10B). The presentation of the US always produces a large, superthreshold EPSP that activates the facilitatory neuron and stimulates the release of the facilitatory transmitter, which in turn activates adenylate cyclase in the sensory neuron (see Fig. 2.3). We have begun to test the computational capabilities of this neural network by attempting to simulate second-order conditioning and blocking.

Simulation of Higher Order Features of Classical Conditioning

Simulations of Second-Order Conditioning. The defining feature of second-order conditioning is that a conditioned stimulus (CS) can come to produce a conditioned response (CR) without ever being paired directly with the US (Pavlov, 1927; Rescorla, 1980, 1988). As illustrated in Fig. 2.7A, the training paradigm for second-order conditioning proceeds in two phases. During phase I, CS 1 is paired with the US, while CS 2 is unpaired with the US. During phase II, the presentation of the US is terminated and CS 2 is paired with CS 1. The critical question is whether the previously conditioned CS 1 will act as a secondary reinforcer for CS 2. If CS 1 can provide reinforcement to CS 2, then CS 2 should undergo associative enhancement during phase II of second-order conditioning. In phase I of the control paradigm (Fig. 2.7B), both CS 1 and CS 2 are unpaired with the US, while in phase II CS 1 and CS 2 are paired as above. Second-order conditioning should only occur if CS 1 has been previously paired with the US, thus during phase II of the control paradigm there should be no associative enhancement of CS 2.

Figure 2.8 illustrates a computer simulation of second-order conditioning. The amplitudes of the EPSPs at the sensory-to-facilitatory neuron synapses (see Fig. 2.6) are plotted in Part A. During training trials 1-5 (phase I in Fig. 2.7A; the intertrial interval is 5 min), the synaptic strength of the CS 1 cell (SN 1) undergoes associative enhancement and increases to approximately 300% of the baseline amplitude. Because the CS 1 cell is paired with the US during phase I, this associative enhancement represents first order conditioning (e.g., compare SN 1 in Fig. 2.8A to the CS+ cell Fig. 2.4). In contrast, the synaptic strength of the CS 2 cell (SN 2) undergoes nonassociative enhancement or sensitization during

Phase I
(Training Trials 1-5)

Phase II
(Training Trials 6-15)

A. Second-order Conditioning

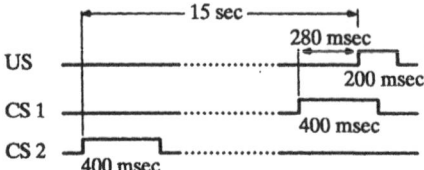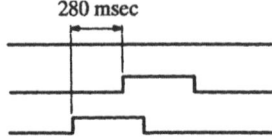

B. Control

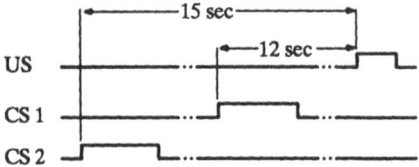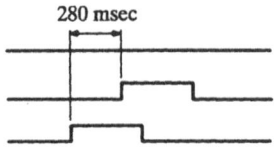

FIG. 2.7. *Second-Order Conditioning Paradigm.* Training paradigms used in the simulations of second-order conditioning and the control for second-order conditioning. **A:** Second-order conditioning. During Phase I of second-order conditioning, CS 1 (spike activity in SN 1) is temporally paired with the US (ISI of 280 msec), whereas CS 2 is presented in a unpaired fashion (ISI of 15 sec). During Phase II, CS 2 is paired with CS 1 in the absence of any US. **B:** Control. The control paradigm for second-order conditioning is similar to the experimental paradigm except that CS 1 is unpaired with the US during Phase I (ISI of 12 sec). Each CS consisted of a 400 msec train of 11 spikes at 25 Hz in the modeled sensory neurons. The US was a 200 msec, superthreshold activation of the facilitatory neuron. The inter-trial interval is 5 min. (Modified from Bounomano, Baxter & Byrne, in press.)

phase I (e.g., compare SN 2 in Fig. 2.8A to the CS- cell in Fig 2.4). During training trials 6-15 (phase II in Fig. 2.8A; the intertrial interval is 5 min), the presentation of the US is terminated and activity in the CS 2 cell is paired with activity in the CS 1 cell. During phase II, the synaptic strength of the CS 2 cell is enhanced above the level produced by nonassociative enhancement. As illustrated by the control paradigm (dotted line), if the CS 1 cell is unpaired with the US during phase I, then there is no enhancement of the CS 2 cell during phase II (see Fig. 2.7B). In other words, the prior conditioning of the CS 1 cell during phase I allows it to act as a secondary reinforcer and to induce an associative enhancement in the synaptic strength of the CS 2 cell. Thus the increase in the synaptic strength of the CS 2 cell during phase II is an example of second-order conditioning.

Details of how the synaptic outputs of the two sensory neurons and the facilitatory neuron change during the simulation of second-order conditioning are shown in Part B of Fig. 2.8. At the start of training (trial 1), the facilitatory neuron is active only during the presentation of the US. An intrinsic property of the facilitatory-to-sensory neuron synapse (see Fig. 2.6) is that the release of transmitter from the facilitatory neuron (FN Output) accommodates rapidly. Thus during the US the FN Output is seen to diminish (recovery occurs within about 30 sec, a time less than the intertrial interval). Initially, the EPSPs at the sensory-to-facilitatory neuron synapse (see Fig. 2.6) for both the CS 1 cell (EPSP 1) and the CS 2 cell (EPSP 2) are weak and do not activate the facilitatory neuron. However, because the CS 1 cell is paired with US, the amplitude of EPSP 1 undergoes significant associative enhancement, and eventually the amplitude of EPSP 1 surpasses the threshold for activation of the facilitatory neuron. As illustrated in trial 5, the facilitatory neuron is active during the presentation of the CS 1 as well as during the presentation of the US. Thus in subsequent trials, the CS 1 cell will be able to stimulate the release of facilitatory transmitter, and function as a secondary reinforcer for the CS 2 cell. This secondary reinforcement is illustrated in trial 6 (Fig. 2.8B). Although the presentation of the US has been terminated during phase II, the facilitatory neuron continues to release transmitter because of activation by the CS 1 cell. The secondary reinforcement provided by the CS 1 to the CS 2 is at an optimum ISI for activity dependent neuromodulation (see Fig. 2.5), and by trial 10, the amplitude of EPSP 2 has undergone significant associative enhancement. In this model, second-order conditioning results directly from the ability of a previously conditioned CS to 'take control' of the facilitatory neuron and thus serve as a secondary reinforcer to another CS in the absence of an US (Buonomano, Baxter, & Byrne, 1989).

Simulations of Blocking.   Blocking emphasizes the predictive value of the CS in relation to the US (Kamin, 1969). A blocking paradigm, like second-order conditioning, consists of two phases of training (Fig. 2.9). During phase I, CS 1 is paired with the US, whereas CS 2 is unpaired with the US. During phase II, CS 1 and CS 2 are activated simultaneously and this compound CS 1/CS 2 is paired with the US. If blocking occurs, then prior conditioning of CS 1 should reduce the magnitude of the associative enhancement of CS 2 during phase II. The expected or control level of associative enhancement of CS 2 is determined by activating both CS 1 and CS 2 in an unpaired fashion with the US during phase I of the control paradigm (Fig. 2.9B) and presenting the compound CS 1/CS 2 in Phase II paired with the US. Any reduction in the magnitude of the associative enhancement of CS 2 that is related to the prior conditioning of CS 1 represents blocking.

Figure 2.10 illustrates a computer simulation of blocking by the neuronal network shown in Fig. 2.6. During phase I, the synaptic strength of the CS 1 cell (SN 1) increases during first-order conditioning, although there is only a

A. Second-order Conditioning (With Low FN Threshold)

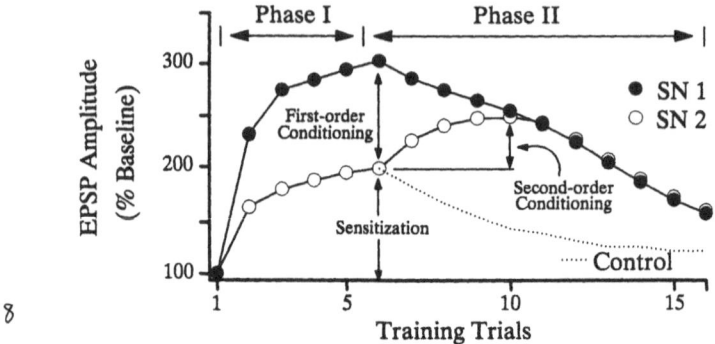

2.8

B. Phase I

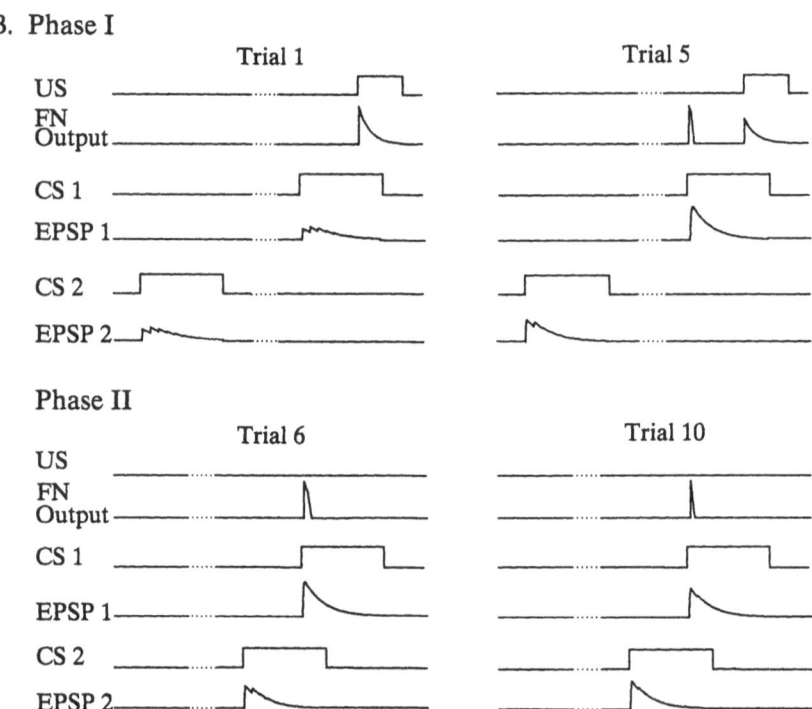

Phase II

FIG. 2.8. *Simulation of Second-Order Conditioning.* **A:** Second-order conditioning of transmitter release. The baseline values of the EPSPs produced by the CS cells were normalized to 100%. During first-order conditioning (Phase I, Trials 1–5), both the CS 1 cell (SN 1) and CS 2 cell (SN 2) show increases in strength. The increase in strength observed in the CS 2 cell is due to a nonassociative effect (sensitization). The CS 1 cell, which was paired with the US, is enhanced to a greater extent. During Phase II (trials 6–15), the EPSPs produced by CS 2 exhibited an increase in strength due to second-order conditioning. During Phase I, of the control paradigm, both CS cells exhibit a nonassociative increase in strength. Phase II of the control paradigm (dotted line) illustrates that

nonassociative enhancement in the synaptic strength of the CS 2 cell (SN 2; Part A). During trials 6–15, the compound CS 1/CS 2 is paired with the US. Although the synaptic strength of the CS 2 cell undergoes some enhancement during phase II, the control paradigm (dotted line) illustrates that the enhancement of the CS 2 cell (SN 2) would have been greater if the CS 1 cell (SN 1) had not been paired previously with the US (see Fig. 2.9). In other words, the prior conditioning of the CS 1 cell partially reduced the amount of associative enhancement seen in the CS 2 cell during phase II. This is referred to as partial blocking since prior conditioning of CS 1 partially "blocked" the ability of the CS 2 cell to gain associative strength.

Details of how the synaptic outputs of the two sensory neurons and the facilitatory neuron change during the simulation of blocking are shown in Part B of Fig. 2.10. Initially (trial 1), EPSP 1 and EPSP 2 are weak, and the facilitatory neuron is activated only by the US. The output of the facilitatory neuron (FN Output) accommodates, but recovers during the 5 min intertrial interval. During phase II (trial 6), simultaneous activity in the CS 1 and CS 2 cells activates the facilitatory neuron. There are two important consequences of this CS-mediated activity in the facilitatory neuron. First, the amplitude of FN Output during the presentation of US is reduced because of incomplete recovery from the accommodation that occurred during the CS-mediated activation of the facilitatory neuron. Second the CS 2 cell receives the majority of its facilitatory input during the compound CS 1/CS 2 at an ISI of 0 msec, rather than at an optimal ISI. As illustrated in Fig. 2.5, an ISI of 0 msec produces about 70% less associative enhancement than does the optimal ISI. In this model, the degree of blocking is intrinsically related to the ISI function and the accommodation of the output of the facilitatory neuron.

It is interesting to note that the single cell model described by Gingrich and Byrne (1985, 1987), when implemented into the circuit suggested by Hawkins and Kandel (1984), can simulate either second-order conditioning *or* partial

---

without preconditioning of CS 1 no associative plasticity is observed in the CS 2 cell. **B:** Synaptic output of network elements during second-order conditioning. The outputs of the sensory neurons and facilitatory neuron are plotted at various points during training. Initially, neither CS cell is strong enough to activate the facilitatory neuron, which is activated only by the US (Trial 1). By the end of first-order conditioning, CS 1 is able to activate the facilitatory neuron (Trial 5). The CS 1 cell can now function as a reinforcing stimulus by activating the facilitatory neuron and condition the CS 2 cell (Trial 6). Towards the end of Phase II, the associative plasticity in both cells begins to extinguish (Trial 10). Note that the duration of the CS outlasts the actual duration of the EPSPs. This is due to synaptic depression. The threshold of the facilitatory neuron was 1100. (Modified from Bounomano, Baxter, & Byrne, in press.)

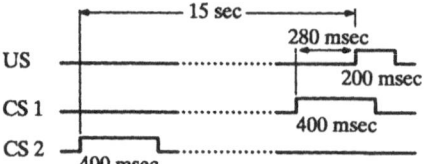

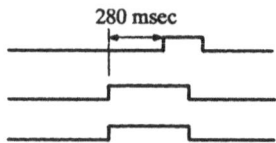

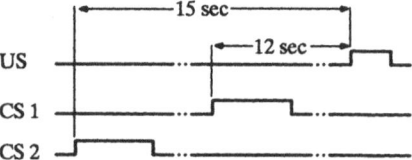

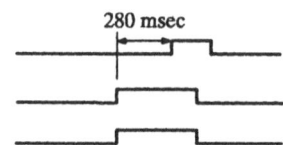

FIG. 2.9. *Blocking Paradigm.* Training paradigms used in the simulations of blocking and the control for blocking. **A:** Blocking. During Phase I of blocking, CS 1 is paired with the US (ISI of 280 msec) and CS 2 is presented in an unpaired fashion with the US (ISI of 15 sec). During Phase II, both CS 1 and CS 2 are presented simultaneously and paired with the US. **B:** Control. The control paradigm for blocking is similar to the experimental paradigm except that during Phase I CS 1 is unpaired with the US (ISI of 12 sec). The inter-trial interval is 5 min. (Modified from Bounomano, Baxter & Byrne, 1989.) (Modified from Bounomano, Baxter, & Byrne, in press.)

blocking, but not both with a single value for the threshold for the facilitatory neuron. To obtain both second-order conditioning and blocking with a three cell network, we found that there must be at least a 100% difference in strength between the CS+ and CS− cells. This can be understood intuitively, since in order to obtain reasonable blocking, the facilitatory neuron must be able to "distinguish" between the control CS−/CS− and experimental CS+/CS− compound stimuli (see Fig. 2.9). If the threshold of the facilitatory neuron is between the strengths of the CS−/CS− and CS+/CS−, it can make the distinction between a control and experimental blocking paradigm. To obtain second-order conditioning, however, the threshold of the facilitatory neuron has to be below the strength of the CS+ cell. Thus, threshold must be above CS−/CS− strength and below CS+ strength. It follows that if the connection of the CS+ cell is not at least twice as strong as that of the CS− cell, it is not possible to obtain both significant second-order conditioning *and* blocking. Thus, the original model of the sensory neuron when incorporated into a three cell network was unable to perform reasonable simulations of blocking, in part, due to the constraints imposed by the empirical data, in which the strength of a CS+ cell was not 100% greater than that of the CS− cell. The importance of factors such as the magnitude of the change in associative strength in our simulations stress the point that if the purpose of a model is to provide insights into biological information processing and to test hypotheses on the mechanisms underlying learning, it may prove essential to maintain network parameters within physiological ranges.

In its present configuration, this network (Fig. 2.6) can not simulate complete blocking (i.e., the prior conditioning of CS 1 preventing all associative enhancement in the CS 2 cell during phase II). The partial associative enhancement of CS 2 results from the ability of the CS 2 cell to undergo some associative conditioning at an ISI of 0 msec (Fig. 2.5), and the incomplete accommodation of the facilitatory neuron during the compound CS 1/CS 2. We are able to simulate complete blocking by either of two methods. We can either alter the properties of the sensory neurons so as to completely prevent associative enhancement of an ISI of 0 msec or we can introduce inhibitory neurons that prevent associative enhancement in the CS 2 cell (Buonomano, Baxter, & Byrne, in press; Byrne, Buonomano, Corcos, Patel, & Baxter, 1988).

In summary, previous work has shown that small neural networks are able to simulate classical conditioning as well as some higher order features of associative learning (Barto, Sutton, & Anderson, 1983; Gluck & Thompson, 1987; Grossberg & Levine, 1987; Hawkins, 1989a, 1989b; Klopf, 1988; Sutton & Barto, 1981). Here we have extended these observations by showing that the same holds true when the construct of the elements better reflect our understanding of real neurons. But from our simulations it is clear that in order for these networks to model second-order conditioning and blocking, certain specific constraints have to be satisfied. Ultimately, these constraints may prove to be useful for testing hypotheses that seek to account for the mechanisms underlying associative learning. For example, in order for this model of blocking to be biologically plausible, the facilitatory neuron would have to inactive within several hundred msec.

Associative Learning: Operant Conditioning

Behavioral and Cellular Analyses.   As previously described in this chapter, classical and operant conditioning can be distinguished in terms of the paradigms that govern the presentation of the reinforcing stimuli (see Mackintosh, 1974). In classical conditioning, presentation of the reinforcing stimulus (the US) is contingent on the presentation of another stimulus (the CS). This contingency is independent of the behavior of the animal. In contrast, during operant conditioning, the presentation of the reinforcement is contingent on the performance of some arbitrary behavior (the operant) by the animal. Thus during operant conditioning, the animal has the opportunity to control the delivery of reinforcement.

A. Blocking (With High FN Threshold)

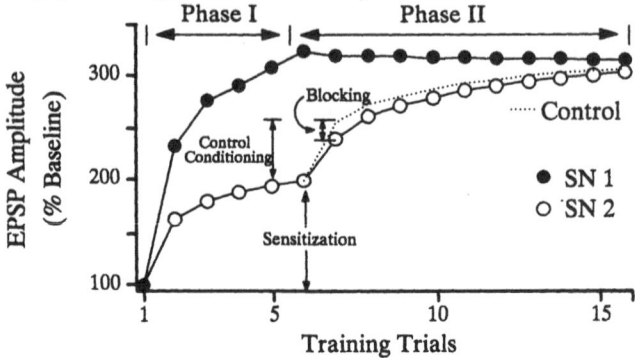

B. Phase I

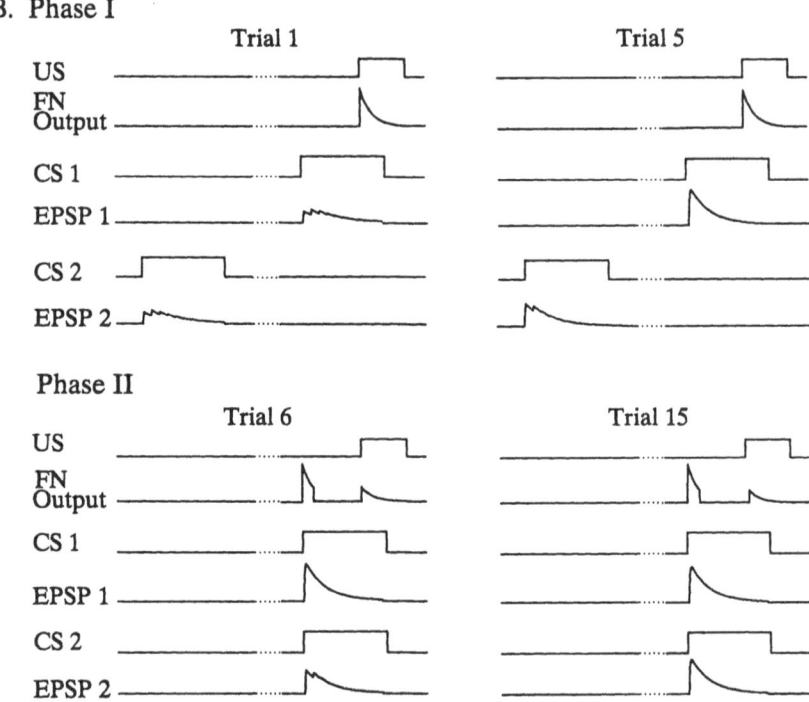

FIG. 2.10. *Simulation of Blocking.* **A**: Partial blocking of associative enhancement of transmitter release. During Phase I (Trials 1–5), the CS 1 cell exhibits first-order conditioning whereas the CS 2 cell exhibits nonassociative plasticity (sensitization). During Phase II (Trials 6–15), the CS 2 cell exhibits slightly less conditioning than the control (dotted line), representing a small degree of partial blocking. **B**: Synaptic output of network elements during blocking. The output of the sensory neurons and facilitatory neuron are shown during different training trials. Initially, neither CS cell is strong enough to activate the facilitatory neuron, which is activated only by the US (Trial 1). After first-order conditioning, the EPSP produced by CS 1 is still subthreshold for activa-

Operant conditioning has been examined in three response systems in *Aplysia*: the gill withdrawal reflex, the feeding system, and the head-waving system.

Operant Conditioning of Gill Withdrawal.   Hawkins, Clark, and Kandel (1985) have reported preliminary evidence that gill withdrawal in *Aplysia* can be operantly modified. Specifically, animals that received an electric shock to the siphon when the gill relaxed beyond a predetermined criterion level (20% full contraction) spent a significantly greater percentage of time with the gill contracted than did animals that received yoked exposure to the shock. Furthermore, in subsequent training sessions, previously conditioned animals exhibited enhanced performance relative to yoked controls.

Conditioned Feeding Behavior.   Susswein and Schwartz (1983) have reported that *Aplysia* can be conditioned to modify their feeding responses in a training protocol that shares a number of features in common with operant conditioning. Normally *Aplysia* will bite and swallow seaweed when it is presented to the mouth. When the food is placed in an inedible plastic net, the animals will initially attempt to consume the netted food, but will eventually expel the material from the buccal cavity because it cannot be swallowed. After several such exposures, the animals learn to avoid biting the netted food. If a hole is placed in the net, allowing animals to eat the netted food, subsequent response to the netted food is enhanced. Moreover, this conditioned response is specific to the taste of the food presented in the net. Finally, preliminary insights into neural pathways mediating aspects of the learning have been obtained: information carried by the esophageal nerves are required for learning to occur since bilateral transsection of these nerves abolishes the conditioned response (Schwarz & Susswein, 1986).

Operant Conditioning of Head-Waving.   Head-waving in *Aplysia* is a naturally occurring behavior, in which animals sweep their heads from side to side in order to probe their environment, for example to find food or to secure a foot-hold (Kupferman & Carew, 1974). Cook and Carew (1986) found that *Aplysia* can be operantly conditioned to avoid head-waving to one side of their body in order to

---

tion of the facilitatory neuron (Trial 5). During Trial 6, both the CS 1 and CS 2 cells are activated simultaneously and their summed output is able to activate the facilitatory neuron and induce its partial accommodation. Thus, the facilitatory neuron can respond only partially when the US is presented. Since the facilitatory neuron output in response to the actual US is decreased and the major of the FN Output is shifted to an ISI of 0 sec, less associative plasticity occurs in the CS 2 cell. The threshold of the facilitatory neuron was 1,500. (Modified from Bounomano, Baxter, & Byrne, in press.)

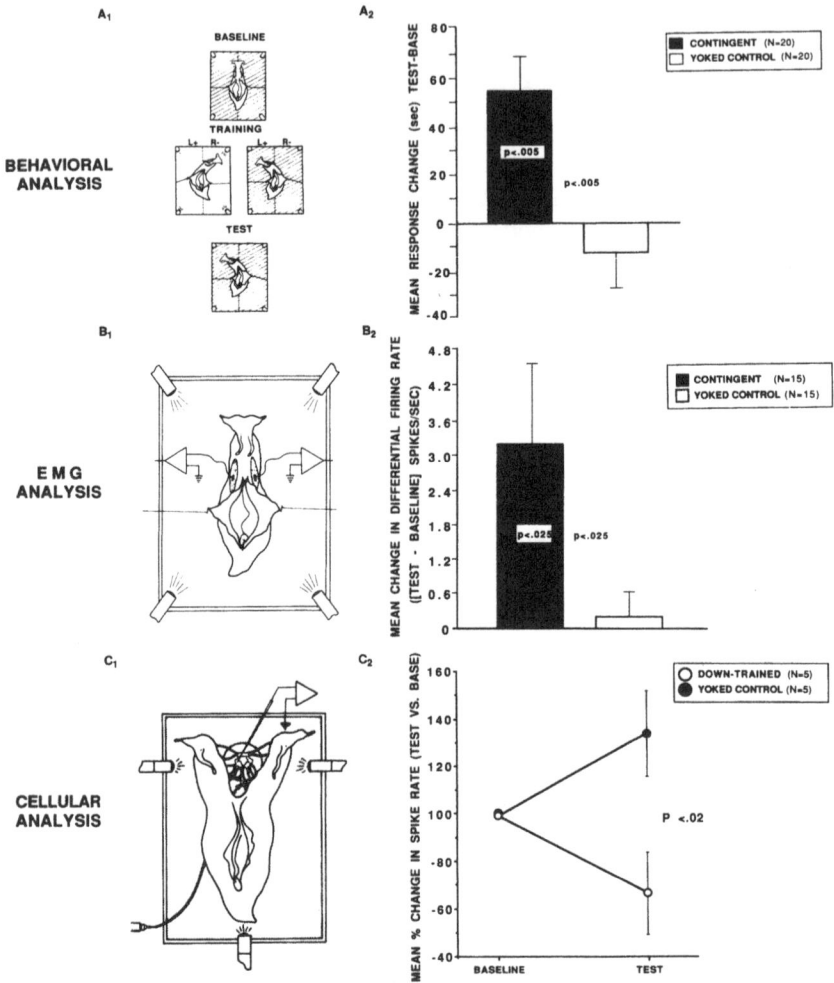

FIG. 2.11. *Operant Conditioning of Head-waving in Aplysia. Behavioral Analysis.* **A₁**: Training procedure has three phases: 5 min baseline; 10 min training where aversive bright light reinforcement occurs when the animal head-waves on the punished (−) side of the body; 5 min test carried out in the absence of reinforcement. **A₂**: Results of operant conditioning experiment. Contingent trained animals show significant learning whereas yoked controls do not. Moreover, contingent trained animals are significantly different from yoked controls.

EMG Analysis. **B₁**: Procedure for operant modification of differential EMG output of LCM muscles. Delivery of reinforcement is contingent solely on the activity of the LCMs, themselves. **B₂**: Outcome of operant training of EMG activity. LCMs of contingently trained animals exhibit significant modification of EMG output, whereas yoked controls do not. Moreover, contingent trained animals are significantly different from yoked controls.

avoid the presentation of bright, whole body illumination, which the animals find aversive.

The basic training protocol is schematically illustrated in Figure 2.11A$_1$. First an animal received a five minute baseline, in which the amount of time it spent head-waving to each side of the body was recorded. Next, a ten minute training period followed, in which head-waving to one (randomly selected) side of the body (the punished side) resulted in exposure to the bright light reinforcement; the light remained on until the animal made a head-waving response back to the safe (unpunished) side. Thus for these animals, reinforcement was contingent upon their performing the appropriate operant response. Control animals received yoked training that was identical to the contingent training described above, except that the reinforcement was not correlated with the behavior of the yoked animals. Finally, animals receiving both types of training were given a five minute test (carried out in the absence of reinforcement) during which the amount of time spent head-waving to each side of the body was recorded. Following training, the contingently trained animals exhibited operant conditioning by significantly increasing the amount of time they spent head-waving on the safe side compared to their baseline performance. In contrast, yoked controls showed no systematic change in responding after training (Fig. 2.11A$_2$).

Additional studies showed that operant conditioning of head-waving behavior is quite sensitive to the contingencies of reinforcement. For example, following training as described earlier, if the contingencies were reversed (the originally safe side now becoming the punished side and *vice versa*) animals rapidly reduced responding to the previously safe side. Moreover, if the original contingencies were subsequently restored, animals once again increased responding to the originally safe side. Finally, if animals that previously received yoked training (and thus exhibited no operant conditioning) subsequently received contingent training, they rapidly acquired the operant response (Cook & Carew, 1986).

The head-waving response in *Aplysia* is quite complex. Thus a cellular analysis of operant conditioning would be greatly facilitated if the conditioning could be expressed in a simple system that retained the essential features of the conditioning on the one hand, and was amenable to cellular analysis on the other. Towards that end Cook and Carew (1989a, 1989b) recently explored whether a restricted group of neck muscles was capable of exhibiting operant modification of their electrical activity under conditions that produce behavioral operant conditioning. In these experiments chronic, bilateral electromyographic (EMG) recordings were

---

Cellular Analysis. C_1: Intracellular recording obtained using the split foot preparation, in which the periphery remains intact during operant conditioning of individual motor neurons firing rate. C_2: Firing rate of motor neurons significantly decreased (compared with yoked cells) when increase in spontaneous firing rate were punished with bright light delivered to the preparation.

obtained from the Lateral Columellar Muscles (LCMs) in freely head-waving animals (Fig. 2.11B$_1$). Animals were trained as previously described, with one exception: reinforcement was not contingent upon the head-waving response, but was contingent solely on the EMG activity of the LCMs themselves. Specifically, during training one LCM was designated as the positive (+) LCM and the other as the negative (−) LCM. Reinforcement was delivered (lights were turned on) when (1) the +LCM decreased its firing rate; (2) the −LCM increased its firing rate; or (3) when both of these events occurred simultaneously. Following training, the differential (L-R) firing rate of the LCMs exhibited contingent modifications: there was a significant increase in the differential rate in the reinforced direction for the contingently trained animals (Fig. 2.11B$_2$), whereas yoked control animals showed no evidence of operant modification (Cook & Carew, 1989a, 1989b).

Cook and Carew (1988) have recently extended the analysis of operant conditioning of head-waving to the cellular level by examining operant modification of the activity of individual motor neurons that innervate the LCMs and other neck muscles. Intracellular recordings were obtained using the split-foot preparation (Fig. 2.11C$_1$) which leaves the entire CNS connected to the periphery. A single motor neuron per animal was trained. During contingent training, as the spontaneous firing rate of the cell increased above its average baseline rate, the preparation was delivered whole field bright light until the firing rate fell below baseline (down training). Following this contingent training the motor neurons showed a decrease in their firing rate (compared to baseline) that was significantly different from yoked control cells, which actually increased their firing rate following noncontingent (yoked) exposure of the preparation bright light (Fig. 2.11C$_2$).

Thus *Aplysia* are capable of expressing operant conditioning of head-waving at the level of intact behavior, restricted muscle groups, and at the level of individual central motor neurons. Much work remains to be done in order to elucidate the elements and connectivity of the neural circuitry that mediates operant conditioning. In parallel with behavioral and cellular studies aimed at addressing these issues, useful insights into the mechanisms of operant conditioning can be obtained by constructing computational models of neural circuits that undergo operant conditioning. Among the interesting questions that can be examined utilizing this combined approach is the theoretically important issue of the relationship between classical and operant conditioning in mechanistic terms.

Simulation of Elementary Features of Operant Conditioning

Despite operational distinctions, it is not known whether the cellular processes underlying classical conditioning and operant conditioning are fundamentally different or whether they may share at least aspects of a common underlying mechanism. Recent theoretical studies indicate that the same artificial neural networks and learning rules that can simulate features of classical conditioning can also simulate elementary forms of operant conditioning (Baxter, Raymond, Buonomano, & Byrne, 1989; Klopf & Morgan, in press; Schwarz, Markovich, & Susswein, 1988; see also Schwarz & Susswein, 1986). Therefore, we were interested in determining whether a small network containing elements with the activity dependent neuromodulation learning rule, which simulates features of classical conditioning, can also simulate features of operant conditioning.

We implemented an artificial neural network that contained three types of elements: pattern generating elements (PGs), associative elements (AEs), and motor neurons (MNs) (Fig. 2.12). The pattern generating elements generate the spontaneous behavior within the network that serves as the target of operant training. Two spontaneously active and mutually inhibitory neurons (PG_A and PG_B) comprise a central pattern generator that drives the network between two

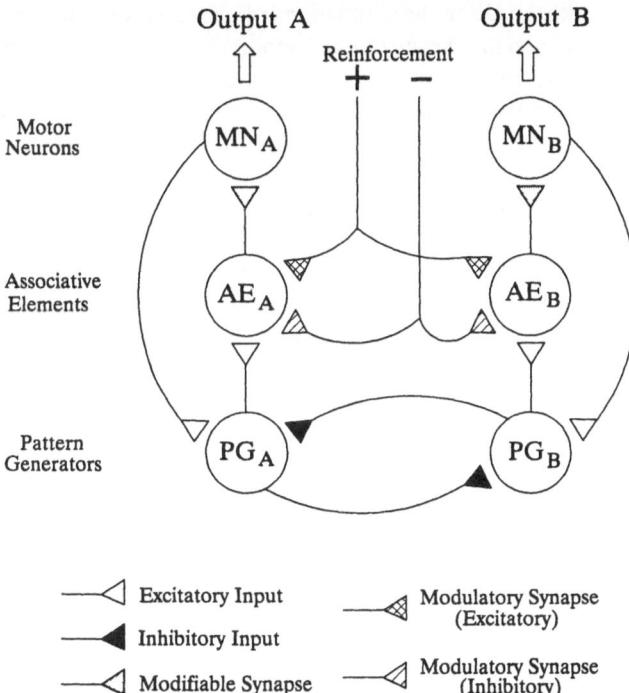

FIG. 2.12. *Network Architecture for Operant Conditioning*. The circuit is driven by a central pattern generator consisting of two pattern generator cells (PG_A and PG_B) connected by inhibitory synapses. The PG cells excite the associative elements (AE_A and AE_B) which in turn excite the motor neurons (MN_A and MN_B). Feedback connections from each MN to the corresponding PG cell tends to increase the duration of activity in that PG. The AE-to-MN synapses are the site of plasticity within the circuit. They can be modified by positive or negative reinforcement that is delivered simultaneously to both AEs.

"behavioral" or output states (Output A or Output B). Each pattern generating neuron makes an excitatory connection onto an associative element (AE). The properties of these associative elements are similar to the sensory neuron illustrated in Fig. 2.3, but they have been modified in order to reduce the transmitter depletion that would otherwise occur during the prolonged periods of activity during these simulations. These associative elements are the only elements in the network that are capable of activity dependent neuromodulation (i.e., associative enhancement of its synaptic strength). The motor elements (MN_A and MN_B) are driven by the associative elements. Activity in the motor elements serves as the measure of network behavior, and it feeds back onto the pattern generating elements. This excitatory feedback from the motor elements contributes to the maintenance of bursts of activity in the pattern generating elements. The neural pathways for reinforcement impinge on both associative elements. Positive reinforcement ($+$), like the US in the simulations of classical conditioning, facilitates the connection between the associative and motor elements. This synaptic facilitation is enhanced by prior activity in the associative element *via* the activity dependent neuromodulation learning rule. In addition, we have incorporated a negative modulator into our model ($-$) (for discussion of inhibitory modulation of sensory neurons see Belardetti & Siegelbaum, 1988). In this chapter, we focus only on the effects of positive reinforcement.

As described earlier, the basic training paradigm consists of three phases (Fig. 2.13): five minutes of baseline measurement, ten minutes of training, and five minutes of testing. During the baseline and the test measurements, no reinforcement is delivered. During training, activity in one of the motor neurons is arbitrarily chosen as the reinforced behavior and the modulator is activated whenever activity in that neuron exceeds a certain criterion level. As a control procedure, we introduced a random reinforcement procedure whereby the modulator is activated during the training period, but its activation is not specifically correlated with the behavior of the network. The results of these procedures are shown in the records of motor neuron activity on the reinforced side (Fig. 2.14). When activation of the modulator is contingent upon occurrence of activity in MN_A, two effects are observed. There is an increase in amplitude of motor neuron activity and there is an increase of the duration of each burst in the motor neuron. Thus during the 5 min test phase, there is an increase in the total time that the network spends in the reinforced behavioral state (Fig. 2.15). In contrast, when the reinforcement is presented randomly during training, there is no significant change in the total time that the network spends with MN_A active (Figs. 2.14B and 2.15B).

To understand these results, we must refer to the circuit diagram. During contingent training, activation of the modulator always coincides with activity (and therefore high Ca^{2+} levels) in AE_A. Because of the mutual inhibition of the cells of the central pattern generator, activation of the modulator will always

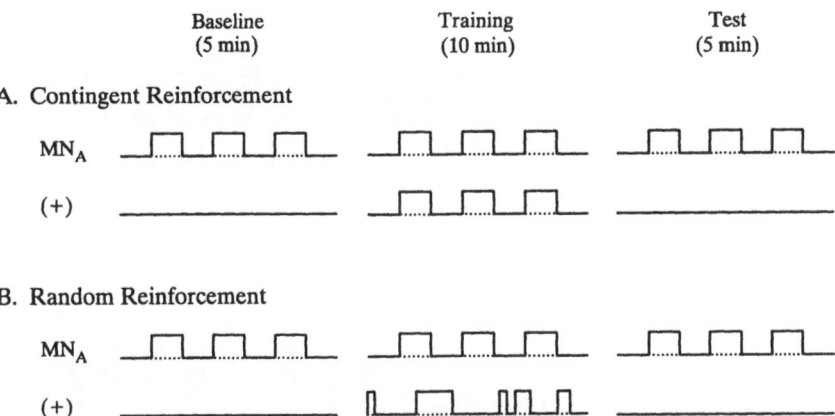

FIG. 2.13. *Operant Conditioning Paradigm.* A three phase test-train-test procedure is used, and a schematic representation of part of each phase is shown. **A:** Contingent reinforcement. During baseline and posttest, activity in MN_A (corresponding to the occurrence of one of the two possible behavioral states) is monitored during oscillation of the network in the absence of any reinforcement. During training, the modulator is activated (+) whenever activity occurs in MN_A. **B:** Random reinforcement control. The protocol is the same as the contingent protocol except that during training, the reinforcement occurs at times that are randomly determined.

occur during a time when AE_B is inactive and contains only the small amount of Ca^{2+} from its last burst that has not yet been buffered. Therefore the modulator will facilitate both AE-to-MN synapses but will enhance AE_A to MN_A to a much greater degree because of the high Ca^{2+} level in AE_A. Consequently, activity in PG_A will result in greater activity in MN_A than in the baseline condition and hence greater positive feedback to PG_A, which results in a longer burst on that side. In the nonreinforced MN, MN_B, activity and feedback to the central pattern generator will not change significantly.

During random reinforcement (Fig. 2.13B, 2.14B, 2.15B), the modulator is sometimes activated during activity in MN_A and is sometimes activated during activity in MN_B. Therefore both AE-to-MN connections will be facilitated to a comparable degree, the activities of both motor elements will increase, feedback to both neurons of the central pattern generator will be increased, and thus bursts in both neurons of central pattern generator will be increased. However, since on average both sides are equally facilitated, there will not no change in the relative amount of time the network spends in behavioral state A.

One feature of operant conditioning is that it extinguishes or decays if the reinforcement of the behavior is discontinued. Accordingly, we find that when we interpose 10 minutes of extinction training (modulator off) between training

A. Contingent Reinforcement

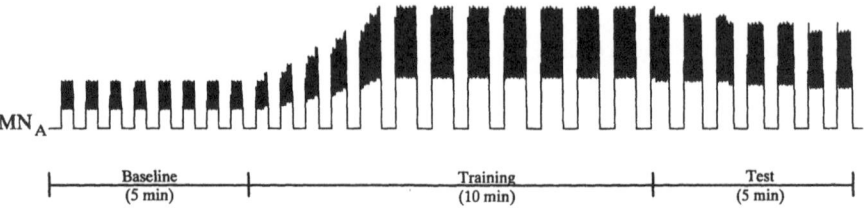

B. Random Reinforcement

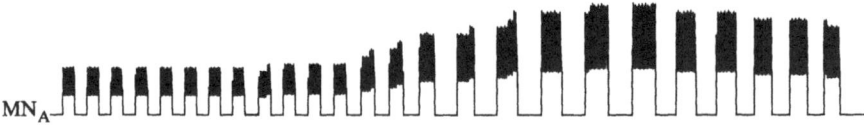

FIG. 2.14. *Network Output During Simulation of Operant Conditioning.* The magnitude and duration of the activity in MN$_A$ is illustrated during the baseline, training and test phases. This activity in MN$_A$ feeds back onto PG$_A$ and contributes to the regulation of the central pattern generator. As illustrated in Fig.2.12, the network generates behavior A whenever MN$_A$ is active. When reinforcement is contingent on activity in MN$_A$, there is an increase in the amplitude of activity in MN$_A$ as well as an increase in the duration of each burst of activity in MN$_A$. In contrast, when reinforcement is delivered randomly, the silent periods increase to the same extent as the active periods. Figure 2. 15 illustrates that contingent reinforcement resulted in an increase in the amount of time that network spent producing behavior A following operant conditioning.

and testing, we do indeed observe decay of the operant response toward baseline level (Fig. 2.16A, 2.17A, 2.18). This extinction results from the gradual decay of the level of cAMP in the AE. Furthermore, if instead of merely removing reinforcement, we reverse the contingency, and now activate the modulator specifically when the previously reinforced behavior, behavior A, is not occurring, we can accelerate the extinction and even decrease the operant behavior to below baseline level (Fig. 2.16B, 2.17B, 2.18).

To summarize, our preliminary simulations using a network containing associative elements with a learning rule derived from empirically observed activity dependent neuromodulation can simulate behavioral data on operant conditioning. Specifically, we are able to demonstrate a change in the time spent performing a reinforced behavior, and this change is dependent upon a contingency between the occurrence of that behavior and the occurrence of reinforcement. Our network also exhibits extinction and reversal learning, two important features of operant conditioning. In future studies, we hope to simulate other

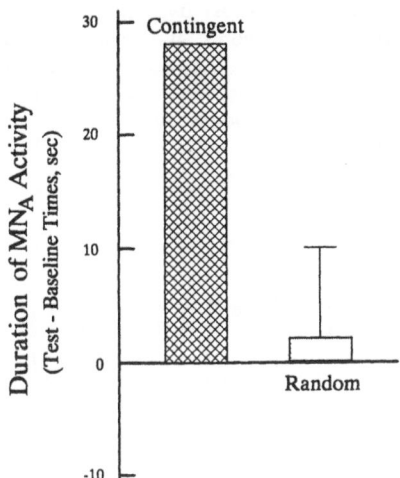

FIG. 2.15. *Simulation of Operant Conditioning.* The degree of operant conditioning is measured as an increase in the total amount of time that the network spends producing the reinforced behavior during the 5 min test phase as compared to the 5 min baseline phase (i.e., test-baseline total times). Contingent reinforcement resulted in an increase in the amount of time that the network spent producing output state A (i.e., activity in MN_A); whereas, random reinforcement did not result in any change in the behavior of the network. The error bar in the random condition is the standard deviation (N=3) that results from using different random patterns of activation of the modulator. These results are similar to those obtained in behavioral experiments (Cook & Carew, 1986; Fig. 2. $11A_1$). Thus, a network that contains associative elements derived from the activity dependent neuromodulation learning rule can simulate features of operant conditioning.

features of operant conditioning, including the effects of various reinforcement schedules on learning.

CONCLUDING REMARKS

In conclusion, our results indicate that the empirically derived model for associative plasticity of a single cell, when incorporated into a appropriate circuit, can simulate higher order features of classical conditioning such as second-order conditioning and blocking as well as the features of another example of associative learning, operant conditioning. Thus, in principle, a simple associative learning rule can be used as a 'building block' to construct more complex forms of learning.

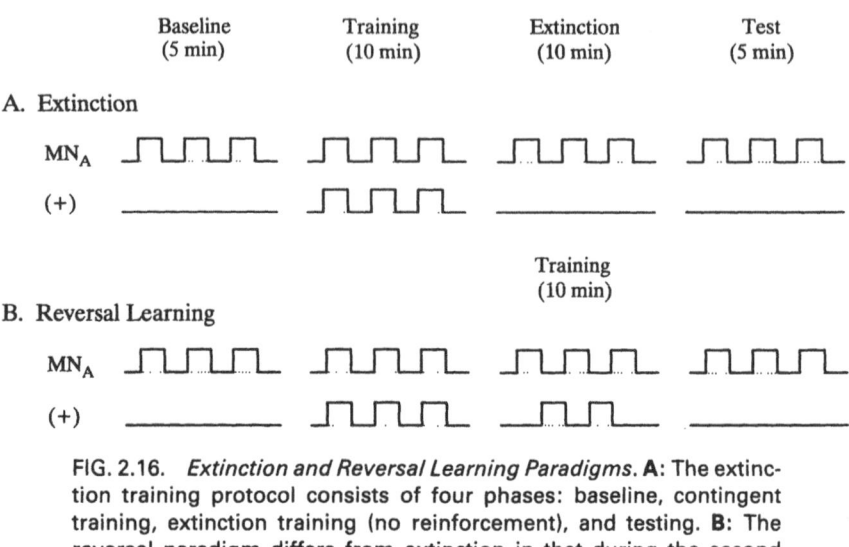

	Baseline (5 min)	Training (10 min)	Extinction (10 min)	Test (5 min)

A. Extinction

B. Reversal Learning

FIG. 2.16. *Extinction and Reversal Learning Paradigms.* **A**: The extinction training protocol consists of four phases: baseline, contingent training, extinction training (no reinforcement), and testing. **B**: The reversal paradigm differs from extinction in that during the second training phase, reinforcement is delivered specifically in response to a lack of activity in MN$_A$; i.e., reserved contingency.

A. Extinction

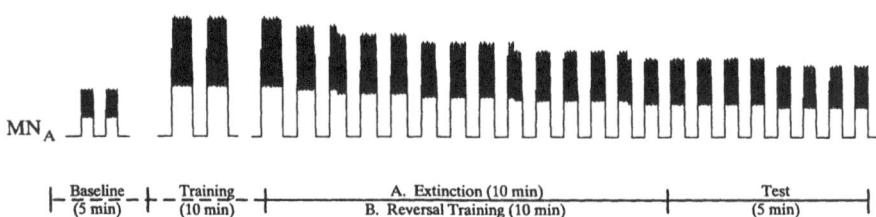

B. Reversal Learning

FIG. 2.17. *Network Output During Simulation of Reversal Learning.* The magnitude and duration of the activity in MN$_A$ is illustrated. Note that the baseline and first training phases are not shown in full; however, these are identical to the baseline and training phases shown in Fig. 2.14. Contingent reinforcement during the first training phase resulted in increases in both the magnitude and duration of activity in MN$_A$ (i.e., increases in the generation of behavior A). During extinction training, the increases in the activity of MN$_A$ are seen to slowly decrease. During reversal training, the increases in the activity in MN$_A$ are seen to decrease more rapidly and to a greater extent. These results are summarized in Fig. 2.18.

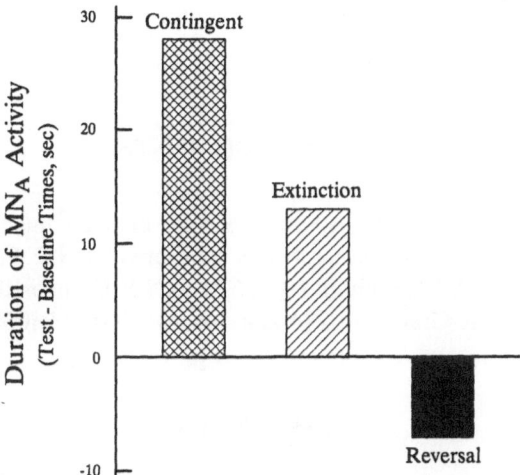

FIG. 2.18. *Simulation of Extinction and Reversal Learning.* The effects of extinction and reversal learning are determined by comparing the total amount that the network spends producing behavior A during the test and baseline phases (i.e., test-baseline times). Following the 10 min of extinction training, the occurrence of the operantly conditioned behavior, is reduced, but has not completely returned to baseline. Thus there is some retention of the operant conditioning during the test phase. Reversed contingency training, however, causes an even greater decrease; reducing the occurrence of behavior A to below baseline level. These results are similar to those obtained experiments (Cook & Carew, 1986).

It should be emphasized that we do not know the properties of the interneurons that may be involved in second-order conditioning and blocking nor do we know whether the basic mechanism for operant conditioning in *Aplysia* is the same as that for classical conditioning. For the near future, major experimental questions to be answered include to what extent are mechanisms for classical conditioning common both within any one animal and between different species? Although there are many common features that are emerging, there seem to be some differences. Thus it will be important to understand the extent to which specific mechanisms are used selectively for one type of conditioning and not another. A related question is what are the mechanistic relationships between different forms of simple learning, such as sensitization, classical conditioning and operant conditioning?

More ambitious questions include the analysis of more complex phenomena such as learning to recognize a face or telephone number and whether such learning involves processes and mechanisms fundamentally different from those underlying simpler forms of learning, such classical conditioning. It may be that the major distinction between simple forms of learning from more complex forms is not the particulars of the learning rules, but rather the particulars of the circuits into which the learning rule is imbedded.

ACKNOWLEDGMENTS

We thank Drs. D. Levine and A. Susswein for their comments on an earlier draft of the manuscript. This research was sponsored by NSF Fellowship RCD-8851871 to J. L. R., AFOSR Grant 89-0362 and NIH Grant 5-R01-MN41083 to T. J. C., and AFOSR Grant 87-0274 and Award 5-K02-MH00649 to J. H. B.

REFERENCES

Abrams, T. W. (1985). Activity dependent presynaptic facilitation: an associative mechanism in *Aplysia. Cellular and Molecular Neurobiology, 5*, 123–145.

Abrams, T. W., Bernier, L., Hawkins, R. D., & Kandel, E. R. (1984). Possible roles of Ca^{++} and cAMP in activity dependent facilitation, a mechanism for associative learning in *Aplysia. Society for Neuroscience Abstracts, 10*, 269.

Abrams, T. W., Elliot, L., Dudai, Y. & Kandel, E. R. (1985). Activation of adenylate cyclase in *Aplysia* neural tissue by Ca^{2+}/calmodulin, a candidate for an associative mechanism during conditioning. *Society for Neuroscience Abstracts, 11*, 797.

Abrams, T. W. & Kandel, E. R. (1988). Is contiguity detection in classical conditioning a system or a cellular property? Learning in *Aplysia* suggest a possible molecular site. *Trends in Neuroscience, 11*, 128–136.

Alkon, D. L., & Woody, C. D. (1986). *Neural mechanisms of conditioning.* New York: Plenum Press.

Anderson, J. A. & Rosenfeld, E. (Eds.) (1988). *Neurocomputing: Foundations of research.* Cambridge, MA: MIT Press.

Aparicio IV, M. (1988). Neural computations for true pavlovian conditioning: control of horizontal propagation by conditioned and unconditioned reflexes. Unpublished PhD Dissertation, University of South Florida.

Bailey, C. H., & Chen, M. (1983). Morphological basis of long term habituation and sensitization in *Aplysia. Science, 220*, 91–93.

Bailey, C. H., & Chen, M. (1988a). Morphological basis of short-term habituation in *Aplysia. Journal of Neuroscience, 8*, 2452–2459.

Bailey, C. H., & Chen, M. (1988b). Long term memory in *Aplysia* modulates the total number of varicosities of single identifies sensory neurons. *Proceedings of the National Academy of Science USA, 85*, 2373–2377.

Barto, A. G., Sutton, R. S., & Anderson, C. W. (1983). Neuronlike adaptive elements that can solve difficult learning control problems. *IEEE Trans System Management and Cybernetics SMC, 13*, 834–846.

Baxter, D. A., & Byrne, J. H. (1987). Modulation of membrane currents and excitability by serotonin and cAMP in pleural sensory neurons of *Aplysia. Society for Neuroscience Abstracts, 13*, 1440.

Baxter, D. A., & Byrne, J. H. (1989). Serotonergic modulation of two potassium currents in the pleural sensory neurons of *Aplysia. Journal of Neurophysiology, 63*, 665–679.

Baxter, D. A., Raymond, J. L., Buonomano, D. V., & Byrne, J. H. (1989). Operant conditioning can be simulated by small networks of neuron-like elements. *Society for Neuroscience Abstracts, 15,* 1263.

Bear, M. F., Cooper, L. N., & Ebner, F. F. (1987). A physiological basis for theory of synaptic modification. *Science, 237,* 42–48.

Belardetti, F., & Siegelbaum, S. A. (1988). Up- and down-modulation of single K^+ channel function by distinct second messenger. *Trends in Neuroscience, 11,* 232–238.

Bernier, L., Castellucci, V. F., Kandel, E. R., Schwartz, J. H. (1982). Facilitatory transmitter causes a selected and prolonged increase in adenosine 3':5'-monophosphate in sensory neurons mediating the gill and siphon withdrawal relfex in *Aplysia. Journal of Neuroscience., 2,* 1682–1691.

Braha, O., Dale, N., Hochner, B., Klein, M., Abrams, T. W. & Kandel, E. R. (1990). Second messengers involved in the two processes of presynaptic facilitation that contribute to sensitization and dishabituation in *Aplysia* sensory neurons. *Proceedings of the National Academy of Science USA, 87,* 2040–2044.

Buonomano, D., Baxter, D. & Byrne J. H. (1990). Small networks in empirically derived adaptive elements simulate some higher order features of classical conditioning. *Neural Networks,* in press.

Buonomano, D. V. & Byrne, J. H. (1990). Long-term synaptic changes produced by a cellular analogue of classical conditioning in *Aplysia. Science,* in press.

Byrne, J. H. (1982). Analysis of the synaptic depression contributing to habituation of gill-withdrawal reflex in *Aplysia. Journal of Neurophysiology, 48,* 431–438.

Byrne, J. H. (1985). Neural and molecular mechanisms underlying information storage in *Aplysia*: Implications for learning and memory. *Trends in Neuroscience, 8,* 478–482.

Byrne, J. H. (1987). Cellular analysis of associative learning. *Physiological Review, 67,* 329–439.

Byrne, J. H. (1990). Learning and memory in *Aplysia* and other invertebrates. In R. P. Kesner & D. S. Olton (Eds.), *Neurobiology of comparative cognition.* Hillsdale, NJ: Lawrence Erlbaum Associates.

Byrne, J. H., Buonomano, D., Corcos, I., Patel, S., & Baxter, D. (1988). Small networks of adaptive elements that reflect the properties of neurons in *Aplysia* exhibit higher order features of classical conditioning. *Society for Neuroscience Abstracts, 14,* 840.

Byrne, J. H., Castellucci, V., & Kandel, E. R. (1978). Contribution of individual mechanoreceptor sensory neurons to defensive gill-withdrawal reflex in *Aplysia. Journal of Neurophysiology, 41,* 418–431.

Byrne, J. H., & Gingrich, K. J. (1989). Mathematical model of cellular and molecular processes contributing to associative and nonassociative learning in *Aplysia*. In J. H. Byrne & W. Berry (Eds.), *Neural models of plasticity: Experimental and theoretical approaches* (pp. 58–73). New York: Academic Press.

Byrne, J. H., Gingrich, K. J., & Baxter, D. A. (1989). Computational capabilities of single neurons: relationship to simple forms of associative and nonassociative learning in Aplysia. In R. D. Hawkins & G. H. Bower (Eds.), *Computational models of learning: Vol. 23: Psychology of Learning and Motivation* pp. 31–63). New York: Academic Press.

Carew, T. J. (1987). Cellular and molecular advances in the study of learning in Aplysia. In J.-P. Changeux & M. Konshi (Eds.), *The neural and molecular cellular bases of learning: Dahlem Konferenzen* (pp. 177–204). Chicester: Wiley.

Carew, T. J., Castellucci, V., & Kandel, E. R. (1971). An analysis of dishabituation and sensitization of the gill-withdrawal reflex in *Aplysia. International Journal of Neuroscience, 2,* 79–98.

Carew, T. J., Hawkins, R. D., & Kandel, E. R. (1983). Differential classical conditioning of a defensive withdrawal reflex in *Aplysia californica. Science, 219,* 397–400.

Carew, T. J., Walters, E. T., & Kandel, E. R. (1981). Classical conditioning in a simple withdrawal reflex in *Aplysia californica. Journal of Neuroscience, 1,* 1426–1437.

Carew, T. J., & Sahley, C. L. (1986). Invertebrate learning and memory: from behavior to molecules. *Annual Review of Neuroscience, 9,* 435.

Castellucci, V. F. & Kandel, E. R. (1974). A quantal analysis of the synaptic depression underlying habituation of the gill-withdrawal reflex in *Aplysia*. *Proceedings of the National Academy of Science USA, 71*, 5004–5008.

Castellucci, V. F., & Kandel, E. R. (1976). Presynaptic facilitation as a mechanism for behavioral sensitization in *Aplysia. Science, 194*, 1176–1178.

Castellucci, V. F., Pinsker, H., Kupfermann, I., & Kandel, E. R. (1970). Neuronal mechanisms of habituation and dishabituation of the gill-withdrawal reflex in *Aplysia. Science, 167*, 1745–1748.

Colwill, R. M., Absher, R. A., & Roberts, M. L. (1988a). Context-US learning in *Aplysia californica. Journal of Neuroscience, 12*, 4434–4439.

Colwill, R. M., Absher, R. A., & Roberts, M. L. (1988b). Conditional discrimination learning in *Aplysia californica. Journal of Neuroscience, 12*, 4440–4444.

Cook, D. G., & Carew, T. J. (1986). Operant conditioning of head waving in *Aplysia. Proceedings of the National Academy of Science USA, 83*, 1120–1124.

Cook, D. G. & Carew, T. J. (1988). Operant conditioning of identical neck muscles and individual motor neurons in *Aplysia. Society for Neuroscience Abstracts, 14*, 607.

Cook, D. G., & Carew, T. J. (1989a). Operant conditioning of head-waving in *Aplysia* I: identified muscles involved in the operant response. *Journal of Neuroscience, 9*, 3097–3106.

Cook, D. G., & Carew, T. J. (1989b). Operant conditioning of head-waving in *Aplysia* II: Contingent modification of electromyographic activity in identified muscles. *Journal of Neuroscience, 9*, 3107–3114.

Cook, D. G., & Carew, T. J. (1989c). Operant conditioning of head-waving in *Aplysia* III: cellular analysis of possible reinforcement pathways. *Journal of Neuroscience, 9*, 3115–3122.

Crow, T. (1988). Cellular and molecular analysis of associative learning and memory in *Hermissenda. Trends in Neuroscience, 11*, 136–142.

Dobbins, A., Zucker, S. W., & Cynader, M. S. (1987). Endstopped neurons in the visual cortex as a substrate for calculating curvatures. *Nature, 329*, 438–442.

Dudai, Y. (1987). The cAMP cascade in the nervous system: molecular sites of action and possible relevance to neuronal plasticity. *CRC Critical Review Biochemistry, 22*, 221–281.

Farley, J., & Alkon, D. L. (1985). Cellular mechanisms of learning, memory and information storage. *Annual Review of Psychology, 36*, 419–494.

Frey, P. W., & Ross, L. E. (1968). Classical conditioning of the rabbit eyelid response as a function of interstimulus interval. *Journal of Comparative and Physiological Psychology, 65*, 246–250.

Fukushima, K., Miyake, S. and Ito, T. (1983). Neocognitron: a neural network model for a mechanism of visual pattern recognition. *IEEE Transactions of System, Man and Cybernetics, 13*, 826–834.

Gingrich, K. J., Baxter, D. A., & Byrne, J. H. (1988). Mathematical model of cellular mechanisms contributing to presynaptic facilitation. *Brain Research Bulletin, 21*, 513–520.

Gingrich, K. J., & Byrne, J. H. (1985). Simulation of synpatic depression, post-tetanic potentiation, and presynaptic facilitation of synaptic potentials from sensory neurons mediating gill-withdrawal reflex in *Aplysia. Journal of Neurophysiology, 53*, 652–669.

Gingrich, K. J., & Byrne, J. H. (1987). Single-cell neuronal model for associative learning. *Journal of Neurophysiology, 57*, 1705–1715.

Gluck, M. A., & Thompson, R. F. (1987). Modeling the neural substrates of associative learning and memory: a computational approach. *Psychological Review, 94*, 176–191.

Grossberg, S. (1971). On the dynamics of operant conditioning. *Journal of Theoretical Biology, 33*, 225–255.

Grossberg, S., & Levine, D. S. (1987). Neural dynamics of attentionally modulated Pavlovian conditioning: blocking, interstimulus interval, and secondary reinforcement. *Applied Optics, 26*, 5015–5030.

Gustafsson, B., & Wigström, H. (1988). Physiological mechanisms underlying long term potentiation. *Trends in Neuroscience, 11*, 156–162.

Hawkins, R. D. (1983). Cellular neurophysiological studies of learning. In J. A. Deutsch (Ed.), *The physiological basis of memory* (pp. 71–120). New York: Academic Press.

Hawkins, R. D. (1984). A cellular mechanism of classical conditioning in *Aplysia*. *Journal of Experimental Biology, 112,* 113–128.

Hawkins, R. D. (1989a). A simple circuit model for higher order features of classical conditioning in *Aplysia*. In J. H. Byrne & W. Berry (Eds.), *Neural models of plasticity: Experimental and theoretical approaches* (pp. 74–93). New York: Academic Press.

Hawkins, R. D. (1989b). A biologically based computational model for several simple forms of learning. In R. D. Hawkins & G. H. Bower (Eds.), *Computational models of learning: Vol. 23: Psychology of Learning and Motivation* (pp. 65–108). New York: Academic Press.

Hawkins, R. D., Abrams, T. W., Carew, T. J., & Kandel, E. R. (1983). A cellular mechanism of classical conditioning in *Aplysia:* Activity-dependent amplification of presynaptic facilitation. *Science, 219,* 400–405.

Hawkins, R. D., Carew, T. J., & Kandel, E. R. (1986). Effects of interstimulus interval and contingency on classical conditioning of the *Aplysia* siphon withdrawal reflex. *Journal of Neuroscience, 6,* 1695–1701.

Hawkins, R. D., Clark, G. A., & Kandel, E. R. (1985). Operant conditioning and differential classical conditioning of gill withdrawal in *Aplysia*. *Society for Neuroscience Abstracts, 11,* 796.

Hawkins, R. D., Clark, G. A., & Kandel, E. R. (1986). Cell biological studies of learning in simple vertebrate and invertebrate systems. In F. Plum (Ed.), *Handbook of physiology, section 1. The nervous system, Vol. 6: Higher functions of the nervous system* (pp. 25–83). Bethesda, MD: American Physiological Society.

Hawkins, R. D., & Kandel, E. R. (1984). Is there a cell biological alphabet for simple forms of learning? *Psychological Review, 91,* 375–391.

Hochner, B., Klein, M. Schacher, S., & Kandel, E. R. (1986). Additional component in the cellular mechanisms of presynaptic facilitation contributing to behavioral dishabituation in *Aplysia*. *Proceedings of the National Academy of Science USA, 83,* 8794–8798.

Hopfield, J. J. (1982). Neural networks and physical systems with emergent collect computational abilities. *Proceedings of the National Academy of Science USA, 79,* 2554–2558.

Hopfield, J. J. (1984). Neurons with graded responses have collective computational properties like those of two state neurons. *Proceedings of the National Academy of Science USA, 81,* 3088–3092.

Hopfield, J. J., & Tank, D. W. (1985). Neural computation of decisions in optimization problems. *Biological Cybernetcis, 52,* 141–152.

Hopfield, J. J., & Tank, D. W. (1986). Computing with neural circuits: a model. *Science, 233,* 625–633.

Hull, C. L. (1943). *Principles of behavior*. New York: Appleton-Century-Crofts.

Hull, C. L. (1952). *A behavior system*. New Haven, CT: Yale University Press.

Kamin, L. J. (1969). Predictability, surprise, attention and conditioning. In R. Church & B. A. Campbell (Eds.), *Punishment and aversive behavior* (pp. 279–296). New York: Appleton-Century-Crofts.

Kandel, E. R., Abrams, T., Bernier, L., Carew, T. J., Hawkins, R. D., & Schwartz, J. H. (1983). Classical conditioning and sensitization share aspects of the same molecular cascade in *Aplysia*. *Cold Spring Harbor Symposia of Quantitative Biology, 48,* 821–830.

Kandel, E. R., & Schwartz, J. H. (1982). Molecular biology of learning: modulation of transmitter release. *Science, 218,* 433–443.

Kelso, S., Ganong, A., & Brown, T. (1986). Hebbian synapses in hippocampus. *Proceedings of the National Academy of Science USA, 83,* 5326–5330.

Klein, M., Camardo, J. & Kandel, E. R. (1982). Serotonin modulates a specific potassium current in the sensory neurons that show presynaptic facilitation in *Aplysia*. *Proceedings of the National Academy of Sciences USA, 79,* 5713–5717.

Klein, M., Shapiro, E., & Kandel, E. R. (1980). Synaptic plasticity and the modulation of the Ca^{2+} current. *Journal of Experimental Biology, 89,* 117–157.

Klopf, A. H. (1988). A neuronal model of classical conditioning. *Psychobiology, 16*, 85–125.

Klopf, A. H. (1989). Classical conditioning phenomena predicted by a drive reinforcement model of neuronal function. In J. H. Byrne & W. O. Berry (Eds.), *Neural model of plasticity: theoretical and empirial approaches* (pp. 104–132). New York: Academic Press.

Klopf, A. H., & Morgan, J. S. (1990). The role of time in natural intelligence: implication of classical and instrumental conditioning for neuronal and neural networking modeling. In J. W. Moore & M. Gabriel (Eds.), *Learning and computational neuroscience* (in press). Cambridge, MA: MIT Press.

Koch, C., & Segev, I. (Eds.) (1989). *Methods in neuronal modeling*. Cambridge, MA: MIT Press.

Kupferman, I. V., & Carew, T. J. (1974). Behavioral patterns of *Aplysia californica* in its natural environment. *Behavioral Biology, 12*, 317–337.

Kupfermann, I. V., Castellucci, V., Pinsker, H., & Kandel, E. R. (1970). Neuronal correlates of habituation and dishabituation of the gill-withdrawal reflex in *Aplysia. Science, 167*, 1743–1745.

Lehky, S. R., & Sejnowski, T. J. (1988). Network model of shape-form shading: neural function arises from both receptive and projective fields. *Nature, 333*, 452–454.

Levy, M., & Susswein, A. J. (1990). Learned changes in rate of respiratory pumping in *Aplysia fasciata. Behavioral and Neural Biology*, in press.

Linsker, R. (1986). From basic network principles to neural architecture: emergence of orientation columns. *Proceedings of the National Academy of Sciences USA, 83*, 8779–8783.

Lippmann, R. P. (1989). Review of neural networks for speech recognition. *Neural Computation, 1*, 1–39.

Lukowiak, K. (1986). *In vitro* classical conditioning of a gill-withdrawal reflex in *Aplysia:* neural correlates and possible neural mechanism. *Journal of Neurobiology, 17*, 83–101.

Lynch, G., McGaugh, J. L., & Weinberger, N. M. (1984). *Neurobiology of learning and memory*. New York: Guilford Press.

Mackintosh, N. J. (1974). *The psychology of animal learning*. New York: Academic Press.

McClelland, J. L., Rumelhart, D.E., & The PDP Research Group (1986). *Parallel distributed processing. Explorations in the microstructure of cognition. Volume II: psychology and biological models*. Cambridge, MA: MIT Press.

Mpitsos, G. J., & Lukowiak, K. (1986). Learning in gastropod molluscs. In A. O. D. Willows (Ed.), *The mullusca volume 8, neurobiology and behavior* (pp. 96–267). New York: Academic Press.

Nicoll, R. A., Kauer, J. A., & Malenka, R. C. (1988). The current excitement in long term potentiation. *Neuron, 1*, 97–103.

Ocorr, K. A., & Byrne, J. H. (1985). Membrane responses and changes in cAMP levels in *Aplysia* sensory neurons produced by serotonin, tryptamine, FMRFamide and small cardioactive peptide B (SCP_B). *Neuroscience Letters, 55*, 113–118.

Ocorr, K. A., Tabata, M., & Byrne, J. H. (1986). Stimuli that produce sensitization lead to elevation of cyclic AMP levels in tail sensory neurons of *Aplysia. Brain Research, 371*, 190–192.

Ocorr, K. A., Walters, E. T., & Byrne, J. H. (1985). Associative conditioning analog selectively increases cAMP levels of tail sensory neurons in *Aplysia. Proceedings of the National Academy of Science USA, 82*, 2548–2552.

Pavlov, I. P. (1927). *Conditioned reflexes*. London: Oxford University Press.

Pearson, J. C., Finkel, L. H., & Edelman, G. M. (1987). Plasticity in the organization of adult cerebral cortical maps: a computer simulation based on neuronal group selection. *Journal of Neuroscience, 7*, 4209–4223.

Pinsker, H., Kupfermann, I., Castellucci, V., & Kandel, E. R. (1970). Habituation and dishabituation of the gill-withdrawal reflex in *Aplysia. Science, 167*, 1740–1742.

Pollock, J. P., Bernier, L., & Camardo, J. S. (1985). Serotonin and cyclic adenosine 3':5'-monophospate modulate the potassium current tail sensory neurons in pleural ganglion of *Aplysia. Journal of Neuroscience, 5*, 1862–1871.

Quinn, W. G. (1984). Work in invertebrates on the mechanisms underlying learning. In P. Marler & H. S. Terrace (Eds.), *The biology of learning* (pp. 197–246). New York: Springer-Verlag.

Rankin, C. H., & Carew T. J. (1988). Dishabituation and sensitization emerge as separate process during development in *Aplysia. Journal of Neuroscience, 8,* 197–211.

Rescorla, R. A. (1980). *Pavlovian second-order conditioning: Studies in associative learning.* Hillsdale, NJ: Lawrence Erlbaum Associates.

Rescorla, R. A. (1988). Behavioral studies of pavlovian conditioning. *Annual Review of Neuroscience, 11,* 329–352.

Rescorla, R. A., & Wagner, A. R. (1972). A theory of Pavlovian conditioning: variations in the effectiveness of reinforcement and nonreinforcement. In A. H. Black & W. F. Prokasy (Eds.), *Classical conditioning II. Current research and theory* (pp. 64–99). New York: Appleton-Century-Crofts.

Rumelhart, D. E., McClelland, J. L., & the PDP Research Group. (1986). *Parallel distributed processing. Explorations in the microstructure of cognition. Volume I: foundations.* Cambridge, MA: MIT Press.

Scholz, K. P., & Byrne, J. H. (1987). Long term sensitization in *Aplysia:* biophysical correlates in tail sensory neurons. *Science, 240,* 1664–1666.

Schwartz, J. H., Bernier, L., Castellucci, V. F., Polazzola, M., Stapleton, A., & Kandel, E. R. (1983). What molecular steps determine the time course of the memory for short term sensitization in *Aplysia: Cold Spring Harbor Symposia of Quantitative Biology, 48,* 811–819.

Schwarz, M., Markovich, S., & Susswein, A. J. (1988). Parametric features of inhibition of feeding in *Aplysia* by associative learning, satiation, and sustained lip stimulation. *Behavioral Neuroscience, 102,* 124–133.

Schwarz, M., & Susswein, A. J. (1986). Identification of the neural pathway for reinforcement of feeding when *Aplysia* learn that food is inedible. *Journal of Neuroscience, 6,* 1528–1536.

Siegelbaum, S. A., Camardo, J. S., & Kandel, E. R. (1982). Serotonin and cyclic AMP close single K$^+$ channels in *Aplysia* sensory neurons. *Nature, 299,* 413–417.

Sejnowski, T. J., Koch, C., & Churchland, P. S. (1988). Computational neuroscience. *Science, 241,* 1299–1306.

Sejnowski, T. J., & Rosenberg, C. R. (1986). NETtalk: A parallel network that learns to read aloud. In *The John Hopkins University Electrical Engineering and Computer Science Technical Report, JHU/EECS-86/01,* 32–41.

Shuster, M. J., Camardo, J. S., Siegelbaum, S. A., & Kandel, E. R. (1985). Cyclic AMP-dependent protein kinase closes serotonin-sensitive K$^+$ channels of *Aplysia* sensory neurons in cell free membrane patches. *Nature, 313,* 392–395.

Skinner, B. F. (1938). *The behavior of organisms.* New York: Appleton-Century.

Smith, M. C., Coleman, S. R., & Gormezano, I. (1969). Classical conditioning of the rabbit's nictitating membrane response at backward, simultaneous, and forward CS-US intervals. *Journal of Comparative Physiological Psychology, 69,* 226–231.

Susswein, A. J., & Schwarz, M. (1983). A learned change of response to inedible food in *Aplysia. Behavioral and Neural Biology, 39,* 1–6.

Susswein, A. J., Schwarz, M., & Feldman, E. (1986). Learned changes of feeding behavior in *Aplysia* in response to edible and inedible foods. *Journal of Neuroscience, 6,* 1513–1527.

Sutton, R. S., & Barto, A. G. (1981). Toward a modern theory of adaptive networks: expectation and prediction. *Psychology Review, 88,* 135–171.

Sutton, R. S., & Barto, A. G. (1990). Time-derivative models of pavlovian reinforcement. In J. W. Moore & M. Gabriel (Eds.), *Learning and computational neuroscience* (in press). Cambridge, MA: MIT Press.

Thompson, R. F. (1986). The neurobiology of learning and memory. *Science, 233,* 941–947.

Thompson, R. F. (1988). The neuronal basis of basic associative learning of discrete behavioral responses. *Trends in Neuroscience, 11,* 152–155.

Thompson, R. F., Berger, T. W., & Madden, J. (1983). Cellular processes of learning and memory in the mammalian CNS. *Annual Review of Neuroscience, 6*, 447–491.

Walters, E. T., & Byrne, J. H. (1983). Associative conditioning of single sensory neurons suggests a cellular mechanism for learning. *Science, 219*, 404–407.

Walters, E. R., Byrne, J. H., Carew, T. J., & Kandel, E. R. (1983). Mechanoafferent neurons innervating the tail of *Aplysia:* II. Modulation of sensitizing stimulation. *Journal of Neurophysiology, 50*, 1543–1559.

Walters, E. R., Carew, T. J., & Kandel, E. R. (1979). Classical conditioning in *Aplysia californica. Proceedings of the National Academy of Sciences USA, 76*, 6675–6679.

Walters, E. R., Carew, T. J., & Kandel, E. R. (1981). Associative learning in *Aplysia:* evidence for conditioned fear in an invertebrate. *Science, 211*, 504–506.

Weinberger, N. M., McGaugh, J. L., & Lynch, G. (1985). *Memory systems of the brain*. New York: Guilford Press.

Woody, C. D. (1986). Understanding the cellular basis of memory mechanisms. *Annual Review of Psychology, 37*, 433–493.

Zipser, D., & Anderson, R. (1988). A back propagation programmed network that simulates response properties of a subset of posterior parietal neurons. *Nature, 331*, 679–684.

3 Adaptive Synaptogenesis Can Complement Associative Potentiation/Depression

William B Levy
Costa M. Colbert
Department of Neurological Surgery
and the Neuroscience Graduate Program
University of Virginia School of Medicine,
Charlottesville

ABSTRACT

Adaptive modifiability is a hallmark of the nervous system and many neural like networks. Consistent with this presumption, there are many network simulations where connection weight modifications store information. However, almost all of these simulations are built with an immutable set of prespecified connections even though natural networks probably can make or break a synapse based on a modest set of rules and on the activity history of each neuron.

In an attempt to relate the experimental literature on synapse formation to such activity dependent rules, we have identified some particularly relevant anatomical studies. These studies imply two simple rules that, together with associative modification of existing synapses, might make up a complete set of rules controlling connectivity of excitatory synapses. In particular, we are inspired by various experimental studies to hypothesize a presynaptic mechanism and a postsynaptic mechanism that together control new synapse formation. We have previously pointed out that such anatomical observations, and the implied growth rules, are consistent with and complement the experimentally observed associative synaptic modification rule of the hippocampal dentate gyrus.

The purpose of this communication is to show, via computer simulations, that these simple microscopic rules can reproduce some of the normal and experimentally induced electrophysiological observations of the developing visual cortex.

As pointed out by Hebb (1949) there is little reason to consider the issues of perceptual development and the issues of learning to be fundamentally different—whether studying them as abstract cognitive processes or studying them at the microscopic, biological level. That is, regardless of how old an animal is, there is the problem of how the brain uses experience to determine its own connectivity. With this perspective and with supporting evidence that implies synaptic modification throughout the lifetime of an animal, we are comfortable interrelating observations from various times in the organism's life to understand adaptive modifications of brain circuitry based on synaptic modification. Within this philosophy one question to consider is—What are the local modification rules used by neurons to control their own connections?

Modification of connectivity falls into two broad classes: modification of existing synapses and synaptogenesis. We have explained previously the evidence for one particular class of associative weight modification rules (Levy, 1982; Levy & Desmond, 1985a, and b; Levy & Steward, 1979). A rule of this class seems also to control the functional characteristics of developing neurons in the kitten visual cortex (Rauschecker & Singer, 1979, 1981). On the other hand, synaptogenesis and the conditions that favor synaptogenesis appear distinct from this associative modification rule based on observations in a number of systems outside of the hippocampus and visual cortex (Lomo & Rosenthal, 1972; Sabel & Schneider, 1988; Wolff, Joo, Dames, & Feher, 1979).

The purpose of this chapter is to describe a pair of synaptogenesis rules consistent with the studies above that: (a) is complementary to the associative weight modification rule; (b) allows a network to usefully self-organize from zero connections; and (c) explains, i.e. reproduces, a set of experimental observations concerning cortical development.

The observations of cortical development of interest here are the ocular dominance characteristics of neurons in visual cortex. Ocular dominance, or just ocularity, refers to the relative ability of a stimulus presented to one eye to fire a neuron compared to the ability of the same stimulus to fire a neuron when presented to the other eye. This chapter presents simulations of the normal development of ocularity of cells in the kitten visual cortex. Additional simulations investigate the effect of monocular deprivation when such deprivation occurs either alone or combined with pharmacological agents that augment or inhibit the processes underlying associative modification and synaptogenesis.

Background

The need for a complementary relationship among the adaptive rules governing synaptic modification, including new synapse formation, has not yet become widely appreciated. In particular, most people seem happy with Hebb's idea that one associative rule should govern both new synapse formation and modification of existing synapses. Hebb's suggestion, however, leads to poorly used neurons. We, on the other hand, take it as axiomatic that the nervous system adaptively wires itself so that with high probability a neuron is not poorly used but is well used. (A poorly used cortical neuron is one that is not firing very much or is firing only in response to one or very few afferents.)

In his 1949 book, Hebb posits his now famous idea that coactivity of pre- and postsynaptic elements forming a synapse leads to a strengthening of this synapse via an associative weight modification rule. However, before such a modification occurs, Hebb posits a synaptogenesis mediated by an associative, electrobiotaxic property that brings neural processes together. That is, coactivity of an axon and a dendrite that are nearby but not in contact leads to a contact (synapse) being formed. However, it is now known that neurons often have many axon collaterals and that sprouting can occur so that single axons seem to easily collateralize. If we consider Hebb's developmental picture plus the capability for such sprouting, there is the immediate inference that a positive feedback situation results. The initial innervation increases the degree of coactivity between the afferent(s) forming the earliest synapse(s) and its postsynaptic targets. With sprouting, more synapses form and correlated coactivity increases even more. Thus, following Hebb's logic of associative biotaxis, a postsynaptic cell will become dominated by one or a small set of inputs that made the earliest synapses.

Likewise, there is the problem of the poorly used cell due to inactivity. Without a certain minimal level of excitatory input, such a cell does not fire, and thus it transmits no information due to its input. This cell is essentially wasted without additional innervation. Unfortunately, new innervation requires coactivity, which is precisely what this uninnervated cell does not have. Thus we see that without other special postulates Hebb's new synapse rule leaves much to be desired in the sense that it cannot guarantee cells are well used. Other authors have considered (Grossberg, 1982) and modeled (Hirai, 1980) similar consequences of specific rules governing new synapse formation.

As we became disenchanted with this one of Hebb's ideas, we began to study the biological literature on new synapse formation. Most importantly, we looked for experimental studies where manipulations of variables local to a presynaptic or postsynaptic cell resulted in changes clearly related to synaptogenesis. In our reading, not all studies were of equal value. Clear and unambiguous manipulation of a functionally related variable was very important. Regarding the measured variables of synaptogenesis experiments, neurophysiologically studied modifications are ambiguous as to whether the observed changes are caused by the modification of existing synapses or by the formation of new synapses. Carefully controlled anatomical work, however, can unambiguously identify the formation of new synapses or even precursor events to new synapse formation.

We now briefly review the features of these studies (see also Levy & Desmond, 1985a) pertinent to the description of the rules governing synaptogenesis. Insight into the postsynaptic rule comes from studies of the peripheral nervous system by two research groups. Lomo and colleagues (see, e.g., Lomo & Rosenthal, 1972; Lomo & Slater, 1978) use the adult neuromuscular system to show that lowered postsynaptic activity leads to increased innervation. Wolff and colleagues (see, e.g., Wolff, Joo, Dames, & Feher, 1979, 1981) use the superior cervical ganglion to show that inhibitory substances, such as γ-aminobutyric acid

(GABA), produce a proliferation of postsynaptic sites, and that these sites are permissive for innervation when a suitable presynaptic nerve is nearby. The most relevant experiments were performed in the superior cervical ganglion, but we first review work in the neuromuscular junction that preceded the ganglion work and foreshadowed the later results.

Normally an adult muscle is innervated by a single nerve. However, if the activity of the muscle is depressed by any one of a variety of manipulations including application of tetrodotoxin or curare, then multiple innervation ensues. When excitatory stimulation is applied, the extra connections are lost via a competitive process that seems no different, in the abstract, from associative potentiation/depression in the dentate gyrus.

The experiments performed in the superior cervical ganglion are more relevant to formulating a rule for the brain for at least two reasons. First, ganglion synapses are more similar than neuromuscular junctions to brain synapses. Second, a normally present inhibitory substance is used in the experiments. In these experiments, bathing the ganglion with the inhibitory substance produces a proliferation of the postsynaptic portion of a synapse; that is, there are more postsynaptic densities without associated presynaptic structures following the treatment. This proliferation of free postsynaptic sites is due to an increase in the number of such sites, not a disengagement of existing presynaptic structures. Two other results from this laboratory extend these observations. First, similar manipulations have similar effects in cerebral cortex, which normally has very few, <1%, free postsynaptic density sites (Balcar, Erdo, Joo, Kasa, & Wolff, 1987). Second, if a cut axon bundle is placed near these free postsynaptic sites in the superior cervical ganglion, then innervation of the free site occurs. However, if the free sites are not induced, then no innervation occurs (Joo, Dames, & Wolff, 1983). Our conclusions from this body of work are that the postsynaptic, target structure limits new synapse formation and that postsynaptic inhibition, or perhaps just lack of postsynaptic excitation, is the trigger for providing sites for additional innervation.

The presynaptic rule comes from work by Schneider and colleagues (see e.g. Sabel & Schneider, 1988) and is based on an observation they call the pruning effect. In gardening, one can prune some portion of a shrub or a tree in order to stimulate growth elsewhere. This is just what they observed in mammalian brain during development. Retinal ganglion cells innervate both the lateral geniculate nucleus and the superior colliculus. If the innervation to the superior colliculus is destroyed, presynaptic terminals in the lateral geniculate proliferate. Sabel and Schneider (1988) describe recent evidence implying that the effect is one of pruning and not of reactive synaptogenesis (i.e., the lesion induced sprouting extensively documented in the hippocampus, see e.g., Steward, 1986).

From studies like these and our own work on associative potentiation/depression (see appendix, and e.g. Levy & Steward, 1979) we devised three adaptive rules (Levy & Desmond, 1985a): An associative modification rule for existing synapses (including both potentiation and depression, or in other words a competition); a presynaptic synaptogenesis rule (avidity); and a postsynaptic synaptogenesis rule controlling the number of suitable sites (receptivity).

We now detail these three rules. Consider a postsynaptic cell j, presynaptic cell i, and the possibility of one or more synapses {k} between i and j with individual weights $w_{i,j}$. Then the version of three adaptive rules from Levy and Desmond (1985b) that are used here are:

I. Weight modification ($\Delta w_{i,j}$) of existing synapses: Long term associative potentiation/depression controls excitatory synapse modification as

$$\Delta w_{i,j} = \varepsilon\, f(y_j)(cx_i - w_{i,j}),$$

where $\Delta w_{i,j}$ is the change in the weight $w_{i,j}$; x_i is the activity of the ith presynaptic cell, y_j is the postsynaptic activity ($= \sum_i \sum_k w_{i,j} x_i$) of the jth cortical cell; $f(y_j)$ is a monotonically increasing function of postsynaptic activity (for simplicity in the simulations a linear function is used though experimental evidence favors a highly nonlinear relationship); ε is a small positive constant that controls the rate of convergence of $w_{i,j}$ to a steady-state value; and c is a positive constant to match units of x_i and $w_{i,j}$. The basis of this rule is found in Levy and Steward (1979) and Levy and Desmond (1985b).

II. New synapse formation—Postsynaptic receptivity (R_j) for new innervation: This variable increases as postsynaptic activity decreases as

$$R_j = \frac{1}{1 + c_r(\overline{y}_j)^q},$$

where $\overline{y}j$ is the average activity of j (in the simulations here $\overline{y}_j$ is averaged over an interval 500 times longer than the interval governing ε); c_r is a constant that determines the value of postsynaptic activity resulting in half-maximal receptivity; and q is a positive constant that determines the steepness of the receptivity curve. This receptivity rule is our quantification of Wolff's and Lomo's research described earlier.

III. New synapse formation—Presynaptic avidity (A_i) for growing into a postsynaptic site. This variable increases as the total weight of a presynaptic cell (i) decreases as

$$A_i = \frac{1}{1 + c_a(\sum_j \sum_k w_{i,j})^p}.$$

where $\sum_j \sum_k w_{i,j}$ is the total synaptic weight of presynaptic cell i; c_a is a constant that determines the total presynaptic weight that results in half-maximal avidity; and p is a constant that determines the steepness of the avidity curve. This avidity rule is one of our quantifications of Schneider's theory of pruning induced terminal proliferation. A more complex rule is the dynamic version found in Levy and Desmond (1985a).

In order to confirm the viability and the compatibility of these three rules, we modeled four sets of observations generated by research on the developing kitten visual cortex. The variable of interest in all these experiments is ocularity (the ability of a stimulus presented to the left eye to fire a cell compared to the ability of the same stimulus presented to the right eye to fire the same cell). The modal result of kitten development in a normal environment is the existence of cells that are roughly equivalent in their responsiveness to the left or right eye although many cells show varying degrees of biased responsiveness to left versus right eye stimuli. One control and three experimental conditions are simulated:

1. Normal environment: This environment leads to normal, binocularly responsive postsynaptic cells (Hubel & Wiesel, 1962).

The other three conditions are manipulations that follow rearing in the normal environment and the ensuing normal innervation. These additional manipulations are made within the critical period during which environmental conditions can alter neuronal eye preference.

2. Monocular deprivation by closing one eye: This condition leads to postsynaptic cells whose activity is dominated by the open eye (Hubel & Wiesel, 1970).

3. Monocular deprivation while the drug 2-amino-5-phosphonovaleric acid (APV) is infused into a region of visual cortex: APV blocks associative modification without blocking either pre- or postsynaptic activity (Kleinschmidt, Bear, & Singer, 1987). When and where this drug is present, monocular deprivation no longer produces ocularity shifts.

4. Monocular deprivation while the GABA agonist muscimol is present in one region of visual cortex. Muscimol inhibits postsynaptic activity and thus blocks associative modification. For the cells outside the region of drug action, however, the environment is the same as in Condition 2 and the result is the same. For cells within the drugged region, the opposite result occurs, namely a shift in ocularity toward the closed eye (Reiter & Stryker, 1988).

The difficulty encountered in explaining all four of these results via the associative synaptic modification rule described earlier is that, under both drug conditions, the associative synaptic modification rule is inoperative. That is, because associative modification depends on excitation, the drug manipulation in Condition 4, which enhances inhibition, does not allow associative modification to occur—yet an ocularity shift occurs under Condition 4 but not Condition 3. Additionally, because the ocularity shift is opposite of the one normally produced, something more than associative modification seems to be involved.

Modeling Development of Ocular Dominance

The network model is a highly simplified version of visual cortex, consisting of two postsynaptic cortical cells that are candidates to receive inputs from 16 lateral geniculate (LGN) cells. Eight of the LGN cells are driven by the right eye, and the other eight are driven by the left eye. Normal activity patterns are stochastically generated with a fair amount of correlated activity between the LGN axons of the two eyes. (The level of correlation was varied in different runs of the simulations.) Monocular deprivation is modeled by reducing the correlation and/or the average level of activity of the LGN cells associated with the deprived eye. Drug effects are modeled by appropriate adjustment of equation parameters. In all cases synaptic weights are regularly updated in parallel. Postsynaptic activity is calculated in parallel and is based on input activity, drug effects, and the relevant synaptic weights.

In order to use rules II and III, new synapse formation is modeled as a stochastic process. Postsynaptic receptivity is equivalent to the number of available, unoccupied postsynaptic sites. Presynaptic avidity quantifies the probing, searching, ameboid-like movements of a cell's axonal branches. This probing growth increases as the presynaptic cell's total synaptic weight $\sum_j \sum_k w_{i,j}$ decreases. Increases of receptivity or avidity then contribute to increases in the probability of an encounter between an axonal branch of i and a vacant postsynaptic site of j to form a new synapse i,j.

In order to perform the simulations, there are variables to set that do not appear in the modification rules. One of these is the weight value of a new synapse. In practice it was rather easy to specify a value that was small, that allowed the simulations to duplicate the biological experiments, and that seemed sensible relative to synaptic strengths of potentiated synapses.

Another variable not appearing in the rules, and one that was modeled in two ways, is the minimum strength of a synapse. [Note that Rule I guarantees non-negativity and only zero asymptotically (Levy & Geman, 1982)] The simulations produced the same results when the minimum was zero or a slightly larger number. However, if the minimum weight was set as non-zero so that the synapse vanishes when its strength falls below this minimum, then it was necessary to specify the initial value of a new synapse to be slightly larger than this minimum.

The receptivity curve R_j was chosen so that (a) synapse formation is low in the steady-state following normal development and so that (b) in the presence of muscimol, receptivity increases as implied by Balcar et al. (1987). Likewise, the avidity curve A_i was chosen so that avidity is low in the steady-state normal environment and increases when synaptic weight is lost due to Rule I induced depression.

Results

Setting the free variables to fit all the experimental observations was readily accomplished. Further work showed that the simulations were relatively insensitive to the exact values of any particular variable.

The network automatically wires the postsynaptic cells to produce binocularity, thus satisfying Condition 1. Figure 3.1 shows an example of how cell activity and the adaptively modifiable variables develop over time in a simulation of normal environmental conditions. The initial innervation is extremely rapid because at the earliest time points both receptivity and avidity are at a maximum for all cells. Note that even in a normal environment there is enough fluctuation in the amount and correlation of input activity to cause ongoing associative synaptic modification as well as occasional increases in avidity. Finally, based on relative weights, each postsynaptic cell is nearly, but not perfectly binocular. Extensive simulations indicate that the mirror symmetry of ocularity between these two cells is not accidental but arises from the modification rules and the statistics of the input environment.

Condition 2, visual deprivation of one eye, was simulated either as a decrease in average activity or as a decrease in the correlation between the afferents of the deprived eye. In either case Rule I causes the synaptic weight(s) of afferents associated with the closed (left) eye to decrease and those associated with the open eye to increase. As a result of these synaptic weight changes, there is a net shift in ocularity toward the open (right) eye for both cortical cells. Figure 3.2 shows the effect of normal development followed, at cycle 10, by the monocular deprivation condition. In this example, monocular deprivation is modeled by greatly decreasing the correlated cell firing of the afferents driven by the deprived eye. Note that the competition effect, i.e. the shift in ocularity towards the open eye, occurs rapidly with the open (right) eye afferents dominating the activity of both postsynaptic cells.

For the two drug manipulations, Conditions 3 and 4, we simulate the experiments as if the drug reaches one cell but the other cell is too distant and remains drug free.

If a drug like APV is used to block Rule I without blocking postsynaptic activity (as in Condition 3), then the ocularity shift favoring the open (right) eye is blocked. Figure 3.3 shows the effect of monocular deprivation with APV blocking Rule I for one of the cortical cells beginning at cycle 10. Where APV blocks Rule I, there is no shift in ocular dominance even though Rules II and III are operating. On the other hand, the untreated cells simultaneously show the normal ocularity shift with monocular deprivation. Because drugs like APV block associative potentiation in the hippocampus (Herron, Lester, Coan, &

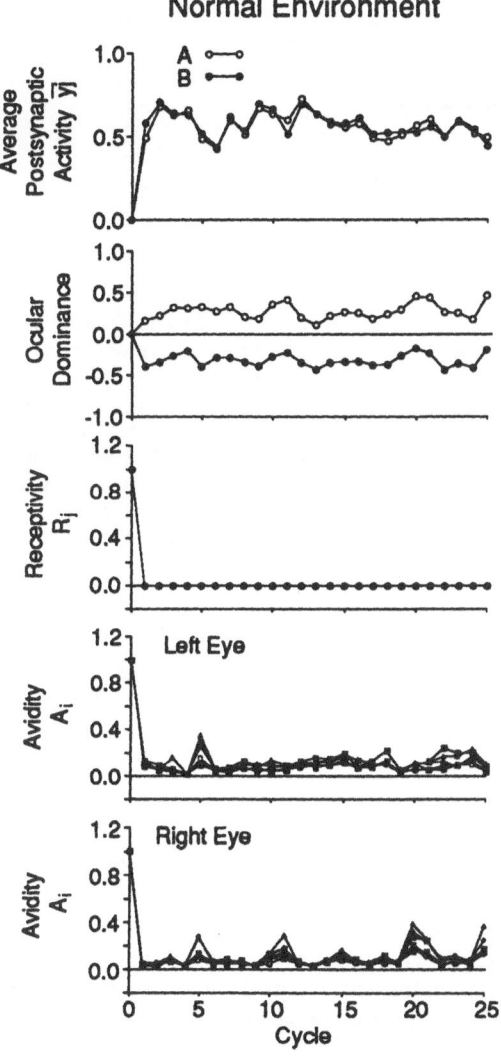

Normal Environment

FIG. 3.1. Development under normal conditions. This figure, and all succeeding figures, illustrate the simulation of a single experiment. Figure 3.1 shows the results of 25 cycles of a normal input environment. The top graph (postsynaptic activity) shows the activity of the two postsynaptic cortical cells, designated A and B, in response to the afferent activity sequence. The next graph (ocularity) shows ocular preference for each of the two cells A and B. Positive numbers indicate left preferring and negative numbers indicate right preferring. The values plotted are the differences between the percentages of total synaptic weight of afferents associated with each eye. Zero is perfectly binocular; plus and minus 100% are perfectly monocular. The middle graph shows the receptivity of cells A and B. The lower two graphs (avidity) shows the avidity of afferents associated with each eye. Each eye commands eight LGN afferents. The values used in this simulation and the other illustrated simulations are:

$$c_a = 200; c_r = 10^5; p = 4; q = 5.$$

A generated stream of 0's and 1's is corrupted by noise (15% of the bits are complemented at random) and converted to frequency by a coarse running-averager. In Condition 1 the average activity of the afferents (i.e. the probability of a particular afferent firing at any given time step in the simulation) is 0.155. The input correlations between left and right eye afferents, before noise corruption are: (.85;{0,0}),(.05;{1,0}),(.05;{0,1}) (.05;{1,1}), where the decimal fraction is the probability of the bivariate binary event in the brackets.

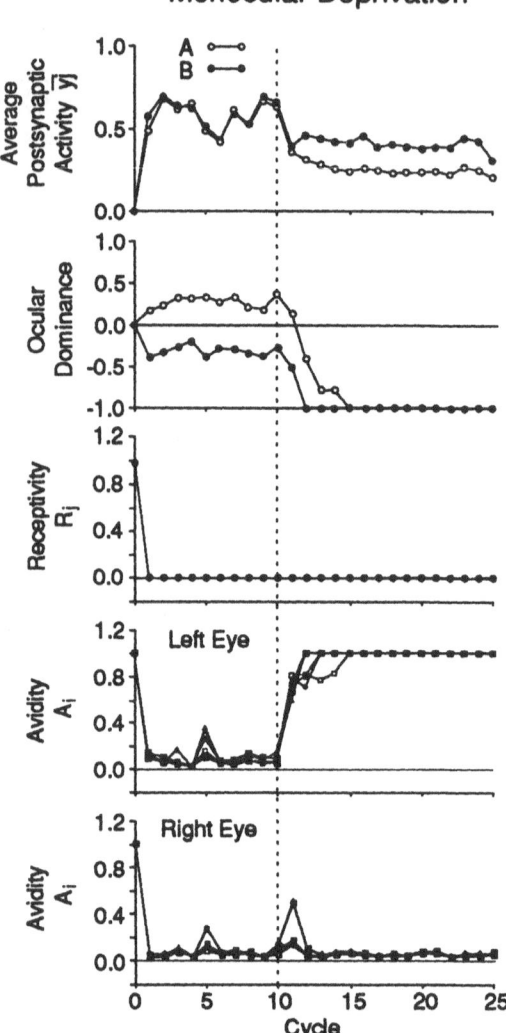

Monocular Deprivation

FIG. 3.2.   The effect of monocular deprivation. After ten cycles of normal activity, the activity pattern is altered to approximate monocular deprivation of the left eye. In this simulation, deprivation is modeled as a decrease in correlated activity associated with the deprived eye. Average frequency remains approximately unchanged. Note that the weights of both postsynaptic cells come to favor the right (open) eye. See Figure 3.1 for more details of the simulation.

Collingridge, 1986), Condition 3 demonstrates the operation of a similar process in developing visual cortex (Artola & Singer, 1987).

Condition 4 produces a different result than Condition 3 because both postsynaptic activity and associative modification are blocked by muscimol application. Figure 3.4 shows the simulation of Condition 4. At cycle 10, monocular deprivation occurs and muscimol blocks Rule I at one of the two cortical cells. Even so, there is an ocularity shift at both cells. In the cortex outside of the drugged region, the situation is identical to Condition 2. That is, presynaptic afferents associated

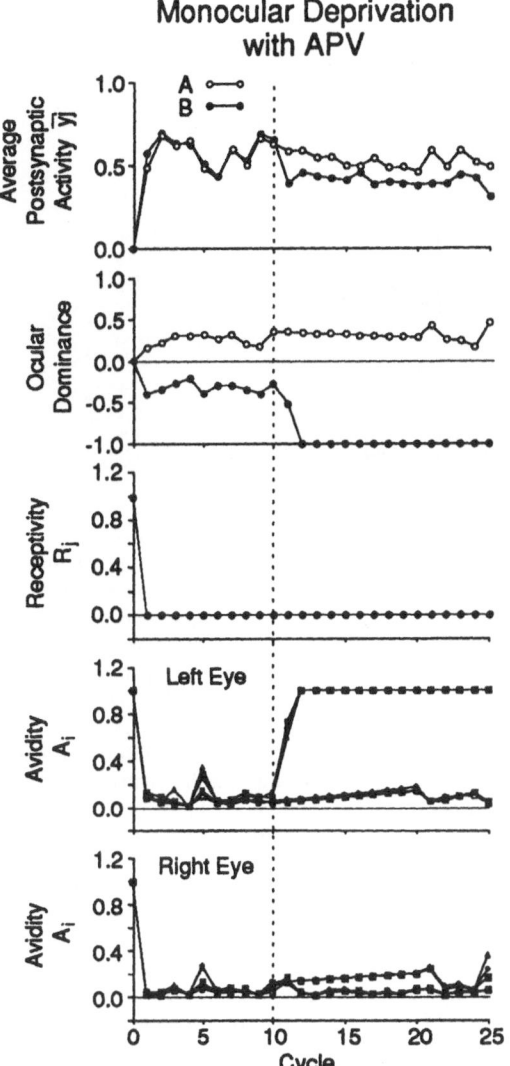

Monocular Deprivation with APV

FIG. 3.3. The effect of APV on monocular deprivation. After ten cycles of normal activity, the activity pattern is altered to approximate monocular deprivation of the left eye as in Figure 3.2. In this simulation, cortical Cell A is in a region affected by the drug APV. APV blocks associative modification but not the activity of Cell A. Cell B is outside the region affected by APV. The effect of APV is simulated by reducing $f(\bar{y}_i)$ 100-fold. Note that, for Cell A, a shift in ocularity does not occur. Conversely, the undrugged Cell B shows an ocularity shift as in Figure 3.2. See Figure 3.1 for more details of the simulation.

with the closed, left eye lose synaptic weight via Rule I while afferents associated with the open eye inputs gain weight.

Now consider the drugged cell. For this cell, Rule I is blocked, and, at the same time, postsynaptic activity is greatly reduced. It is this reduction of postsynaptic activity that causes the receptivity of this cell to rise. If there are presynaptic afferents of high avidity, new synapses will form. Since afferents driven by the deprived eye are losing the competition for synaptic weight on the undrugged postsynaptic

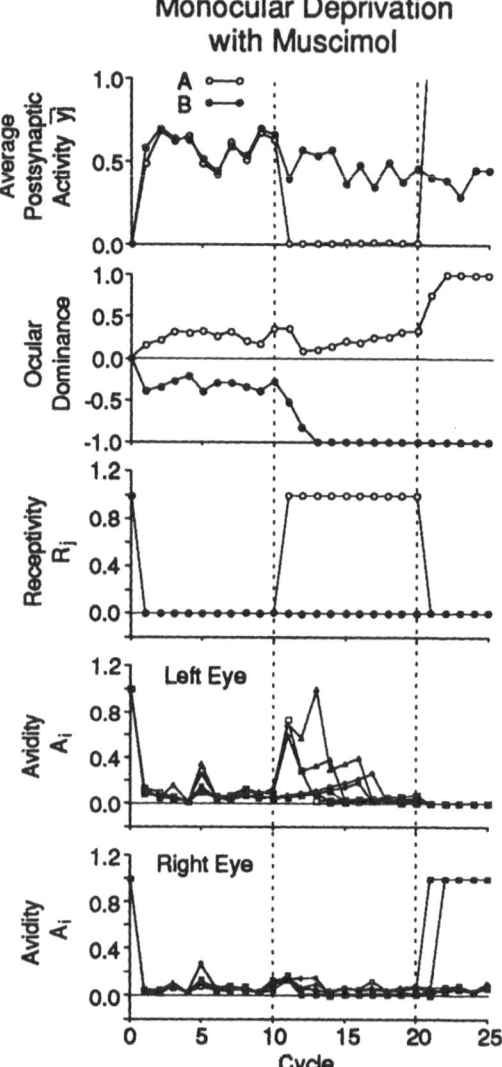

FIG. 3.4. The effect of muscimol on monocular deprivation.
After ten cycles of normal activity, the activity pattern is altered
to approximate monocular deprivation of the left eye as in Figure 3.3. In this simulation, cortical Cell A is in a region affected
by muscimol, which blocks postsynaptic activity, and thus associative modification. Cell B is
outside the affected region.
Muscimol is simulated by reduc-
ing $\bar{y}_i$ 100-fold. Note that Cell A
shows an ocularity shift toward
the left (deprived) eye whereas
Cell B shows a shift toward the
right (open) eye as in Figure 3.2.
See Figure 3.1 for more details
of the simulation.

cell, their avidity is higher. As a result, muscimol plus deprivation induces many
new synapses between high avidity afferents of the deprived (left) eye and the drug-
inhibited, high receptivity cortical cell. Thus, the net effect for the drugged cortical
cell is a shift in ocularity toward the closed eye.

Discussion

This research demonstrates how a set of three microscopic, locally sensing
and locally operating modification rules can be implemented to produce a rather
simple, but biologically relevant, self-assembling neural network. What is inter-

esting about these simulations compared to some others in the literature is that the simulation concerns neocortical development although the three rules are based on experimental results generated outside of neocortex and in adult systems. In terms of the attractiveness of these rules it is important to note that each of these rules depends only on parameters readily available to each neuron.

Even though this model successfully simulates cortical development across the four conditions set forth here, the problem of modeling this development process is *not* uniquely solved by our system of equations. We note that Bear and Cooper (1990) have been able to fit these same developmental observations using a rather different rule. In fact, it was their work that pointed out to us the interesting set of experimental developmental observations modeled here, and we are grateful for their communicating their own results to us prior to publication.

It is also interesting that the rules presented here share some of the characteristics, though not the form, of the rule advocated by Cooper's lab, see e.g., Bienenstock, Cooper, and Munro (1982). In particular, both their and our approaches depend on average postsynaptic activity where the time constant for these averaging processes must be much slower than the time constant controlling weight modification. A second similarity is the appearance of nonlinearities in the slow modification process(es). There are, however, significant differences, particularly in the associative modification rule itself (see Munro, 1983). Most notably, Cooper's synaptic weight modification rule requires presynaptic activity, whereas the weight modification rule presented here requires postsynaptic activity for modification to occur. Second, Cooper's rule requires an adaptively modified threshold. Our rules will work with or without a modifiable threshold.

In contrast to Hebb's (1949) original suggestion, the three rules used here are complementary because they help guarantee that neurons will be well-used. This well-used characteristic can be interpreted more generally than its relation to visual cortex. The postsynaptic receptivity rule prevents individual neurons from being underutilized, in that the activity levels of all neurons will converge toward the same average value and thus the same probability of firing and the same Shannon entropy. This, in turn, would guarantee that each neuron carries its share of the representational information of the network (see Levy, 1989). In a similar vein, the avidity rule prevents all postsynaptic cells from being dominated by the same set of well-correlated inputs by biasing against synapse formation with presynaptic cells that have made numerous synapses. More specifically, because of the avidity rule, there should be less statistical dependence between postsynaptic cells which, in effect, helps maximize the actual representational entropy of a set of such cells.

Though modeled as developmental processes, we believe these same rules govern connectivity in adult animals. In new environments, or more specifically in environments the network perceives as new, postsynaptic activity will tend to decrease because the old correlations, which set up the network originally, will no longer be present. The set of rules used here allows for the establishment of new connections that can then encode new correlations. However, implementation of such modification processes is likely a delicate affair in terms of losing previously learned environments. It may be necessary to consider the existence of permissive hormonal events or mismatch detectors to fully account for control of synaptic modification in the adult animal.

ACKNOWLEDGMENTS

Supported by NIH RO1 NS15488 and NIMH RSDA MH00622 to WBL and by the Department of Neurological Surgery, Dr. John A. Jane, Chairman. CMC was supported MSTP Training Grant NIH 5T32 GM0726713.

APPENDIX

The associative modification equation used in our models is the one implied by experiments in the hippocampal dentate gyrus. In these experiments there are two inputs: a powerfully depolarizing ipsilateral input that controls the postsynaptic term $f(y_j)$ and a very weakly excitatory contralateral input whose activity corresponds to the presynaptic term x_i.

Although these experiments are crude field potential studies, only qualitative questions need to be answered in order to eliminate many of the proposed synaptic modification equations found in the literature. The underlying assumption is that increases of intensity or frequency at a stimulating electrode do not decrease, and sometimes increase, the immediately evoked monosynaptic response.

The initially important observation is that both potentiation and depression can occur simultaneously on the same cells. (This observation is based on an anatomically based argument.) These increases and decreases of synaptic strength are a function of (a) the strength of ipsilaterally evoked postsynaptic excitation; (b) the presynaptic frequency or number of pulses; and (c) the previous history of modification.

Observation (1) includes experiments that varied the ipsilateral frequency of activation or varied its intensity ($\equiv$ the number of active inputs). From these manipulations two conclusions are clear: The postsynaptic requirement is highly nonlinear, probably threshold-like; and the same postsynaptic decision process is permissive for both potentiation or depression.

If a properly timed postsynaptic permissive event is provided in an experiment, then potentiation, depression or no change is a function of both the activity of the test input, x_i, and a function of how much potentiation or depression has been induced previously. Higher frequencies of x_i cause potentiation, and lower frequencies cause depression. The more a synapse has been potentiated without depression, the less it can be potentiated in the future. The equivalent result holds for depression. The simplest way to express this behavior is the equation given in Rule I where the parenthetical term takes on value zero when $cx_i = w_{ij}$ and is either positive or negative otherwise. The functions $f(y_j)$ and ε are, by definition, always non-negative.

Our final comment concerns the temporal dependence of associativity. In order to get potentiation the permissive input must follow the weak input (i.e., the one being potentiated) although too large a time window leads to depression of the weak input.

REFERENCES

Artola, A. & Singer, W. (1987). Long-term potentiation and NMDA receptors in rat visual cortex. *Nature, 330,* 649–652.

Balcar, V. J., Erdo, S. L., Joo, F., Kasa, P., & Wolff, J. R. (1987). Neurochemistry of GABAergic system in cerebral cortex chronically exposed to bromide in vivo, *Journal of Neurochemistry, 48,* 167–169.

Bear, M. F., & Cooper, L. N. (1990). Molecular mechanisms for synaptic modification in the visual cortex: interaction between theory and experiment. In M. Gluck & D. Rumelhart (Eds.), *Neuroscience and connectionist models* (pp. 65–93). Hillsdale, NJ: Lawrence Erlbaum Associates.

Bienenstock, E. L., Cooper, L. N., & Munro, P. W. (1982). Theory for the development of neuron selectivity: Orientation specificity and binocular interaction in visual cortex. *Journal of Neuroscience, 2,* 32–48.

Grossberg, S. (1982) *Studies of mind and brain.* Dordrecht, Holland: D. Reidel.

Hebb, D. O. (1949). *The organization of behavior. A neuropsychological theory.* New York: Wiley.

Herron, C. E., Lester, R. A. J., Coan, E. J., & Collingridge, G. L. (1986). Frequency-dependent involvement of NMDA receptors in the hippocampus: A novel synaptic mechanism. *Nature, 322,* 265–268.

Hirai, Y. (1980) A new hypothesis for synaptic modification: an interactive process between postsynaptic competition and presynaptic regulation. *Biological Cybernetics, 36,* 41–50.

Hubel, D. H., & Wiesel, T. N. (1962). Receptive fields, binocular interactions and functional architecture in the cat's visual cortex. *Journal of Physiology (Lond.), 160,* 105–154.

Hubel, D. H., & Wiesel, T. N. (1970). The period of susceptibility to the physiological effects of unilateral eye closure in kittens. *Journal of Physiology, 206,* 419–436.

Joo, F., Dames, W., Parducz, A., & Wolff, J. R. (1983). Axonal sprouts of the hypoglossal nerve implanted in the superior cervical ganglion of adult rats establish synaptic contacts under long-lasting GABA effect. A degeneration study, *Acta Biologica Hungarica, 34,* 177–185.

Kleinschmidt, A., Bear, M. F., & Singer, W., (1987). Blockade of "NMDA" receptors disrupts experience-dependent modifications of kitten striate cortex. *Science, 238,* 355–358.

Levy, W. B. (1982). Associative encoding of synapses. *Proceedings of the Fourth Annual Conference Cognitive Science Society,* 135–136.

Levy, W. B. (1989). A computational approach to hippocampal function. In R. D. Hawkins & G. H. Bower (Eds.), *Computational models of learning in simple neural systems,* (pp. 243–305). Orlando, FL: Academic Press.

Levy, W. B. & Desmond, N. L. (1985a). The rules of elemental synaptic plasticity. In W. B. Levy, J. Anderson, & S. Lehmkuhle (Eds.), *Synaptic modification, neuron selectivity and nervous system organization* (pp. 105–121). Hillsdale, NJ: Lawrence Erlbaum Associates.

Levy, W. B. & Desmond, N. L., (1985b). Associative potentiation/depression in the hippocampal dentate gyrus. In G. Buzsaki & C. H. Vanderwolf (Eds.), *Electrical activity of the archicortex,* (pp. 359–373). Budapest, Hungary: Akademiai Kiado.

Levy, W. B. & Geman, S. (1982). Limit behavior of experimentally derived synaptic modification rules. *Reports in Pattern Analysis No. 121,* Technical Report, Div. Applied Mathematics, Brown University, Providence, RI.

Levy, W. B. & Steward, O., (1979). Synapses as associative memory elements in the hippocampal formation. *Brain Research, 175,* 233–245.

Lomo, T. & Rosenthal, J. (1972). Control of ACh sensitivity by muscle activity in the rat. *Journal of Physiology (London), 221,* 493–513.

Lomo, T. & Slater, C. R. (1978). Control of acetylcholine sensitivity and synapse formation by muscle activity. *Journal of Physiology (London), 275,* 391–402.

Munro, P. W. (1983). *Neural plasticity: Single neuron models for discrimination and generalization and an experimental ensemble approach.* Unpublished doctoral dissertation, Department of Physics, Brown University.

Rauschecker, J. P., & Singer, W. (1979). Changes in the circuitry of the kitten visual cortex are gated by postsynaptic activity. *Nature, 280,* 58–60.

Rauschecker, J. P., & Singer, W. (1981). The effects of early visual experience on the cat's visual cortex and their possible explanation by Hebb synapses. *Journal of Physiology, 310,* 215–239.

Reiter, H. O., & Stryker, M. P. (1988). Neural plasticity without action potentials: Less active inputs become dominant when kitten visual cortical cells are pharmacologically inhibited. *Proceedings of the National Academy of Science, USA, 85,* 3623–3627.

Sabel, B. A., & Schneider, G. E. (1988). "The principle of total axonal arborizations": massive compensatory sprouting in the hamster subcortical visual system after early tectal lesions. *Experimental Brain Research, 73,* 505–515.

Steward, O. (1986). Lesion-induced synapse growth in the hippocampus. In search of cellular and molecular mechanisms. In R. L. Isaacson & K. H. Pribram (Eds.), *The hippocampus (Vol. 3, pp. 65–111).* New York: Plenum Press.

Wolff, J. R., Joo, F., Dames, W., & Feher, O. (1979). Induction and maintenance of free postsynaptic membrane thickenings in the adult superior cervical ganglion. *Journal of Neurocytology, 8,* 549–563.

Wolff, J. R., Joo, R., Dames, W., & Feher, O. (1981). Neuroplasticity in the superior cervical ganglion as a consequence of long-lasting inhibition. In F. Joo & O. Feher (Eds.), *Cellular Analogues of Conditioning and Neural Plasticity* (pp. 1–9). New York: Pergamon Press.

4

A Neural Network Architecture for Pavlovian Conditioning: Reinforcement, Attention, Forgetting, Timing

Stephen Grossberg
Boston University

ABSTRACT

A real-time neural network model is described, in which reinforcement helps to focus attention on and organize learning of those environmental events and contingencies that have predicted behavioral success in the past. The same mechanisms control selective forgetting of memories that are no longer predictive and adaptive timing of behavioral responses.

Computer simulations of the model reproduce properties of attentional blocking, inverted-U in learning as a function of interstimulus interval, primary and secondary excitatory and inhibitory conditioning, anticipatory conditioned responses, attentional focussing by conditioned motivational feedback, and limited capacity short term memory processing. Qualitative explanations are offered of why conditioned responses are forgotten, or extinguish, when a conditioned excitor is presented alone, but do not extinguish when a conditioned inhibitor is presented alone. These explanations invoke associative learning between sensory representations and drive, or emotional, representations, between sensory representations and learned expectations of future sensory events, and between sensory representations and learned motor commands. Drive representations are organized in opponent positive and negative pairs (e.g., fear and relief), which are modelled by gated dipole opponent processes. The module wherein associative learning interacts with a recurrent gated dipole at a drive representation is called a READ circuit.

Cognitive modulation of conditioning is regulated by adaptive resonance theory, or ART, circuits that control the learning and matching of expectations, and the reset of sensory short term memory in response to disconfirmed expectations. Disconfirmed expectations interact with gated dipole opponent mechanisms to regulate selective forgetting of associative memory in a READ circuit. Unless expectations are disconfirmed, memory does not passively decay, yet does not saturate due to cumulative learning over successive learning trials. To achieve selective forgetting, the read-in and read-out of learned memories are functionally dissociated within the opponent READ circuit. Such dissociation may be achieved by using dendritic spines as a site of associative learning. The selective forgetting mechanism is called *opponent extinction*.

The learning mechanism from sensory representations to drive representations may be refined to derive a neural network model that controls behavioral timing. This Spectral Timing Model controls the type of timing whereby an animal or robot can learn to wait for an expected goal by discounting expected nonoccurrences of a goal object until the expected time of arrival of the goal. Activation of a drive representation by the Spectral Timing Model inhibits output signals from the orienting subsystem of the ART network and thereby prevents resets of short term memory, forgetting, and orienting responses from being caused by events that mismatch the goal expectation prior to the expected arrival time of the goal. If the goal object does not then materialize, the model can respond to unexpected nonoccurrences of the goal with changes appropriate to a disconfirmed expectation, including short term memory reset, associative forgetting, emotional frustration, and exploratory behavior. The timing model is a gated dipole generalized to include a spectrum of cellular response rates within a large population of cells. The model's population output signal generates accurate learned timing properties that collectively provide a good quantitative fit to animal learning data. Simulated data properties include the inverted U in learning as a function of the interstimulus interval (ISI) that occurs between onset of the conditioned stimulus (CS) and the unconditioned stimulus (US); correlations of peak time, standard deviation, Weber fraction, and peak amplitude of the conditioned response as a function of the ISI; increase of conditioned response amplitude, but not its timing, with US intensity; multiple timing peaks in response to learning conditions using multiple ISI's; and conditioned timing of cell activation within the hippocampus and of the contingent negative variation (CNV) event-related potential.

1. INTRODUCTION

A central problem in biological theories of intelligence concerns the manner in which external events interact with internal organismic requirements to trigger learning processes capable of focussing attention on and generating appropriate actions towards motivationally desired goals. The results reported herein further develop a neural theory of learning and memory (Grossberg, 1982, 1987) in which sensory-cognitive and cognitive-reinforcement circuits help to learn goal-oriented behaviors that engage those environmental events that predict behavioral success, and regulate selective forgetting of reinforcement contingencies that no longer predict behavioral success. The present chapter considers these phenomena within the context of the Pavlovian conditioning paradigm.

The first set of results (Grossberg & Levine, 1987) describe computer simulations that show how the model circuits reproduce properties of attentional blocking, inverted-U in learning as a function of interstimulus interval, anticipatory conditioned responses, secondary reinforcement, attentional focussing by conditioned motivational feedback, and limited capacity short-term memory processing. To explain these phenomena, the model utilizes three types of internal representations (Grossberg, 1971, 1972a): *Sensory representations*, at which properties of external events are encoded in short term memory; *drive representations*, at which sensory, reinforcement, and internal drive, or homeostatic, inputs converge to generate motivational decisions; and *motor representations*, at which action commands are generated. Conditioning occurs from sensory representations to drive representations ("conditioned reinforcer" learning), from drive representations to sensory representations ("incentive motivational" learning), and from sensory representations to motor representations ("habit" learning). The conditionable pathways contain long-term memory traces that obey a non-Hebbian associative law. This neural model embodies a solution of two key design problems of conditioning, the synchronization problem (Grossberg, 1971) and the persistence problem (Grossberg, 1975) of Pavlovian conditioning. Predictions derived from this model of vertebrate learning have been supported by data about invertebrate learning, including data from *Aplysia* about facilitator neurons (see Byrne, chap. 2 in this volume) and data from *Hermissenda* about voltage-dependent Ca^{++} currents (see Alkon, chap. 1 in this volume).

In the second set of results (Grossberg & Schmajuk, 1987), drive representations are expanded to include positive and negative opponent drive representations, as in the opponency between fear and relief, using a neural model of opponent processing called a *gated dipole* (Grossberg, 1972a, 1972b). This expanded real-time neural network model is developed to explain data about the learning and selective forgetting, or acquisition and extinction, of both conditioned excitors and conditioned inhibitors. The expanded model of conditioning from sensory representations to opponent drive representations is called a READ circuit (REcurrent Associative gated Dipole). READ circuit properties clarify how positive and negative reinforcers are learned and forgotten during primary and secondary conditioning. Habituating chemical transmitters within a gated dipole determine an affective adaptation level, or context, against which later events are evaluated. Neutral *CS*'s can become reinforcers by being associated either with direct activations or with antagonistic rebounds triggered within a previously habituated dipole. This circuit provides a real-time mechanistic realization, rationale, and generalization of the type of transient reactions relative to baseline comparisons that the Rescorla-Wagner model and Time-Derivative models (Sutton & Barto, 1989) of Pavlovian conditioning have modelled.

The same mechanisms permit an explanation of how learning can be selectively forgotten, by a process called *opponent extinction*, even though no passive memory decay occurs. The opponent extinction process makes precise the idea that disconfirmed expectations actively regulate the process of selective forgetting. This type of active regulation is achieved within a larger neural architecture that combines READ circuit mechanisms of reinforcement learning with ART (Adaptive Resonance Theory) circuit mechanisms of recognition learning. The ART circuit includes mechanisms for activating and storing internal representations of sensory cues in a limited capacity short term memory (STM); for learning, matching, and mismatching sensory expectations, leading to the enhancement or updating of STM; and for shifting the focus of attention toward sensory representations whose reinforcement history is consistent with momentary appetitive requirements. This architecture has been used to explain conditioning and extinction of a conditioned excitor; conditioning and extinction of a conditioned inhibitor; properties of conditioned inhibition as a "slave" process and as a "comparator" process, including effects of pretest deflation or inflation of the conditioning context, of familiar or novel training or test contexts, of weak or strong shocks, and of preconditioning US-alone exposures (Grossberg & Schmajuk, 1987). The same mechanisms have previously been used (Grossberg, 1982, 1987) to explain phenomena such as unblocking, overshadowing, latent inhibition, superconditioning, partial reinforcement acquisition effect, learned helplessness, and vicious-circle behavior.

A simple generalization of the gated dipole circuitry used for learning from sensory representations to drive representations generates a competence for conditioned timing, including the inhibition of unappropriate behavioral reactions due to expected nonoccurrences of a goal object during the interstimulus (ISI) interval between CS and US; the timed release of conditioned responses to the expected occurrence of the goal object during the ISI interval; and the release of attention-shifting, forgetting, and exploratory reactions to the unexpected nonoccurrence of the goal object subsequent to the ISI interval.

Finally, the chapter includes a comparison of the present theory with data and models of other laboratories, such as those of Alkon, Barto and Sutton, Berger and Thompson, Byrne, Church, Hawkins and Kandel, Levy, Moore, and Staddon.

2. MACROCIRCUITS FOR SENSORY-COGNITIVE AND COGNITIVE-REINFORCEMENT PROCESSING

Two types of macrocircuits control learning within the theory.

Sensory-Cognitive Circuit.   Sensory-cognitive interactions in the theory are carried out by an Adaptive Resonance Theory (ART) circuit (Carpenter & Grossberg, 1987a, 1987b, 1988, 1989; Grossberg, 1976a, 1976b, 1987). The ART architecture suggests how internal representations of sensory events, including conditioned stimuli (CS) and unconditioned stimuli (US), can be learned in stable fashion (Figure 4.1). Among the mechanisms used for stable self-organization of sensory recognition codes are top-down expectations that are matched against bottom-up sensory signals. When a mismatch occurs, a STM reset wave in the form of a nonspecific arousal burst acts to reset the sensory representations of all cues that are currently being stored in STM. In particular, representations with high STM activation tend to become less active, representations with low STM activation tend to become more active, and the novel event that caused the mismatch tends to be more actively stored than it would have been had it been expected. Thus the effect of STM reset is to shift the focus of attention towards sensory representations that may better predict environmental contingencies.

The STM reset wave also triggers orienting responses, such as the activation of motor reactions to orient towards the unexpected event.

Cognitive-Reinforcement Circuit.   Cognitive-reinforcer interactions in the theory are carried out in the type of circuit described in Figures 4.1b and 4.2. In this circuit, there exist cell populations called drive representations, at which sensory, reinforcement, and drive signals converge to regulate motivational decisions (Grossberg, 1972a, 1972b, 1987). Repeated pairing of a *CS* sensory representation, S_{cs}, with activation of a drive representation, D, by a reinforcer causes the modifiable synapses connecting S_{cs} with D to become strengthened. Incentive motivation pathways from the drive representations to the sensory representations are also assumed to be conditionable. Activation of these $S \rightarrow D \rightarrow S$ feedback pathways can shift the attentional focus to the set of previously reinforced sensory cues that are motivationally consistent with D (Figure 4.2). This shift of attention occurs because the sensory representations, in addition to emitting conditioned reinforcer signals and receiving incentive motivation signals, compete among themselves for a limited capacity short-term memory (STM) via a shunting on-center off-surround anatomy. When strong incentive motivational feedback signals are received at the sensory representational field, these signals can bias the competition for STM activity towards the set of motivationally preferred cues.

3. ATTENTIONAL BLOCKING AND INVERTED U IN CONDITIONING

The attentional modulation of Pavlovian conditioning is part of the general problem of how an information processing system can selectively process those environmental inputs that are most important to the current goals of the system. A typical example is the blocking paradigm studied by Kamin (1969) (Figure 4.3). First, a stimulus CS_1, such as a tone, is presented several times, followed at a given time interval by an unconditioned stimulus *US*, such as electric shock, until a conditioned response, such as fear, develops. Then CS_1 and another stimulus CS_2, such as a light, are presented simultaneously, followed at the same time interval by the *US*. Finally, CS_2 is presented alone, not followed by a *US*, and no conditioned response occurs. Intuitively, the CS_1 "blocks" conditioning of the simultaneously presented CS_2 because the CS_1, by itself, perfectly predicts

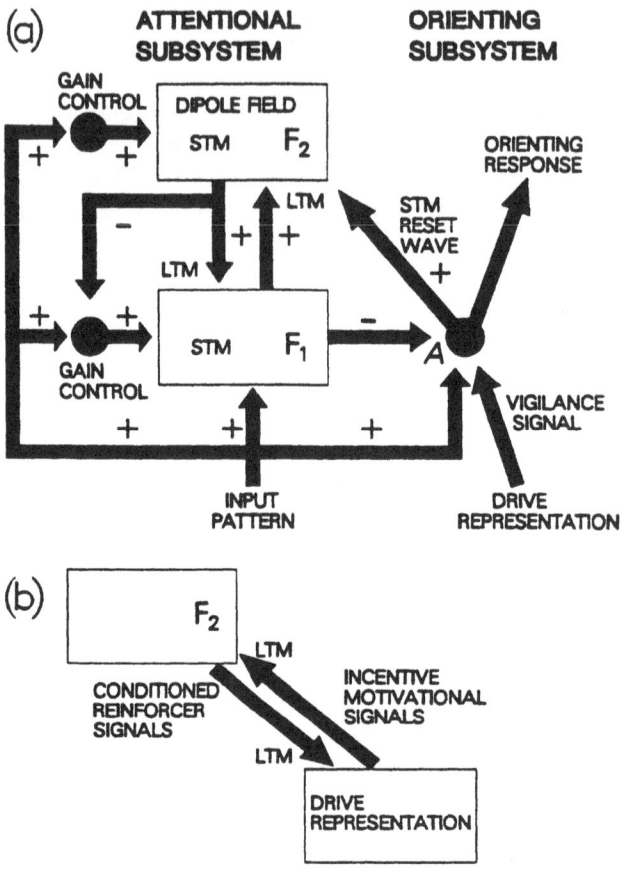

FIG. 4.1.   Anatomy of an adaptive resonance theory (ART) circuit: (a) Interactions between the attentional and orienting subsystems. Code learning takes place at the long term memory (LTM) traces within the bottom-up and top-down pathways between levels F_1 and F_2. The top-down pathways can read-out learned expectations, or templates, that are matched against bottom-up input patterns at F_1. Mismatches activate the orienting subsystem A, thereby resetting short term memory (STM) at F_2 and initiating search for another recognition code. Subsystem A can also activate an orienting response. Sensitivity to mismatch at F_1 is modulated by vigilance signals from drive representations. (b) Trainable pathways exist between level F_2 and the drive representations. Learning from F_2 to a drive representations endows a recognition category with conditioned reinforcer properties. Learning from a drive representation to F_2 associates the drive representation with a set of motivationally compatible categories. (Adapted from Carpenter & Grossberg, 1987c.)

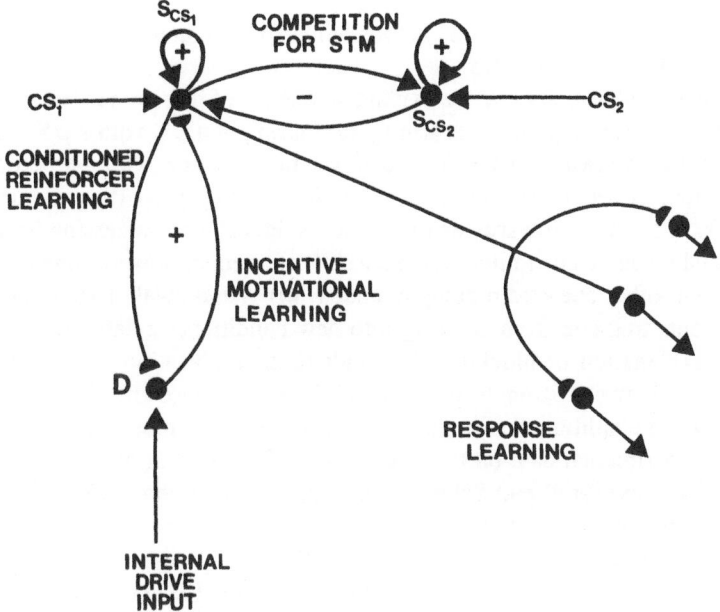

FIG. 4.2. Schematic conditioning circuit: Conditioned stimuli (CS_i) activate sensory representations (S_{csi}) that compete among themselves for limited capacity short term memory activation and storage. The activated S_{csi} elicit conditioned signals to drive representations and motor command representations. Learning from an S_{csi} to a drive representation D is called conditioned reinforcer learning. Learning from D to S_{csi} is called incentive motivational learning. Signals for D to S_{csi} are elicited when the combination of external sensory plus internal drive inputs is sufficiently large. In the simulations reported herein, the drive level is assumed to be large and constant.

1. CS_1 —— US

 CS_1 ⟶ CR

2. $CS_1 + CS_2$ —— US

 CS_2 ↛ CR

FIG. 4.3. A blocking paradigm. The two stages of the experiment are discussed in the text.

its consequence, the *US;* the CS_2 is thus redundant and unpredictive, hence does not get conditioned to the *US.*

The blocking property may be explained in terms of four properties of selective information processing during learning: (a) Pairing of a CS_1 with a *US* in the first phase of the blocking experiment endows the CS_1 cue with properties of a conditioned, or secondary, reinforcer. (b) Reinforcing properties of a cue shift the focus of attention towards its own processing. (c) The processing capacity of attentional resources is limited, so that a shift of attention towards one set of cues can prevent other cues from being attended. (d) Withdrawal of attention from a cue prevents that cue from entering into new conditioned relationships.

This explanation of blocking also leads to an explanation of the inverted-U relationship between strength of the conditioned response and the time interval (ISI) between conditioned stimulus and unconditioned stimulus. Figure 4.4 summarizes experimental data on the effects of ISI from studies of Smith, Coleman, and Gormezano (1969) and Schneiderman and Gormenzano (1964) of the rabbit nictitating membrane response.

A unified explanation of blocking and ISI data is suggested later. Such an explanation is noteworthy because Sutton and Barto (1981) claimed that the ISI data pose a difficulty for any network with associative synapses; that is, synapses whose efficacy changes as a function of the correlation between presynaptic and

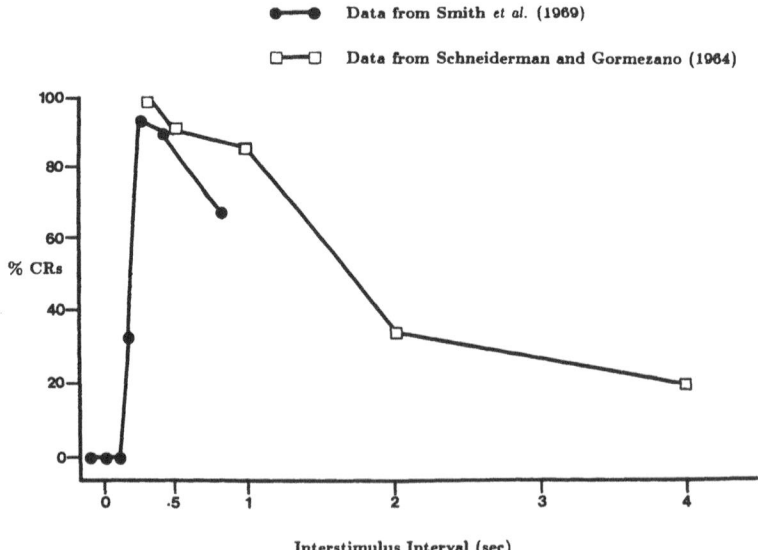

FIG. 4.4. Experimental relationship between conditioned response strength (measured by percentage of trials that response occurs) and interstimulus interval in the rabbit nictitating membrane response. (Reprinted with permission from Sutton & Barto, 1981.)

postsynaptic activities. They argued that a network with associative synapses should, to a first approximation, have an optimal ISI of zero because cross-correlation between two stimulus traces is strongest when the two stimuli occur simultaneously. In order to reproduce the ISI property, other modellers had introduced a delay in the *CS* pathway that was equal to the optimal ISI. But such a delay would delay the *CR* by an equal amount, and hence is incompatible with the anticipatory *CR* that occurs before *US* onset. To overcome this difficulty, Sutton and Barto (1981) suggested that a different type of learning law was needed that utilizes an approximation to a time-derivative operation. Although such transient reactions do play an important role in conditioning when they are elaborated in a gated dipole opponent process, they are not needed to explain the coexistence of the ISI effect and the anticipatory *CR*.

Our computer simulations contradict the Sutton and Barto (1981) claim by showing how the Grossberg (1975) conditioning model can reproduce the ISI data and the anticipatory *CR* without invoking a delay in the *CS* pathway. Poor conditioning with a small ISI is, moreover, simulated using the same mechanisms that were used to explain attentional blocking in Grossberg (1975). In the blocking paradigm, a conditioned reinforcer CS_1 blocks conditioning of a simultaneously presented CS_2. In the ISI paradigm, a *US* blocks conditioning of a simultaneously presented *CS*. In both cases, the stimulus with more motivational significance inhibits the processing of the stimulus with less motivational significance. Poor conditioning with *CS* and *US* at a large ISI occurs because, by the time the *US* arrives, the *CS* representation had decayed in short-term memory to a level that is below the threshold for affecting efficacy of the appropriate synapses.

The properties (a) to (d), listed above, arise within a network that include bidirectional associative links between sensory representations and drive representations and competitive links between different sensory representations (Figure 4.2). Associative learning within these pathways does not obey the classical Hebb (1949) postulate that requires associative strength to increase with every conditioning trial. Instead, as illustrated in equations (13) and (14) discussed later, associative learning is balanced by a process of gated memory decay, which permits synaptic strength to either increase or decrease due to pairing of presynaptic and postsynaptic activities when an activity-dependent learning gate opens (Grossberg, 1968, 1969a, 1982). Such a gated associative law has recently received direct neurophysiological support from data collected in hippocampus (Levy, Brassel, & Moore, 1983; Levy & Desmond, 1985) and visual cortex (Rauschecker & Singer, 1979; Singer, 1983). It is thus an early neural network prediction that has received impressive experimental support after a delay of many years.

The drive representations (Figure 4.2) are network loci where sensory, reinforcement, and homeostatic drive signals converge to regulate the network's motivational decisions. Such a representation was introduced into the neural network modeling literature in Grossberg (1971). Grossberg (1971, 1972a,

1972b) used drive representations to analyse a variety of data about Pavlovian and instrumental conditioning, and showed that their properties could be derived as part of the solution of a fundamental problem, called the *synchronization problem* of conditioning; that is, of how a stable conditioned response can develop even if variable time lags occur between the *CS* and the *US*, as they typically do in real-time learning environment. Subsequently, a number of authors have invoked representations that play a role analogous to the role of drive representations. Bower has called them *emotion nodes* (Bower, 1981; Bower, Gilligan, & Monteiro, 1981) and Barto, Sutton, and Anderson (1983) have called them *adaptive critic elements*.

In the model, conditioned reinforcer learning from a sensory representation to a drive representation works as follows. A *US* unconditionally activates its drive representation if the drive input to the node is sufficiently large. Repeated pairing of a *CS* with, for example, a food *US* causes correlated activation of the *CS* sensory representation, denoted S_{CS}, with that of the drive representation corresponding to hunger, denoted D_H. As a result, associative learning occurs in the synapses of pathway $S_{CS} \rightarrow D_H$, thereby enabling subsequent activation of S_{CS} to activate D_H if the drive input to D_H is also large. This is how a *CS* becomes a conditioned, or secondary, reinforcer, which clarifies the first of the four properties needed to analyse blocking.

Property (b) concerns the manner in which reinforcing properties of a sensory event shift the focus of attention towards its own processing. This property utilizes the fact that synapses in the incentive motivational pathway $D_H \rightarrow S_{CS}$ are also strengthened by conditioning. When the sensory representation S_{CS_1} is activated by the conditioned reinforcing cue CS_1, it enhances its own activation via the positive feedback loop $S_{CS_1} \rightarrow D_H \rightarrow S_{CS_1}$. Other sensory representations, such as S_{CS_2}, are placed at a competitive disadvantage because the synapses in their feedback pathways $S_{CS_2} \rightarrow D_H \rightarrow S_{CS_2}$ are weak. The enhanced activation of S_{CS_1} causes the suppression of activation at representations such as S_{CS_2} because the competition between sensory representations limits the possible total activation, or capacity, across all the sensory representations. The same competitive property attenuates conditioning of a *CS* to a *US* if the ISI is small, because the *US* activates a strong feedback pathway $S_{US} \rightarrow D \rightarrow S_{US}$ corresponding to its own sensory representation S_{US}, and inhibits activation of S_{CS} before strong conditioning in the pathways $S_{CS} \rightarrow D \rightarrow S_{US}$ can occur.

The limited capacity of short-term memory, which is needed to achieve property (c), is a property of shunting on-center off-surround feedback networks. The limited capacity property follows from more basic capabilities of this class of networks: their ability to process and store in STM spatially distributed input patterns without distorting these patterns due to either noise or saturation (Ellias & Grossberg, 1975; Grossberg, 1982; Grossberg & Levine, 1975). Figure 4.2 schematizes a network with conditionable sensory-to-drive and drive-to-sensory pathways, and on-center off-surround feedback interactions between sensory representations.

Our computer simulations, reported more completely in Grossberg and Levine (1987), modelled the behavior of the network illustrated in Figure 4.5, which is a variant of the network in Figure 4.2. Figure 4.5 includes three sensory representations, CS_1, CS_2, and US, and one drive representation, D. The $US \rightarrow D$ and $D \rightarrow US$ synapses are fixed at high value. The $CS_i \rightarrow D$ and $D \rightarrow CS_i$ synapses are strengthened by appearance of the US while the CS_i short term memory representation is active. In this variant of the network, sensory representations are divided into two successive stages. The activity of x_{i1} of the first stage of S_{CS_i} can activate conditioned reinforcer pathways $S_{CS_i} \rightarrow D$. The activity x_{i2} of the second stage of S_{CS_i} receives conditioned incentive motivational signals $D \rightarrow S_{CS_i}$ from D, which can thereupon activate x_{i1} to enhance the activation of S_{CS_i}. Activation of x_{i2} can also generate output signals to conditionable motor control pathways.

The same set of network parameters was used to simulate both the ISI inverted-U curve and attentional blocking. In both cases, the CR anticipated the US.

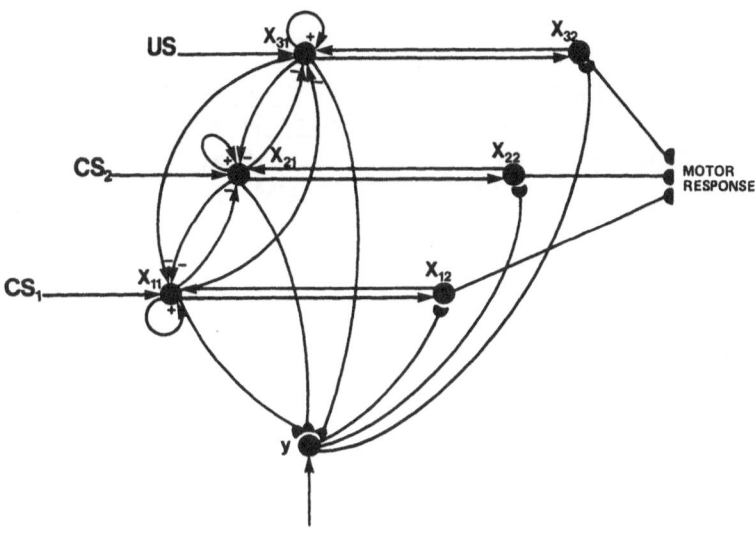

FIG. 4.5. Simulated network: Each sensory representation possesses two stages with STM activities x_{i1} and x_{i2}. A CS or US input activates its corresponding x_{i1}. Activation of x_{i1} elicits unconditionable signals to x_{i2} and conditioned reinforcer signals to D, whose activity is denoted by y. Incentive motivational feedback signals from D activate the second stage potentials x_{i2}, which then send feedback signals to x_{i1}. Conditionable long-term memory traces are designated by hemi-disks.

The simulated ISI curves (Figure 4.6) are qualitatively compatible with experimental data on the rabbit's conditioned nictitating membrane response shown in Figure 4.4. For ISI's of fewer than two time units in the numerical algorithm, competition for the US representation prevented CS activity from staying above the $S_{CS} \rightarrow D$ pathway's threshold long enough to appreciably increase the pathway's strength during the time interval when D was activated by the US. At long ISI's, the prior decay of the CS's short term memory trace prevented the $S_{CS} \rightarrow D$ pathway from sensing the later activation of D by the US.

In the blocking simulation (Figures 4.7a–4.7d), pairing of CS_1 with a delayed US enabled the long term memory trace of the $CS_1 \rightarrow D$ pathway to achieve an S-shaped cumulative learning curve (Figure 4.7c). After CS_1 became a conditioned reinforcer, it enhanced its own short-term memory storage (Figure 4.7a) by generating a large $S_{CS_1} \rightarrow D \rightarrow S_{CS_1}$ feedback signal. As a result, when CS_1 and CS_2 were simultaneously presented, the short term memory activity of S_{CS_2} was quickly suppressed by competition from CS_1 (Figure 4.7b). Consequently, the long term memory $S_{CS_2} \rightarrow D$ pathway did not grow in strength (Figure 4.7d), thereby "blocking" CS_2 and preventing it from becoming a conditioned reinforcer or eliciting a CR.

4. EXPERIMENTAL SUPPORT FOR LEARNING AT DRIVE AND MOTOR REPRESENTATIONS: FACILITATOR NEURON, HIPPOCAMPAL PYRAMID, AND CEREBELLUM

Striking neurophysiological support for the existence of drive representations has been derived from the invertebrate learning literature. Hawkins et al. (1983) and Walters and Byrne (1983) have described a *facilitator neuron* in *Aplysia* that possesses many properties of a drive representation (see Byrne et al. chap. 2 in this volume). In addition, Byrne et al. have recently modeled their *Aplysia*

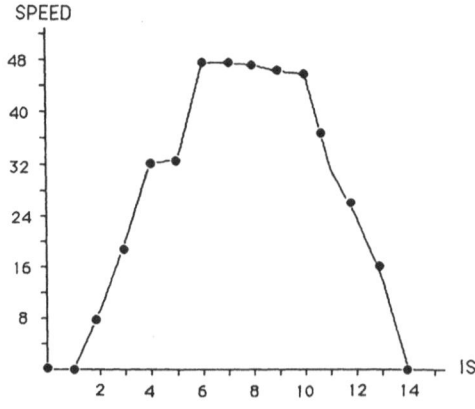

FIG. 4.6. Plot of *CR* acquisition speed as a function of ISI. This speed was computed by the formula 100 x (number of time units per trial)/(number of time units to first *CR*).

conditioning data using essentially the same model as in Figure 4.2. It is remarkable that the type of neural network learning circuit shown in Figures 4.2 and 4.5, which was first introduced to explain vertebrate behavioral conditioning data (Grossberg, 1971, 1975), has received such striking experimental support from invertebrate neurobiological experiments after a delay of two decades. The existence of such similar circuits across species seems to illustrate an evolutionary invariant of associative learning (Grossberg, 1984), which is in keeping with the derivation of the circuit from general, species-independent constraints such as the synchronization problem and the persistence problem of classical conditioning.

Vertebrate neurobiological learning data have also provided experimental support for the theory's prediction that learning occurs from a sensory representation to a drive representation. In Grossberg (1971, 1975, 1978), it was predicted that this type of learning would be found to be conditioned reinforcer learning from a sensory representation, built from thalamo-cortical circuitry, to a drive representation that should have a final common path in the hippocampus. In studies of how the rabbit nictitating membrane response is conditioned, Berger and Thompson (1978) reported hippocampal learning, but first identified their results as the discovery of a general neural "engram." Subsequent experiments considered the effects of selective ablations on learning both in hippocampus and in cerebellum. Their results led to the conclusion that the hippocampal learning is indeed a variant of the predicted conditioned reinforcer learning, whereas the cerebellum carries out a type of motor learning (Thompson et al., 1984). These experiments provided strong support for the predicted distinction between Sensory-to-Drive learning and Sensory-to-Motor learning that is summarized in Figure 4.2, and which Thompson et al. (1984, p.82) accepted when they wrote that "learning theorists have not necessarily specified that the process of 'conditioned fear' occurs first and is essential for subsequent development of the second process—learning of the discrete adaptive response."

Cerebellar motor learning was described by Thompson et al. as "the discrete adaptive response." In a broader context, it may be compared with the cerebellar motor learning that occurs during eye movements that Robinson and his colleagues have described (Optican & Robinson, 1980; Ron & Robinson, 1973). This type of cerebellar motor learning has been modeled as a specialized type of adaptive gain control by Grossberg and Kuperstein (1986, 1989). It remains to test whether the cerebellar motor learning that controls the rabbit nictitating membrane response is also this type of adaptive gain control.

5. CAN BLOCKING BE DUE TO HABITUATION AT THE DRIVE REPRESENTATION?

An alternative explanation of blocking was suggested by Hawkins and Kandel (1984). Based on data from the invertebrate *Aplysia*, these authors developed a model whereby each *US* activates a *facilitator neuron* that modulates learning by *CS* pathways. The data supporting the existence of a facilitator neuron, its

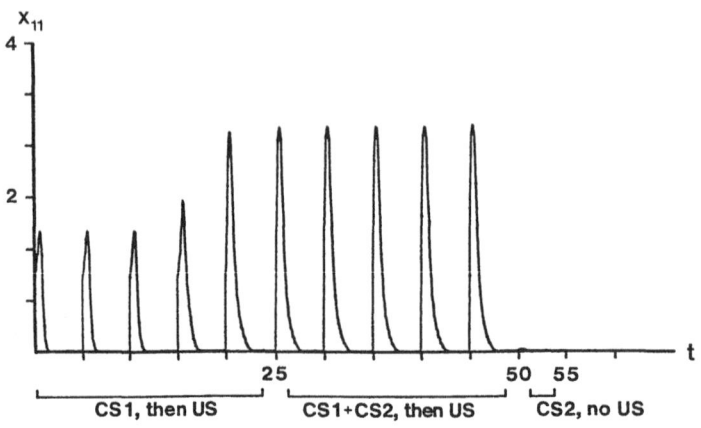

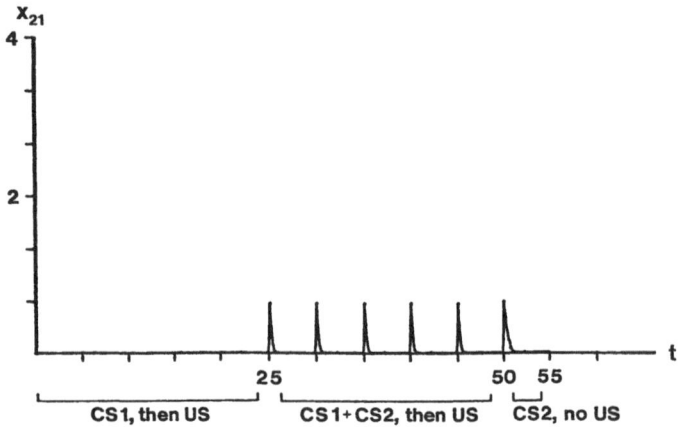

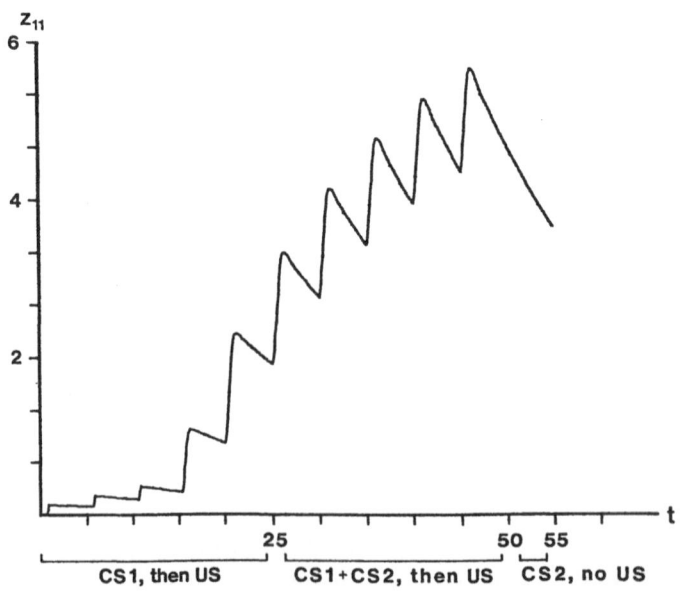

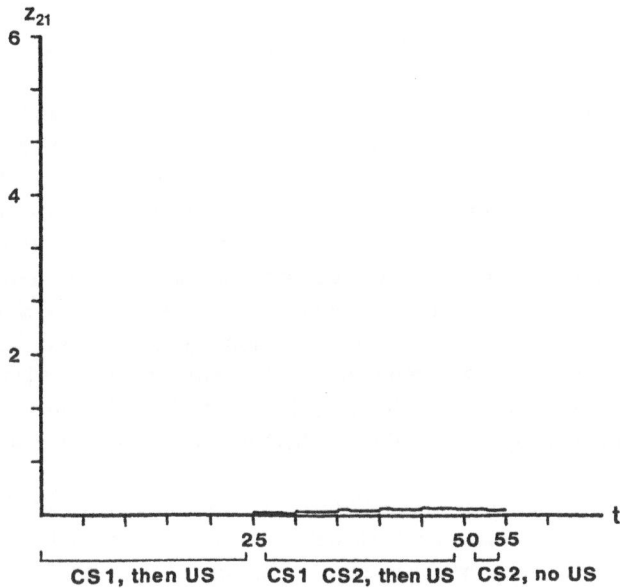

FIG. 4.7. Blocking simulation: In (a)-(d), the ISI = 6 between CS_1 and US onset. Five trials of $CS_1 - US$ pairing are followed by five trials of $(CS_1 + CS_2) - US$ pairing. Then CS_2 is presented alone for one trial. (a) Activity x_{11} of S_{cs_1} through time; (b) Activity x_{21} of S_{cs_2} through time; (c) LTM trace z_{11} from S_{cs_1} to D through time; (d) LTM trace z_{21} from S_{cs_2} to D through time.

anatomical connections with sensory representations, and its qualitative role in conditioning (Hawkins, Abrams, Carew, & Kandel, 1983; Walters & Byrne, 1983) have provided strong qualitative support for our model's predictions concerning drive representations (Grossberg, 1971, 1972a, 1972b, 1975). On the other hand, the quantitative learning mechanism posited by Hawkins and Kandel (1984) was not based on data and cannot be correct as stated. In particular, they analysed blocking using a mechanism whereby the sensitivity of the facilitator neuron habituates with use. Hawkins and Kandel (1984, p.385) wrote that "the output of the facilitator neurons decreases when they are stimulated continuously." Thus after a CS_1 is paired with a US on a number of trials, subsequent presentation of a compound stimulus $CS_1 + CS_2$ with a US does not condition CS_2 because the facilitator neuron cannot fire adequately.

Hawkins and Kandel's explanation, however, is incompatible with the fact (Kamin, 1969) that blocking can be overcome ("unblocked") if $CS_1 + CS_2$ is paired with either a higher or lower intensity of shock that CS_1 alone. A model that depends on the rate of habituation and recovery of the facilitator neuron cannot, in principle, explain unblocking because the times of stimulus occurrence in the blocking and unblocking paradigms are identical. Only the US intensity differs. Moreover, *US* intensity cannot, in itself, be used to alter the rate of facilitator recovery, because either a decrease or an increase in *US* intensity can cause unblocking.

A model that depends upon habituation of the facilitator neuron also faces difficulties in explaining secondary conditioning. After a *CS* becomes well enough conditioned to act as a *US*, it may not be able to activate the facilitator neuron during a subsequent time interval because the facilitator neuron has become depressed due to previous activation by the *US*. As a result, pairing a new CS_2 with the conditioned reinforcer CS_1 does not enable the CS_2 to become conditioned; hence, secondary conditioning does not occur. The blocking model described by Byrne and his colleagues (chap. 2 in this volume) uses facilitator habituation as a blocking mechanism. A careful choice of parameters was needed to simulate secondary conditioning using this model. This problem has more recently led Byrne and his colleagues (see this chapter) to use the type of model shown in Figure 4.2.

In our alternative model, competition between sensory representations interacts with conditioned positive feedback between sensory representations and drive representations to account for blocking. This model does not merely attenuate learning rate during blocking. It also suggests how reinforcing cues come to be better attended as a result of reinforcement learning, and offers explanations of secondary conditioning and unblocking that are parametrically robust. Most interestingly, the model used an analysis of vertebrate learning behavior to predict the existence of neural structures, such as the facilitator neuron, which also control invertebrate learning. The design constraints that were used to derive the model are not peculiar to vertebrate behavioral conditions (Grossberg, 1971, 1975; see Grossberg, 1982, 1987 for a review). The existence of facilitator neurons in invertebrates recommends a comparative neurobiological analysis to test whether similar conditioning circuits exist in a large number of animal species.

6. RELATING GATED DIPOLE OPPONENT
PROCESSES TO TEMPORAL DIFFERENCE MODELS

Our explanation of unblocking uses gated dipole opponent processes that link together positive and negative drive representations (Figure 4.8). These positive and negative opponent channels carry out comparisons in real time between current and expected levels of positive or negative reinforcement. See Grossberg (1982, 1987) for this explanation of unblocking.

In the next section, some computer simulation results of Grossberg and Schmajuk (1987) using gated dipoles are summarized. Gated dipoles are needed because, in the cognitive reinforcement circuit, *CS*'s may be conditioned to either the onset or the offset of a reinforcer. In order to explain how the offset of a positive (or negative) reinforcer can generate an antagonistic rebound in an opponent negative

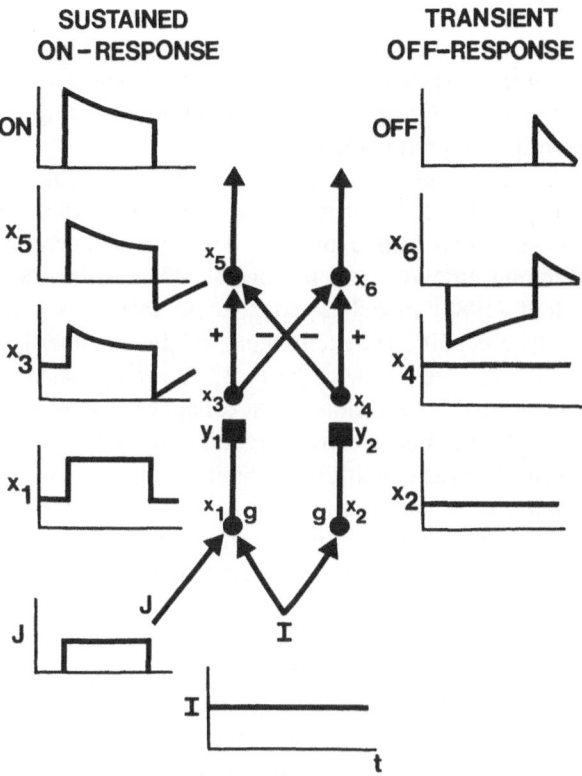

FIG. 4.8. Example of a feedforward gated dipole: A sustained habituating on-response (top left) and a transient off-rebound (top right) are elicited in response to onset and offset, respectively, of a phasic input J (bottom left) when tonic arousal I (bottom center) and opponent processing (diagonal pathways) supplement the slow gating actions (square synapses). See text for details.

(or positive) channel to which a simultaneous *CS* can be conditioned, gated dipoles were introduced in Grossberg (1972a, 1972b). A gated dipole is a minimal neural network that is capable of generating a sustained, but habituative, on-response to onset of a cue, as well as a transient off-response, or antagonistic rebound, to offset of the cue in the opponent channel (Figure 4.8).

Conditioned reinforcer learning from a sensory representation to the habituating on-responses and transient off-rebounds that occur at a gated dipole pair of opponent drive representations explicates the types of transient properties that time-derivative models of Pavlovian conditioning have strived to capture (e.g., Klopf, 1988; Moore et al., 1986; Sutton & Barto, 1981, 1989). The earlier time-derivative models (e.g., Sutton & Barto, 1981) explicitly computed the difference between two successive activity levels at each synapse. Such a model could not capture more gradual transient effects, such as the habituative properties of a gated dipole (Figure 4.8). It is also too sensitive to noise fluctuations. The modification in Sutton and Barto (1989) of the Sutton and Barto (1981) model, called the *Temporal-Difference* (TD) *model,* uses computations that are more similar to those that occur during conditioned reinforcer learning at a gated dipole pair of opponent drive representations.

In the time-derivative models, only a single drive representation, or adaptive critic element, is postulated, rather than opponent pairs of such representations. This prevents these models from explaining how opponent drive representations can activate different emotions and responses. When a time-derivative model such as the Klopf (1988) model is extended to include a pair of opponent representations, then conditioned excitors and inhibitors either both extinguish, or neither extinguish, in conflict with the data. Thus either such a model cannot represent opponent affective reactions, or it cannot explain selective forgetting.

Our model's explanation of selective forgetting, in particular how conditioned excitors extinguish whereas conditioned inhibitors do not (Grossberg & Schmajuk, 1987), is summarized later. This explanation uses the same mechanisms that had earlier been used to explain unblocking (Grossberg, 1975, 1982). Both explanations analyse how the match or mismatch of a learned sensory expectation by sensory data leads to memory persistence or to selective forgetting, respectively. Mechanisms for learning and matching sensory expectations have not yet been included in the time-derivative models.

The need to understand how sensory expectations are learned and matched for the purpose of controlling blocking and unblocking (Grossberg, 1975) was one of the major problems that led to the introduction of Adaptive Resonance Theory (Grossberg, 1976a, 1976b). This analysis has since clarified how processes of recognition learning modulate reinforcement learning, and how reinforcement learning may, in turn, modify recognition learning to generate internal representations that can better predict behavioral success. A large body of interdisciplinary data have by now supported both the main principles and key mechanisms of this theory (Grossberg, 1987, 1988). The strongest data sets for analysing the process of recognition learning are not, however, usually studied as part of the Pavlovian conditioning literature. They form part of the cognitive science literature. This sociological fact may explain why an analysis of recognition learning has not yet played a greater role in the design of time-derivative conditioning models.

7. THE READ CIRCUIT: A SYNTHESIS OF OPPONENT PROCESSING AND ASSOCIATIVE LEARNING MECHANISMS

Several versions of a gated dipole circuit can be used to model associative learning between a CS and the onset or the offset of a reinforcer. A gated dipole needs to contain internal feedback pathways, however, to explain secondary inhibitory conditioning (Grossberg, 1975, 1987). Secondary inhibitory conditioning consists of two phases. In phase one, CS_1 becomes an excitatory conditioned reinforcer (e.g., a source of conditioned fear) by being paired with a US (e.g., a shock). In phase two, the offset of CS_1 generates an off-response that is used to condition a subsequent CS_2 to become an inhibitory conditioned reinforcer (e.g., a source of conditioned relief). It was to explain how this happens that gated dipole circuits with internal feedback pathways—that is, recurrent dipoles—were introduced and joined to a mechanism of associative learning from sensory representations to dipole-organized drive representations. Grossberg and Schmajuk (1987) have further developed these properties into a model called a READ circuit, as a mnemonic for REcurrent Associative gated Dipole.

The equations for a typical READ circuit, depicted in Figure 4.9, are as follows:

Arousal + US + Feedback On-Activation:

The reinforcing US input J, a tonically active arousal input I, and a positive feedback signal $T(x_7)$ activate the on-potential x_1, as in equation (1):

$$\frac{d}{dt}x_1 = -A_1x_1 + I + J + T(x_7) \tag{1}$$

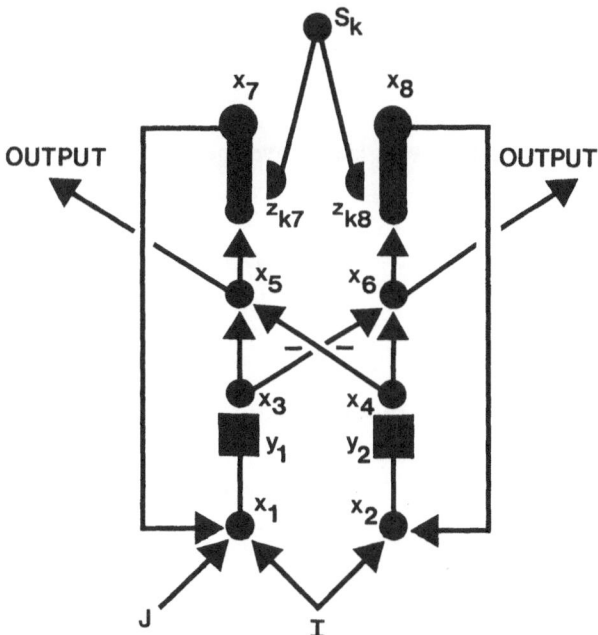

FIG. 4.9.   A READ I circuit: This circuit joins together a recurrent gated dipole with an associative learning mechanism. Learning is driven by signals S_k, which activate long term memory (LTM) traces z_{k7} and z_{k8} that sample activation levels at the on-channel and off-channel, respectively, of the gate dipole. See text for details.

Arousal + Feedback Off-Activation:

A similar law governs the off-potential x_2:

$$\frac{d}{dt}x_2 = -A_2x_2 + I + T(x_8) \tag{2}$$

On-Transmitter:

The on-transmitter gate y_1 accumulates at rate $B(1 - y_1)$ and is inactivated at rate $Cg(x_1)y_1$, which leads to a net level of transmitter habituation:

$$\frac{d}{dt}y_1 = B(1 - y_1) - Cg(x_1)y_1 \tag{3}$$

Off-Transmitter:

A similar law governs the off-transmitter gate y_2:

$$\frac{d}{dt}y_2 = B(1 - y_2) - Cg(x_2)y_2 \tag{4}$$

Gated On-Activation:

The gated output signal $Dg(x_1)y_1$ is time averaged by the on-potential x_3 at the next processing stage:

$$\frac{d}{dt}x_3 = -A_3x_3 + Dg(x_1)y_1 \tag{5}$$

Gated Off-Activation:

A similar law governs the off-potential x_4:

$$\frac{d}{dt}x_4 = -A_4x_4 + Dg(x_2)y_2 \tag{6}$$

Normalized Opponent On-Activation:

Opponent inputs are time-averaged by the on-potential x_5 at the next processing stage:

$$\frac{d}{dt}x_5 = -A_5x_5 + (E - x_5)x_3 - (x_5 + F)x_4 \tag{7}$$

Normalized Opponent Off-Activation:

Opponent inputs of opposite sign are time-averaged by the off-potential x_6:

$$\frac{d}{dt}x_6 = -A_6x_6 + (E - x_6)x_4 - (x_6 + F)x_3 \tag{8}$$

Total On-Activation:

The net on-signal $G[x_5]^+$ after opponent processing is added to the total input from the sensory representations and then time-averaged by on-potential x_7:

$$\frac{d}{dt}x_7 = -A_7x_7 + G[x_5]^+ + L \sum_{k=1}^{n} S_k z_{k7} \tag{9}$$

Total Off-Activation

A similar process occurs at the off-potential x_8:

$$\frac{d}{dt}x_8 = -A_8x_8 + G[x_6]^+ + L \sum_{k=1}^{n} S_k z_{k8} \tag{10}$$

On-Conditioned Reinforcer Association:

The associative learning trace z_{k7} from the kth sensory representation to the on-channel is activated when the signal S_k from the kth sensory representation becomes positive; it then time averages the net output signal $K[x_5]^+$ from the opponent process:

$$\frac{d}{dt}z_{k7} = S_k[-Hz_{k7} + K[x_5]^+], \tag{11}$$

where the notation $[x_i]^+$ denotes a linear signal above the threshold value zero; that is, max $(x_i,0)$.

Off-Conditioned Reinforcer Association:

A similar law governs the learning traces z_{k8} to the off-channel:

$$\frac{d}{dt}z_{k8} = S_k[-Hz_{k8} + K[x_6]^+] \tag{12}$$

On-Output Signal:

The on-channel output signal from the drive representation to other network circuits is generated by the opponent process:

$$O_1 = [x_5]^+ \tag{13}$$

Off-Output Signal:

A similar law governs the output signal from the off-channel to other network circuits:

$$O_2 = [x_6]^+, \tag{14}$$

Parameters A,B,C,D,E,F,G,H,K, and L were kept constant for all simulations. When $E = F$, x_5 and x_6 in equations (7) and (8) compute an opponent process and a ratio scale. For example, at equilibrium, equation (7) then implies

$$x_5 = \frac{E(x_3 - x_4)}{A_5 + x_3 + x_4}. \tag{15}$$

This property of the READ circuit enables the circuit to achieve the property of *associative averaging*, rather than of simple summation. The opponent ratio scale described in equation (17) defines the baseline value against that transient reactions in the on-channel and off-channel are compared to generate net output signals, as in (15) and (16), from the gated dipole.

8.   OPPONENT EXTINCTION BY DISSOCIATING LONG TERM MEMORY READ-IN AND READ-OUT

The READ circuit controls selective forgetting through a mechanism that we have called *opponent extinction*. Passive memory decay does not occur in the parameter ranges that we simulated. A network wherein passive memory decay occurs cannot achieve a large memory capacity, because all associations must be actively practiced before they are forgotten, but only a limited number of events can be practiced within a fixed amount of time.

An active process of selective forgetting is achieved by opponent extinction. In particular, when the net signals in the on- and off-channels are balanced, then by equation (15), $x_5 = 0 = x_6$. Consequently, by equations (11) and (12), the conditioned reinforcer associations z_{k7} and z_{k8} from sensory representation S_k to the on- and off-drive representations approach 0. By a similar argument, it can be seen that these associations continually readjust themselves to the *net imbalance* of activation between the on- and off-channels. Opponent extinction hereby avoids the possibility that both associative traces z_{k7} and z_{k8} could saturate at maximal values, and thereby terminate the ability of the kth sensory representation to engage in further learning.

Opponent extinction requires the third key property of the READ circuit for its realization. This property is a dissociation between read-in and read-out of associative memories. For example, in the on-channel, associative read-out is proportional to $[x_7]^+$, whereas associative read-in is proportional to $[x_5]^+$. The read-out signals $\sum_k S_k z_{k7}$ and $\sum_k S_k z_{k8}$ in equations (9) and (10) are fed back through the dipole via the signal pathways $x_7 \rightarrow x_1 \rightarrow x_3 \rightarrow \{x_5, x_6\}$ and $x_8 \rightarrow x_2 \rightarrow$

$x_4 \rightarrow \{x_6, x_5\}$ to compete via opponent interactions. These interactions generate a consensus, or competitive decision, as in equation (15), which provides the data $\{x_5, x_6\}$ that are read-in to associative memory via equations (11) and (12).

9. ROLE OF DENDRITIC SPINES AND Ca^{++} CURRENTS IN DISSOCIATION OF LONG TERM MEMORY READ-IN AND READ-OUT

Grossberg (1975) predicted that such a dissociation between associative read-in and read-out could be physiologically implemented by assuming that synaptic plasticity occurs at the dendritic spines of neural cells, notably those of hippocampal and cortical pyramidal cells. In Figure 4.10, signal $[x_5]^+$ causes a global potential change that invades all the spines, thereby inducing learned changes at all synapses throughout the dendritic column, as in equation (11). However, due to the geometry and electrical properties of the dendritic tree, a sensory input S_k that activates a particular dendritic branch may not be influenced by inputs that activate different dendritic branches. Activation at a particular dendritic branch

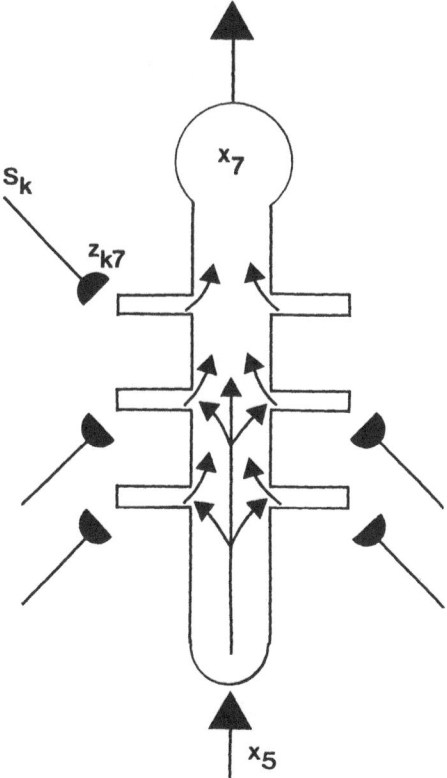

FIG. 4.10. A possible microarchitecture for dissociation of LTM read-in and read-out: Individual LTM-gated sensory signals $S_k z_{k7}$ are read-out into local potentials that are summed by the total cell body potential x_7 without significantly influencing each other's learned read-in. In contrast, the input signal x_5 triggers a massive global cell activation that drives learned read-in at all active LTM traces abutting the cell surface. Signal x_5 also activates the cell body potential x_7.

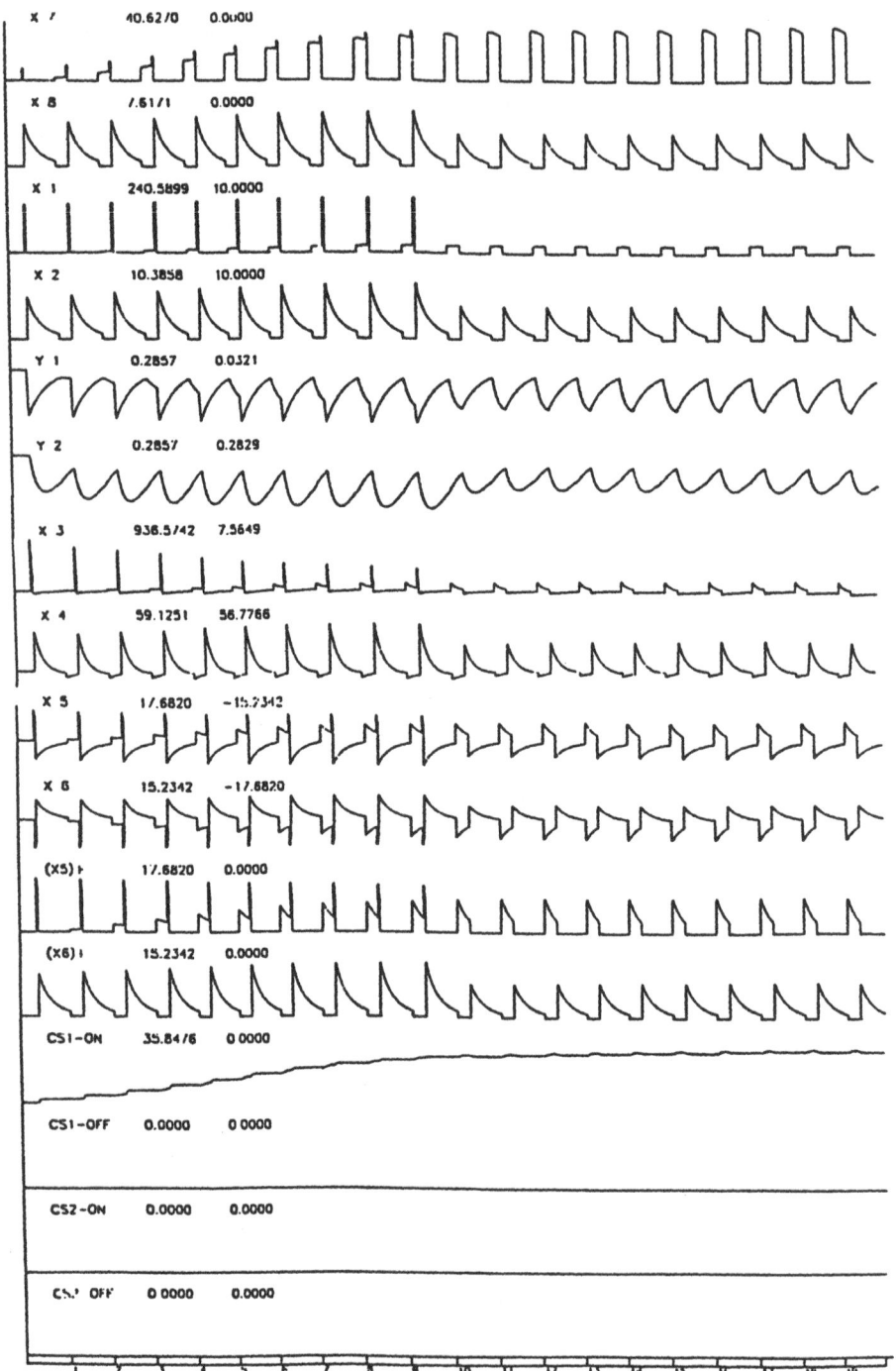

would produce local potentials that do not, in themselves, cause learning. These local potentials propagate to the cell body where they influence axonal firing via potential x_7 in equation (9). Potential x_7 activates x_5 via the competitive feedback loop, and x_5 can influence learning, as in equations (15) and (11).

An early analysis of possible biochemical substrates of associative learning laws such as (11) and (12) further predicted that associative learning may be controlled by an inward Ca^{++} current interacting synergetically with an inward Na^+ and/or outward K^+ current (Grossberg, 1968, 1969b; see also Grossberg, 1982, chap. 3). Given the fragmentary biochemical evidence that was available in the 1960's, the prediction emphasized the possible synergy between inward Ca^{++} and Na^+ currents to augment cell depolarization. Recent data have suggested that this effect may be achieved by a synergy between an inward Ca^{++} current and an outward K^+ current.

The work of Alkon and his colleagues (chap. 1 in this volume) is relevant to both predictions. They have reported that Ca^{++} moves cytoplasmic C kinase into a membrane channel where it alters an outward K^+ current, and that such an effect seems to occur at the dendrites of CA1 hippocampal pyramidal cells.

10.   COMPUTER SIMULATIONS OF PRIMARY AND SECONDARY CONDITIONING

This section summarizes computer simulations by Grossberg and Schmajuk (1987) of several classical conditioning paradigms. Although the simulations show the competence of the READ circuit in these paradigms, additional neural machinery (such as the ART circuit in Figure 4.1) is also necessary to explain conditioning data that involves attentional regulation.

Excitatory primary conditioning.   Because the CS is presented in the presence of the US, it becomes associated with the on-response. Variable CS_1-ON describes conditioning of the LTM trace z_{17} within the pathway from the sensory representation of CS_1 to the READ on-channel. After 10 acquisition trials, presentations of CS_1 alone do not cause extinction of the CS_1-ON association (Figure 4.11). As noted in Section 8, forgetting of CS_1-ON associations is due to the acquisition of CS_1-OFF associations.

FIG. 4.11. Computer simulation of primary excitatory conditioning and extinction with slow habituation and large feedback in a READ I circuit: CS_1 is paired with the US during the first 10 simulated trials, and CS_1 is presented in the absence of the US in the next 10 simulated trials. The numbers above each plot are the maximum and minimum values of the plot. Parameters are $A = 1$, $B = .005$, $C = .00125$, $D = 20$, $E = 20$, $F = 20$, $G = .5$, $H = .005$, $K = .025$, $L = 20$, $M = .05$.

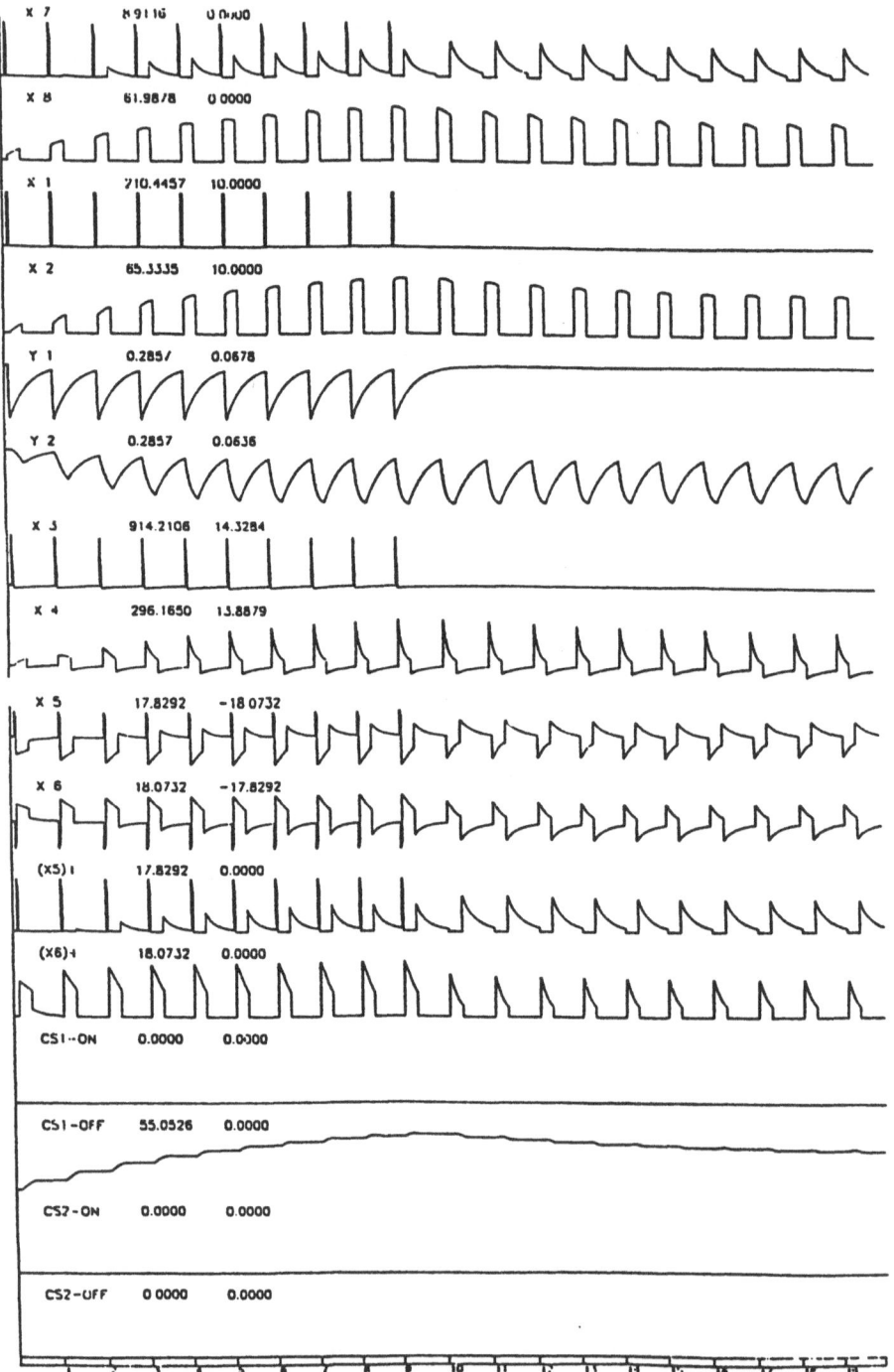

Inhibitory primary conditioning.   Because the *CS* is presented after *US* offset, it becomes associated with the off-response that is generated by an antagonistic rebound within the READ circuit. Variable CS_1-OFF describes conditioning of the LTM trace z_{18} within the pathway from the sensory representation of CS_1 to the off-channel. After 10 acquisition trials, presentations of CS_1 alone cause the CS_1-OFF association to relax to a persistent remembered value (Figure 4.12). As noted in Section 8, forgetting of the CS_1-OFF association is due to the acquisition of CS_1-ON associations.

In Grossberg and Schmajuk (1987), the following additional types of secondary conditioning phenomena were also simulated:

Excitatory secondary conditioning.   The LTM trace CS_1-ON grows during the first 10 trials and is then used to induce the growth of the LTM trace CS_2-ON during the next 10 trials.

Inhibitory secondary conditioning.   The LTM trace CS_1-ON grows during the first 10 trials and is then used, by presenting a CS_2 after CS_1 offset, to induce the growth of the LTM trace CS_2-OFF during the next 10 trials.

11.   DISCONFIRMED EXPECTATIONS CONTROL SELECTIVE FORGETTING

This section presents qualitative explanations for some difficult conditioning data that require the additional neural machinery of an ART circuit. These explanations clarify how disconfirmed sensory expectations regulate selective forgetting by interacting with the opponent extinction mechanism.

Excitatory conditioning and extinction.   When a *CS* is paired with an aversive *US* on successive conditioning trials, the sensory representation S_1 of CS_1 is conditioned to the drive representation D_{on} corresponding to the fear reaction, both through its conditioned reinforcer path $S_1 \rightarrow D_{on}$ and through its incentive motivational path $D_{on} \rightarrow S_1$. As a result, later presentations of CS_1 tend to generate an amplified STM activation of S_1, and thus CS_1 is preferentially attended. Due to the limited capacity of STM, less salient cues tend to be attentionally blocked when CS_1 is presented.

FIG. 4.12.   Computer simulation of primary inhibitory conditioning and extinction with slow habituation and large feedback in a READ I circuit: CS_1 is presented after the *US* offset during the first 10 simulated trials, and CS_1 is presented in the absence of the *US* in the next 10 simulated trials. The same parameters were used as in Figure 4.11.

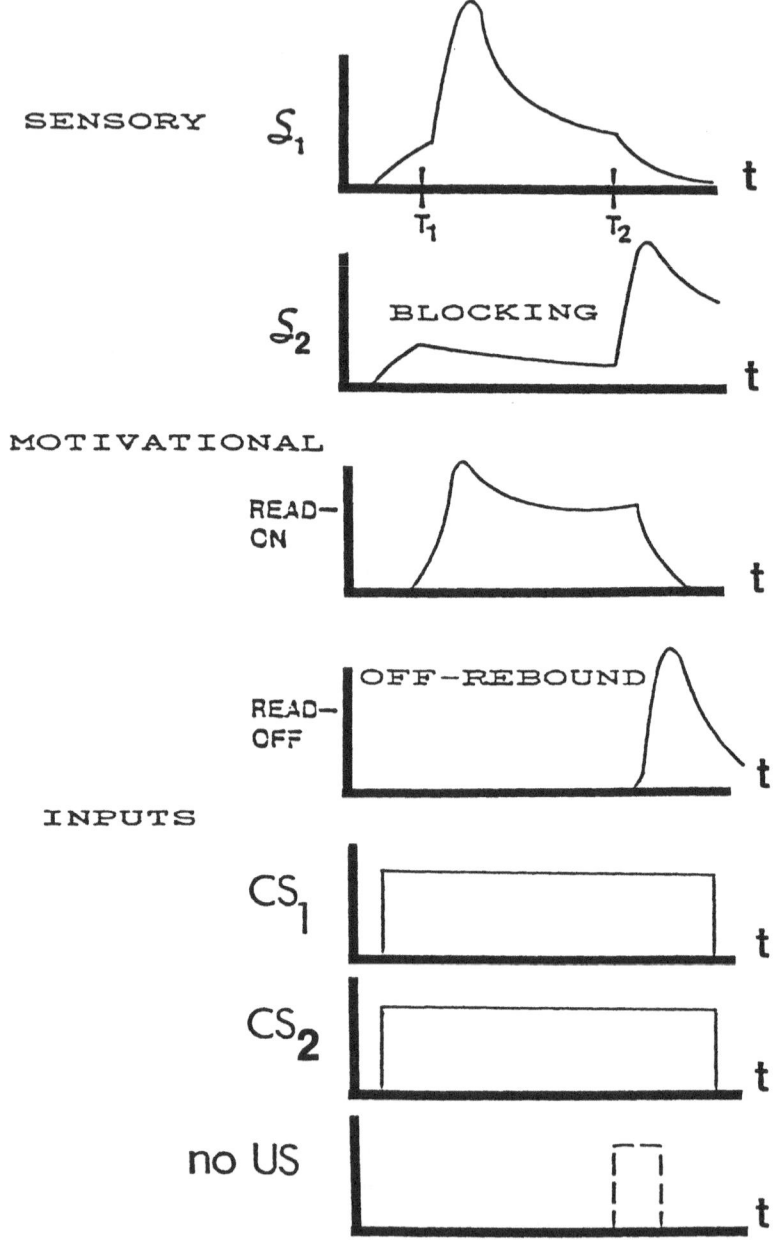

FIG. 4.13. Presentation of CS_1 and CS_2 when CS_1 has become a conditioned excitor and the compound stimulus is followed by no-shock: During the no-shock interval between times T_1 and T_2, S_1 is actively amplified by positive feedback and S_2 is blocked. Nonoccurrence of the expected shock causes both S_1 and S_2 to be reset. S_1's STM activity decreases and S_2's STM activity increases. Due to S_1's increase, D_{on}

As the cognitive-emotional feedback loop $S_1 \rightarrow D_{on} \rightarrow S_1$ is strengthened during conditioning trials, S_1 is also associated to a sensory expectation of the shock within an ART circuit. During extinction, S_1 is presented on unshocked trials. Parameters of the READ circuit are chosen to prevent passive decay of LTM traces from occurring on these trials. However, when the expected shock does not occur, a mismatch occurs with the learned sensory expectation of shock that is read-out by S_1. The mismatch causes STM to be reset (Grossberg, 1982, 1987). In particular, the STM activity of S_1 is reduced by the STM reset. This reduction in S_1 causes a reduction in the total input to D_{on}, as in equation (9). An antagonistic rebound consequently occurs in the off-channel D_{off} of the READ circuit. As a result, S_1 becomes associated with the antagonistic rebound at D_{off}. Because S_1 is smaller after reset than before, $S_1 \rightarrow D_{off}$ associations take place at a slower rate than during conditioning. After several learning trials, however, the pathway $S_1 \rightarrow D_{off}$ can become as strong as the $S_1 \rightarrow D_{on}$ pathway. Due to opponent extinction, the associative weights to both D_{on} and D_{off} are selectively forgotten. Thus a conditioned excitor extinguishes.

Inhibitory conditioning and non-extinction. Suppose that CS_1 has become a conditioned excitor by being paired with the US, and that CS_1 and CS_2 are then presented together in absence of the US. When CS_1 and CS_2 are simultaneously presented (Figure 4.13), S_1's activity is amplified by positive feedback through the strong conditioned $S_1 \rightarrow D_{on} \rightarrow S_1$ pathway. As a result of the limited capacity of STM, the STM activity of S_2 is blocked at time T_1, as in Figure 4.13. When the expected US does not occur at time T_2, the mismatch with S_1's sensory expectation causes both S_1 and S_2 to be reset, and S_1's STM activity decreases whereas S_2's STM activity increases. Due to S_1's decrease, a rebound occurs at D_{off}. Consequently, the unexpected nonoccurrence of the shock enables S_2 to become associated with D_{off} in both the pathways $S_2 \rightarrow D_{off}$ and $D_{off} \rightarrow S_2$. These learned changes turn CS_2 into a conditioned inhibitor. In addition, S_2 is conditioned to a sensory expectation of situational cues that do not include the occurrence of shock.

Due to properties of the READ circuit, when CS_2 is subsequent presented alone, the conditioned value of $S_2 \rightarrow D_{off}$ does not passively decay. In addition, S_2 activates a sensory expectation that predicts the absence of the US. Since this sensory expectation is not disconfirmed, S_2's STM activity is not reset. Consequently, no antagonistic rebound occurs within the READ circuit, and no opponent extinction occurs. Thus a conditioned inhibitor does not extinguish.

---

also decreases, thereby causing a rebound at D_{off}. This rebound becomes associated with the increased activity of S_2.

12. TIMING THE BALANCE BETWEEN EXPLORATION
FOR NOVEL REWARDS AND CONSUMMATION
OF EXPECTED REWARDS

The remainder of this chapter summarizes a model of a neural circuit that controls behavioral timing (Grossberg & Schmajuk, 1989). Several different types of brain processes organize the temporal unfolding of serial order in behavior. The present model concerns one type of timing circuit, and outlines how this circuit may be embedded in larger neural systems that regulate several different types of temporal organization.

Many goal objects may be delayed subsequent to the actions that elicit them, or the environmental events that signal their subsequent arrival. Humans and many animal species can learn to wait for the anticipated arrival of a delayed goal object. Such behavioral timing is important in the lives of animals that are capable of exploring their environments for novel sources of gratification. If an animal could not inhibit its exploratory behavior, then it could starve to death by restlessly moving from place to place, unable to remain in one place long enough to carry out the consummatory behaviors needed to acquire food there. On the other hand, if an animal inhibited its exploratory behavior for too long, waiting for an expected source of food to materialize, then it could starve to death if food was not, after all, forthcoming.

Thus the animal's task is to accurately time the *expected* delay of a goal object based upon its previous experiences in a given situation. It needs to regulate the balance between its exploratory behavior aimed at searching for novel sources of reward, and its consummatory behavior aimed at acquiring expected sources of reward. To effectively control this balance, the animal needs to be able to suppress its exploratory behavior and focus its attention upon an expected source of reward at around the time that the expected delay transpires for acquiring the reward.

13. DISTINGUISHING EXPECTED
NONOCCURRENCES FROM UNEXPECTED
NONOCCURRENCES: INHIBITING THE NEGATIVE
CONSEQUENCES OF EXPECTED NONOCCURRENCES

This type of timed behavior calibrates the delay of a single behavioral act, rather than organizing a correctly timed and speed controlled sequence of acts. For example, suppose that after pushing a lever an animal typically receives a food pellet from a food magazine two seconds later. Suppose that the animal orients to the food magazine right after pushing the lever. When the animal inspects the food magazine, it perceives the nonoccurrence of food during the subsequent two seconds. These nonoccurrences disconfirm the sensory expectation that food will appear in the magazine. Because the perceptual processing cycle that processes this sensory information occurs at a much faster rate than two seconds, it can compute this sensory disconfirmation many times before the two second delay has elapsed.

The central issue is: What spares the animal from erroneously reacting to these *expected nonoccurrences* of food during the first two seconds as predictive failures? Why does not the animal immediately become frustrated by the nonoccurrence of food and release exploratory behavior aimed at searching for food in another place? On the other hand, if food does not appear after two seconds have elapsed, why does the animal then react to the *unexpected nonoccurrence* of food by becoming frustrated and releasing exploratory behavior?

Grossberg and Schmajuk (1989) argued that the primary role of the timing mechanism is to inhibit, or *gate,* the process whereby sensory mismatches trigger the orienting and reinforcing mechanisms that would otherwise reset the animals' attentional focus, negatively reinforce its previous consummatory behavior, and release its exploratory behavior. The process of registering these sensory mismatches or matches is not inhibited, because if the food happened to appear earlier than expected, the animal could perceive it and eat. Thus the sensory matching process is not inhibited by the timing mechanism. Instead, the effects of sensory mismatches upon processes of sensory reset and reinforcement are inhibited.

This inhibitory action is assumed to be part of a more general competition that occurs between the motivational, or arousal, sources that energize different types of behavior. The posited inhibitory action is from the motivational sources of consummatory behaviors to the motivational sources of orienting and exploratory behaviors (Grossberg, 1975). The consummatory motivational sources are also assumed to be in mutual competition, enabling only the strongest combinations of sensory, reinforcing, and homeostatic signals to control observable behaviors (Grossberg, 1982, chap. 6; Staddon, 1983, Staddon & Zhang, chap. 11 in this volume). Thus the posited competition is a special case of the general hypothesis that the output signals from all motivational sources compete for the control of observable behaviors.

14. SPECTRAL TIMING MODEL: AN APPLICATION OF GATED DIPOLE THEORY

The timing model is a specialized design for an opponent processing network of gated dipole type. It is called the Spectral Timing Model for reasons described later. The circuit diagram of the Spectral Timing model is schematized in Figure 4.14. The CS activates a population of cells whose members react at different rates, according to a spectrum of rates α_i. Neural populations whose elements are distributed along a temporal or spatial parameter are familiar throughout the nervous system. Two examples are the *size principle,* which governs variable

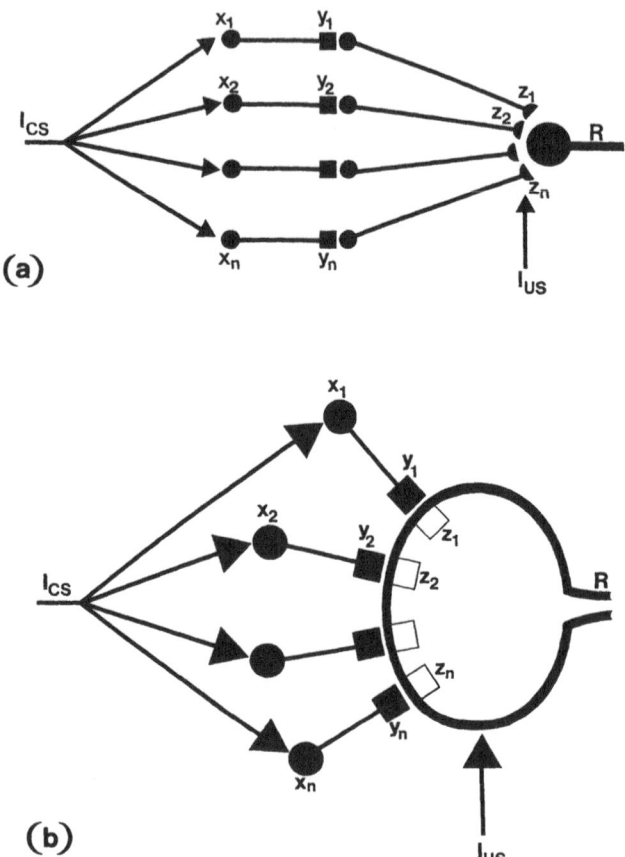

FIG. 4.14.   Circuit diagram of the Spectral Timing Model. The function $I_{cs}(t)$ denotes a step function input that is proportional to the CS intensity and stays on after the CS offset; x_i denote cell activities with different growth rates a_i; z_i denote adaptive long term memory traces; and $R(t)$ denotes the total circuit output. In version (a) of the model, the z_i are computed in terminals of the presynaptic pathways converging upon the output neuron, and the I_{us} activates them presynaptically. In (b), the z_i are computed as part of the postynaptic membrane of the output neuron, and the I_{us} activates them via a postsynaptic route.

rates of responding in spinal motor centers (Henneman, 1957, 1985), and the spatial frequency-tuned cells of the visual cortex, which also react at different rates (Jones & Kreck, 1978; Musselwhite & Jeffreys, 1985; Parker & Salzen, 1977a, 1977b; Parker, Salzen, & Lishman, 1982a, 1982b; Plant, Zimmerin, & Durden, 1983; Skrandies, 1984; Vassilev & Strashimirov, 1979; Vassilev, Manahilov, & Mitov, 1983; Williamson, Kaufman, & Brenner, 1978).

The model equations will first be listed and then explained.

SPECTRAL TIMING EQUATIONS

Spectral Activation

$$\frac{d}{dt}x_i = \alpha_i[-Ax_i + (1 - Bx_i)I_{CS}(t)\,]; \tag{16}$$

Transmitter Gate

$$\frac{d}{dt}y_i = C(1 - y_i) - Df(x_i)y_i, \tag{17}$$

where $f(x_i)$ is a sigmoid signal function of the form

$$f(x_i) = \frac{x_i^n}{\beta^n + x_i^n}; \tag{18}$$

Associative Learning (LTM Trace)

$$\frac{d}{dt}z_i = Ef(x_i)y_i[-z_i + I_{US}(t)]; \tag{19}$$

Output Signal

$$R = [\sum_i f(x_i)y_iz_i - F]^+; \tag{20}$$

where

$$[w]^+ = \begin{cases} w \text{ if } w > 0 \\ 0 \text{ if } w \leqslant 0. \end{cases} \tag{21}$$

A. The Activation Spectrum

The function $I_{CS}(t)$ is assumed to be a step function whose amplitude is proportional to the CS intensity, and which stays on for a fixed time after CS offset because it is internally stored in short term memory (STM). Figure 4.15 depicts a typical relationship between CS, $I_{CS}(t)$, and the US input $I_{US}(t)$. Input $I_{CS}(t)$ activates all potentials x_i in (16) of the cells in its target population. The potentials x_i respond at rates proportional to α_i, $i = 1, 2, \ldots, n$.

Each potential x_i generates the output signal $f(x_i)$. Figure 4.16a depicts the results of a computer simulation, in which $f(x_i(t))$ is plotted as a function of time t for values of α_i ranging from .2 ("fast cells") to .0025 ("slow cells").

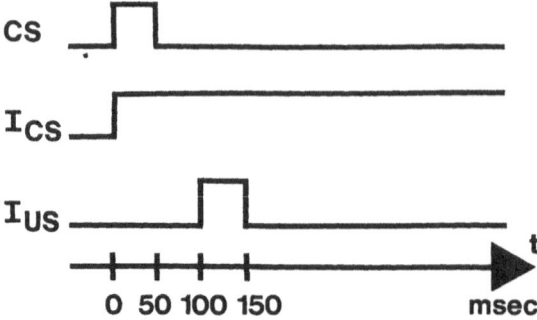

FIG. 4.15. Temporal arrangement of a 50-msec CS and a 50-msec US separated by a 100-msec ISI. I_{cs} is the step function activated by the CS that inputs to the Spectral Timing Model.

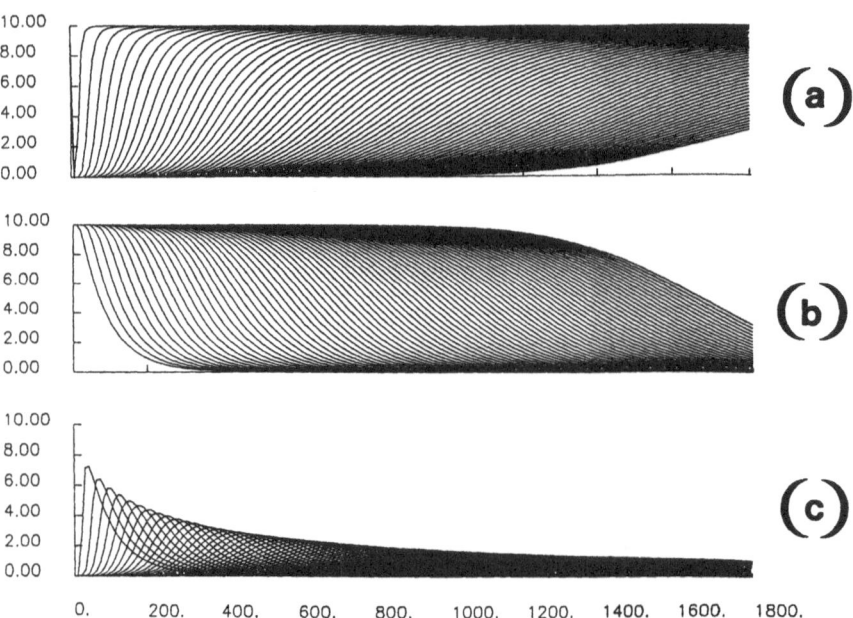

FIG. 4.16. The spectrum of reactions to a step input I_{cs}: (a) Eighty signal functions $f(x_i(t))$, $i = 1,2,...,80$, are plotted as a function of t. (b) The corresponding eighty habituative transmitter gates $y_i(t)$ are plotted as a function of t. (c) The corresponding gated signals $g_i(t) = f(x_i(t))y_i(t)$ are plotted as a function of t. Parameters are: $α_i = .2i^{-1}$ for $i = 1,2,...,80$; $A = 0$; $B = 1$; $C = .0001$; $D = .125$; $β = .8$; $n = 8$; $I_{cs}(t) = 1$ for $t > 0$. In all simulations, one time step represents 1 msec and all $f(x_i(0)) = 0$ and $y_i(0) = 1$.

B. The Habituation Spectrum

Each output signal $f(x_i)$ activates a neurotransmitter y_i. According to equation (17), process y_i accumulates to a constant target level 1, via term $C(1 - y_i)$, and is inactivated, or *habituates*, due to a mass action interaction with signal $f(x_i)$, via term $-Df(x_i)y_i$. The different rates α_i at which each x_i is activated causes the corresponding y_i to become habituated at a different rate. A habituation spectrum is thereby generated. The signal functions $f(x_i(t))$ in Figure 4.16a generate the habituation spectrum of $y_i(t)$ curves in Figure 4.16b.

C. The Gated Signal Spectrum

Each signal $f(x_i)$ interacts with y_i via mass action. This process is also called *gating* of $f(x_i)$ by y_i to yield a net signal g_i proportional to $f(x_i)y_i$. Each gated signal $g_i(t) \equiv f(x_i(t))y_i(t)$ has a different rate of growth and decay, thereby generating the gated signal spectrum shown in Figure 4.16c. In these curves, each function $g_i(t)$ is a unimodal function of time, where function $g_i(t)$ achieves its maximum value M_i at time T_i; T_i is an increasing function of i; and M_i is a decreasing function of i.

D. Temporally Selective Associative Learning

Each *long term memory (LTM) trace* z_i in (19) is activated by its own sampling signal g_i. The sampling signal g_i turns on the learning process, and causes z_i to approach I_{US} during the sampling interval at a rate proportional to g_i. Each z_i grows by an amount that reflects the degree to which the curves $g_i(t)$ and $I_{US}(t)$ have simultaneously large values through time.

The time interval between CS onset and US onset is called the *interstimulus interval*, or ISI. The individual LTM traces differ in their ability to learn at different values of the ISI. This is the basis of the network's timing properties.

Figure 4.17 illustrates how six different LTM traces z_i, $i = 1, \ldots, 6$, learn during a simulated learning experiment. The CS and US are paired during 4 learning trials, after which the CS is presented alone on a single performance trial.

E. The Doubly Gated Signal Spectrum

The CS input $I_{CS}(t)$ remains on and constant throughout the duration of each learning trial. The US input $I_{US}(t)$ is presented after an ISI of 500 msec. unit and remains on for 50 msec. The upper panel in each part of the figure depicts the gated signal function $g_i(t)$ with α_i chosen at progressively slower rates. The middle panel plots the corresponding LTM trace $z_i(t)$, and the lower panel plots the doubly gated signal $h_i(t) = f(x_i(t))y_i(t)z_i(t)$. Each doubly gated signal function $h_i(t)$ registers how well the timing of CS and US is registered by the ith processing channel. Note that in Figure 4.17c, whose gated signal $g_i(t)$ peaks at approxi-

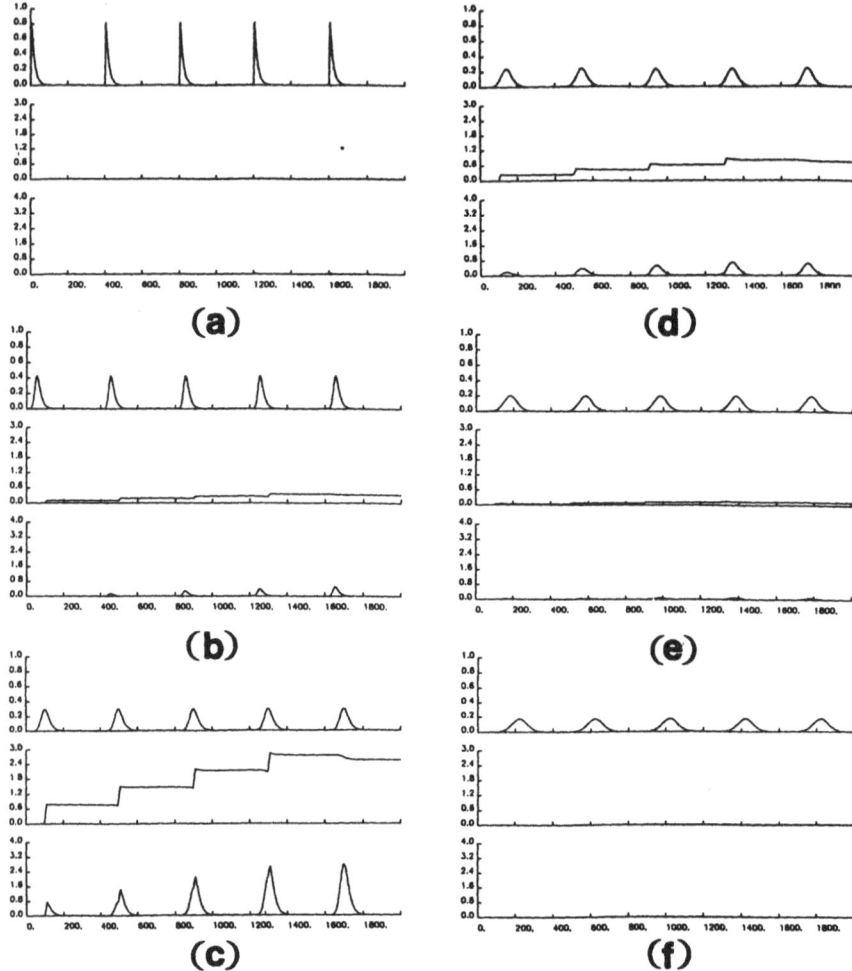

FIG. 4.17. Selective learning within different spectral populations at a fixed ISI = 500 msec. Each three-image panel from (a) to (f) represents the gated signal $g_i(t)$ [top], long term memory trace $z_i(t)$ [middle], and doubly gated signal $h_i(t) = g_i(t) z_i(t)$ [bottom], at a different value of i. Each i labels a different rate a_i of spectral activation in equation (18), such that $a_i = .2i^{-1}$. In (a), $i = 1$; in (b), $i = 10$; in (c), $i = 20$; in (d), $i = 30$; in (e), $i = 40$; in (f), $i = 50$. The same parameters as in Figure 4.16 were used. In addition, $E = .01$ and $I_{US}(t) = 10$ for $t \in (500,550)$ and $= 0$ otherwise.

mately 500 msec., the LTM trace $z_i(t)$ exhibits maximum learning. The doubly gated signal $h_i(t)$ also shows a maximal enhancement due to learning, and exhibits peaks of activation at approximately 500 msec. after onset of the CS on each trial. This behavior is also generated on the fifth trial, during which only the CS is presented.

F. The Output Signal

The output signal $R(t)$ defined in equation (20) is the sum of all the doubly gated signal functions $h_i(t)$ minus a threshold F. The output signal computes the cumulative learned reaction of all the cells to the input pattern.

Figure 4.18a plots the output signal generated in a computer experiment through time across all five trials, using an ISI of 400 msec. In Figure 4.18b, successive responses in Figure 4.18a are superimposed to show how they are aligned with respect to the ISI and increase due to learning on successive trials. Figure 4.18c plots all of the doubly gated signal functions $h_i(t)$ that are summated to form R(t) on the fifth trial. Figure 4.18d plots all the gated signal functions $g_i(t)$ whose multiplication by $z_i(t)$ generates the $h_i(t)$ curves. Together these Figures illustrate how function $R(t)$ generates an accurately timed response from the cumulative partial learning of all cells in the population spectrum.

15. EFFECT OF INCREASING ISI AND US INTENSITY

Figures 4.19a–4.19c plot the curves that are generated by ISI's of 0, 500, and 1000 msec. In every case, the learned cumulative response $R(t)$ is accurately centered at the correct ISI.

Figure 4.20 plots the functions $R(t)$ that are generated by different ISI's in a series of learning experiments. These are the $R(t)$ functions generated on the fifth trial of each experiment in response to a CS alone, after four trials of prior learning, with all time axes synchronized with CS onset. In Figure 4.20a, the $I_{US}(t)$ was chosen twice as large as in Figure 4.20b. Halving $I_{US}(t)$ amplitude reduces the $R(t)$ amplitudes without changing their timing or overall shape. Note that the envelope of the $R(t)$ functions increases and then decreases through time, and that the individual $R(t)$ functions corresponding to larger ISI's are broader.

16. COMPARISON WITH NICTITATING MEMBRANE CONDITIONING DATA: WEBER LAW PROPERTY

The computer simulations summarized in Figure 4.20 are strikingly similar to the data of Smith (1968) summarized in Figure 4.21. Smith (1968) studied the effect of manipulating the CS-US interval and the US intensity on the acquisition of the classically conditioned nictitating membrane response. The CS was a 50 msec

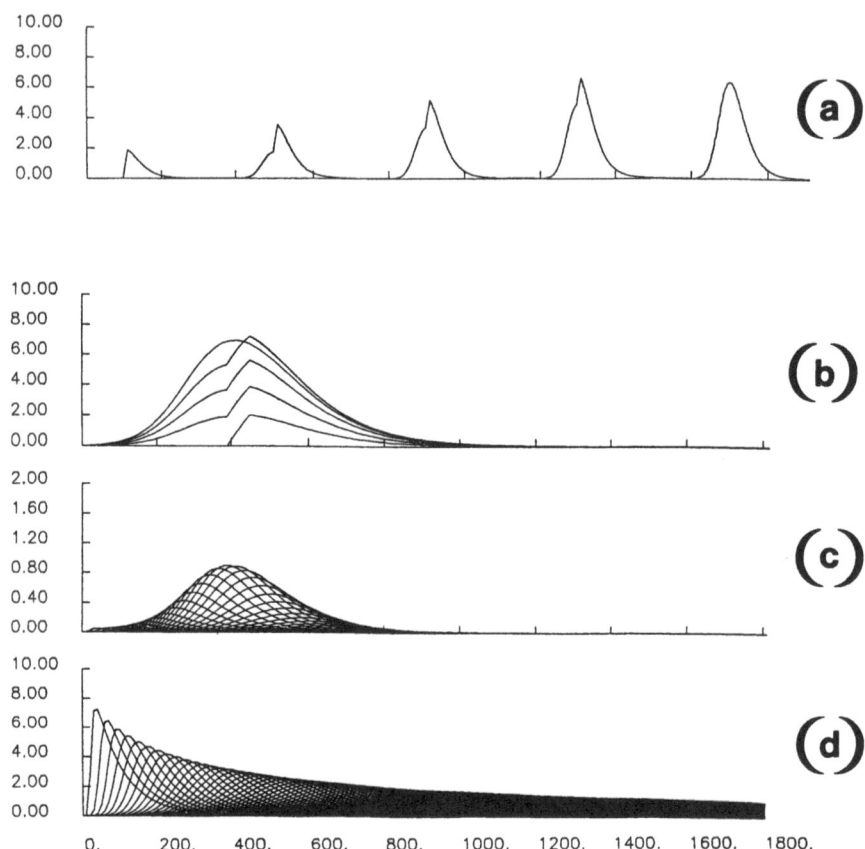

FIG. 4.18. Generation of the response function $R(t)$. The CS was presented at the beginning of each learning trial. The US was presented 400 msec later (thus the ISI=400) and kept on for 50 msec during 4 learning trials, which were followed by one test trial during which only the CS was presented. (a) Graph of the output signal $R(t)$ through time on all five trials. (b) After each trial, the time scale was reset to $t = 0$ to superimpose the output signal with a common initial time. The sudden jump in four of the five curves is due to the I_{US}. All the output curves are centered at the ISI because the output threshold $F = 0$ in (22). If F is chosen positive, the successive output curves move progressively backwards in time and become progressively better centered at the ISI as learning proceeds. (c) All the doubly gated signals $h_i(t) = f_i(x_i(t))y_i(t)$ $z_i(t)$, $1 = 1,2,...,80$, are plotted through time on the fifth trial. (d) All the gated signals $g_i(t) = f_i(x_i(t))y_i(t), i = 1,2,...,80$, are plotted through time on the fifth trial. Parameters are chosen as in Figure 4.4.

106

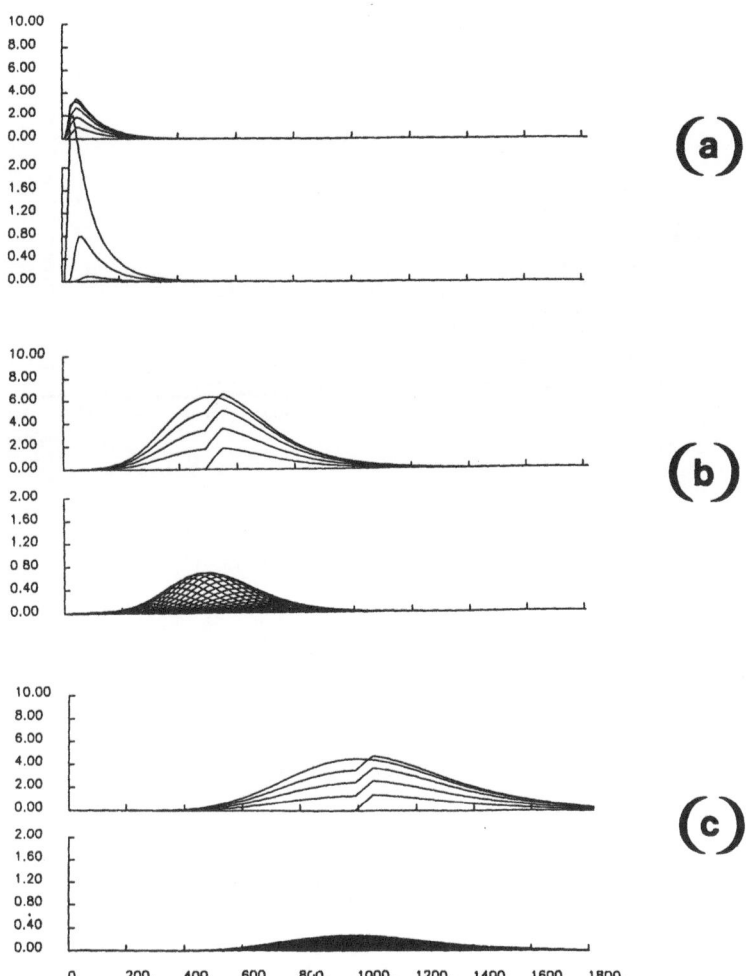

FIG. 4.19.   As in Figure 4.5b, superimposed plots of the output signal *R(t)* on four successive learning trials and one performance trial are shown, along with plots of all the doubly gated signals *h_i(t)*, *i* = 1,2,. . .,80, on the fifth trial. Each panel displays the results at a different ISI: (a) ISI = 0 msec; (b) ISI = 500 msec; and (c) ISI = 1,000 msec.

tone and the US was a 50 msec electric shock. The ISI values were 125, 250, 500, and 1,000 msec. The fact that conditioning occurred at ISI's much larger than CS duration implies that an internal trace of the CS, which we have called I_{CS}, is stored in short-term memory subsequent to CS offset, as in Figure 4.15. The US intensities were 1, 2, and 4 mA.

Smith (1968) found that the conditioned response, measured as percentage of

$$\sum_i f(x_i)y_iz_i$$

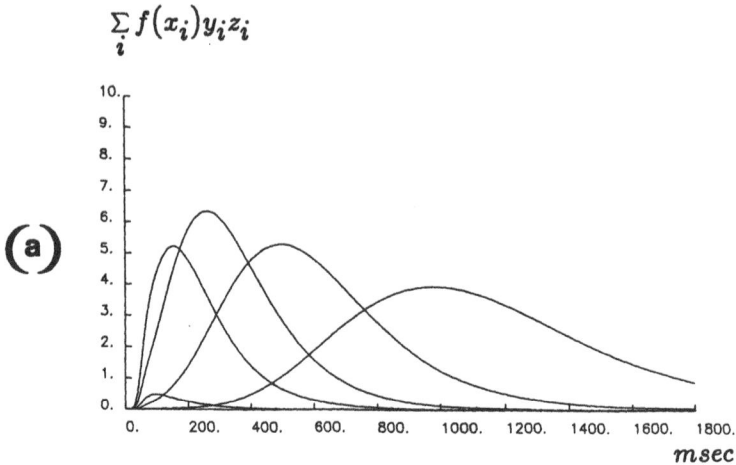

(a)

msec.

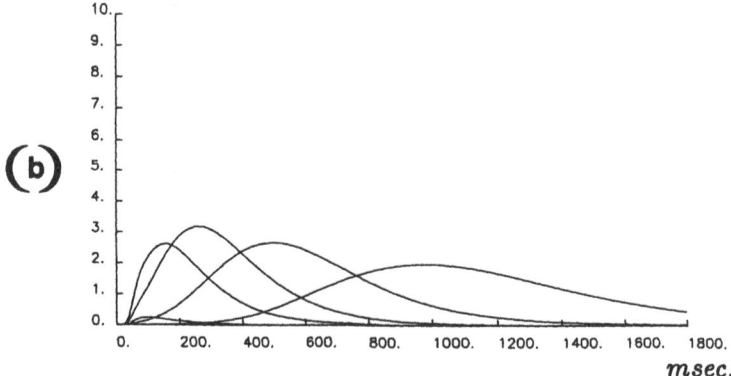

(b)

msec.

FIG. 4.20.   Inverted U in learning as a function of ISI. The output signal functions $R(t)$ are plotted on a test trial, in response to the CS alone, subsequent to 10 prior learning trials with CS-US separated by different ISI's. Successive curves from left to right were generated by ISI's of 0 (the lowest amplitude curve), 125, 250, 500, and 1,000 msec using a US duration of 50 msec. Two different I_{US} intensities were used in (a) and (b), respectively. In (a), $I_{US} = 10$. In (b), $I_{US} = 5$. All other parameters were chosen as in previous figures.

responses and response amplitude, was determined by both ISI and US intensity, whereas response onset rate and peak time were determined by the ISI essentially independently of US intensity. An increase in the mean of the peak response time correlated with an increase in the variance of the response curve, for each ISI. We call this fundamental property the Weber Law property.

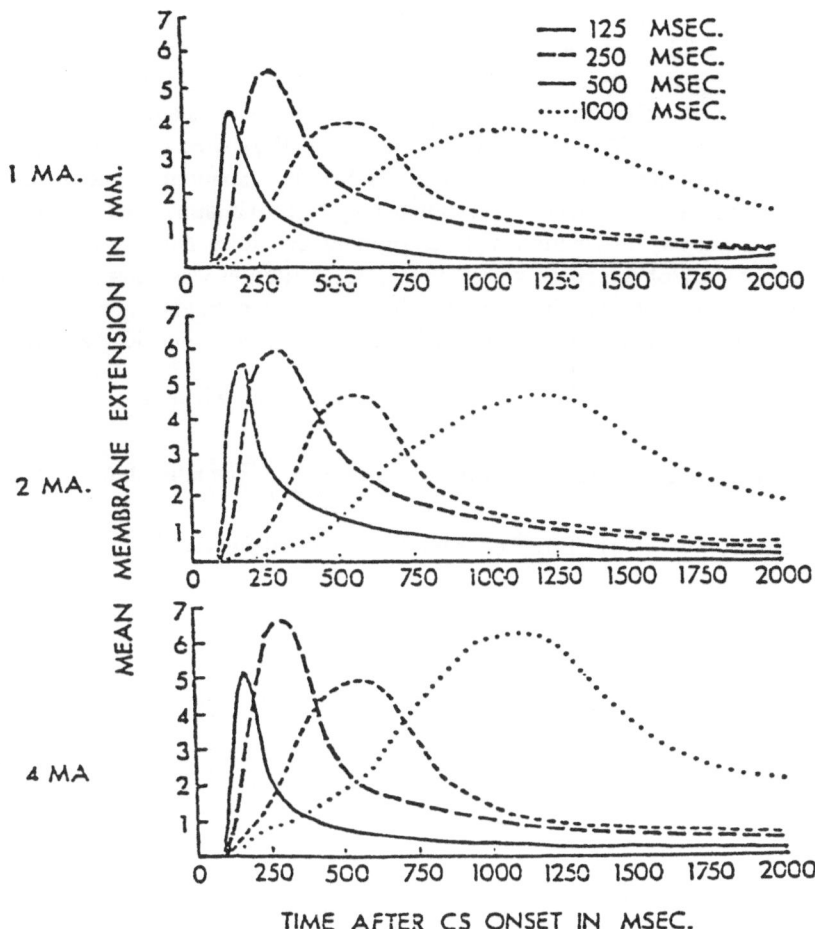

FIG. 4.21. Conditioning data from a nictitating membrane learning paradigm. Mean topograph of nictitating membrane response after learning trial 10 with a 50 msec CS, ISI's of 125, 250, 500, and 1,000 msec, and different (1,2,4 MAmp) intensities of the shock US in each subsequent panel. Reprinted from Smith (1968) with permission.

All of these properties are evident in the computer simulation of Figure 4.20. The absolute sizes of the empirically measured responses increase slower-than-linearly in Figure 4.21 as a function of shock intensity, rather than linearly as in the computer simulations in Figure 4.20. This fact suggests that shock intensity is transformed by a slower-than-linear signal function *in vivo* (cf. Grossberg, 1972b, Equation 21), rather than the linear signal function that we used to engage the activation spectrum of the model.

The qualitative similarities between the data in the top panel of Figure 4.21

and the computer simulation in Figure 4.20a are quantified in Figure 4.22 and 4.23. Figure 4.22 plots data points and computer simulations together. Figure 4.23 plots four measures of data and simulation at ISI values of 0, 125, 250, 500, and 1,000 msec. The four measures are peak time (μ), standard deviation (σ), Weber fraction (W), and peak amplitude (A). Peak time (μ) was defined as the time that the response amplitude reached its maximum value at each ISI. Standard deviation (σ) was estimated by approximating each response curve by a normal distribution and determining the times that the amplitude was equal to .61 of the curve's peak value. This criterion was chosen because the interval between the times at which response amplitude equals .61 of its peak value is approximately 2σ in length. To see this, consider a normal

distribution $\dfrac{1}{\sqrt{2\pi}\sigma} \exp\left[-\dfrac{(t-\mu)^2}{2\sigma^2}\right]$. Its amplitude when $|t-\mu| = \sigma$

is $\dfrac{1}{\sqrt{2\pi}\sigma} \exp\left(-\dfrac{1}{2}\right)$. Its amplitude when $t = \mu$ is $\dfrac{1}{\sqrt{2\pi}\sigma}$. The ratio of these ampli-

tudes is $\exp\left(-\dfrac{1}{2}\right) \cong .61$. The Weber fraction W was defined as $W = \dfrac{\sigma}{\mu}$.

$$\sum_i f(x_i) y_i z_i$$

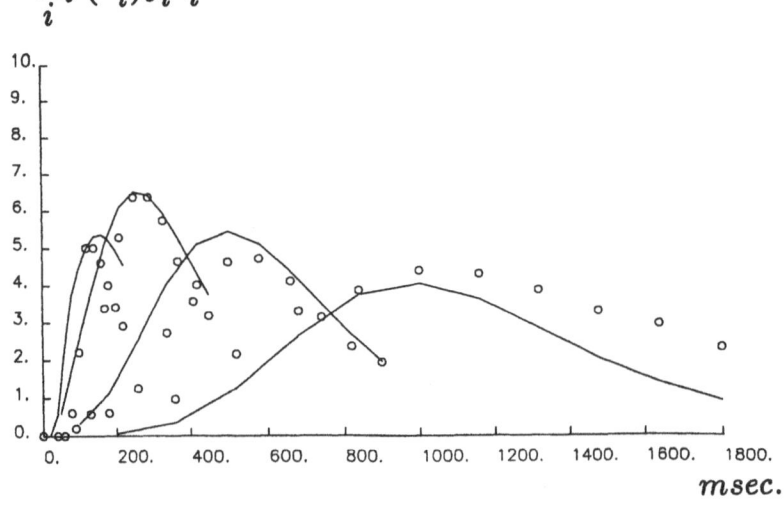

FIG. 4.22. Comparison of experimental data from Figure 4.21 (US intensity equal 1 MAmp) with computer simulation from Figure 4.20a (I_{us} intensity equal 10). Simulated values are computed as curves interpolated through 10 values around the time of US presentation separated by .16 × ISI. Open circles represent experimental data computed at the same times corrected to make the peak values of the experimental and simulated 250 msec ISI curves coincide. The correlation between simulated and experimental points equals $r = .835$.

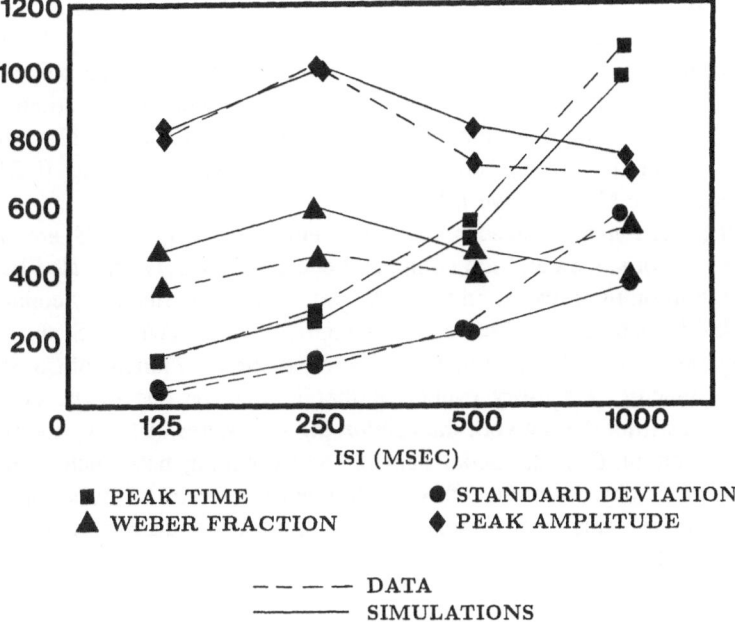

FIG. 4.23. Comparison between experimental and simulated peak time (μ), standard deviation (σ), Weber fraction (W), and peak amplitude (A). See text for details. The correlation between simulated and experimental points for μ is $r = .9996(p < .001)$, for σ is $r = .9761(p < .005)$, and for A is $r = .9666(p < .01)$.

Figure 4.23 demonstrates a good fit between experimental and simulated values of all the parameters μ, σ, W, and A at all the reported ISI's. Of particular interest is the Weber Law property; viz, the approximately constant value of the Weber fraction W as a function of ISI, in particular its tendency to approach a positive asymptote with increasing values of the ISI (Killeen & Weiss, 1987).

17.   INVERTED U IN LEARNING AS A FUNCTION OF ISI

A property of both the simulated response functions $R(t)$ in Figure 4.20 and the data summarized in Figure 4.21 is an inverted U in learning as a function of the ISI. In other words, there exists a positive ISI that is optimal for learning. In Figure 4.20, this optimal ISI is approximately 250 msec. Learning is weaker at both smaller and larger values of the ISI.

Although the Spectral Timing Model successfully generates such a positive optimal ISI, this circuit is not the only one subserving the optimal ISI that is behaviorally observed. To see this, consider once again the blocking paradigm (Section 4.3). Here two CS's are employed; call them CS_1 and CS_2. Let CS_1 be conditioned with a US until CS_1 can elicit some of the reinforcing properties of the US. Then present the two CS's simultaneously as a compound stimulus CS_1 + CS_2 (zero ISI). The conditioning of CS_2 to the new reinforcer CS_1 is attenuated, or blocked, relative to the conditioning that would have occurred if CS_1 was presented before CS_2 (positive ISI).

On the other hand, consider an experiment where CS_1 and CS_2 are equally salient to the organism and the compound cue CS_1 + CS_2 is presented before a US on conditioning trials. Then both CS_1 and CS_2 can be effectively conditioned to the US. Thus blocking of the conditioning of CS_2 to CS_1 when CS$_1$ and CS_2 are simultaneously presented cannot be due merely to the *simultaneity* of CS_1 and CS_2 in their capacity as sensory events. Rather it must be due to the effects of reading-out within the network the *reinforcing properties* of CS_2 by the sensory representation of CS_2. A model capable of explaining how such attentional blocking of CS_2 by a simultaneous conditioned reinforcer CS_1 was outlined in Section 4.3. How this model is related to the Spectral Timing Model is described in Section 4.19.

18.   MULTIPLE TIMING PEAKS

A single CS can read out responses at a series of learned delays. This multiple timing property provides strong indirect evidence that each CS sends signals to a complete activation spectrum, rather than to a single tunable delay.

Figure 4.24 depicts the outcome of a computer simulation in which a CS is paired with a US whose ISI is chosen on alternate trials at two different values. When the CS is subsequently activated on a recall trial, the response function $R(t)$ generates two peaks, with each peak centered at one of the ISI's. The parameters used in the simulation of Figure 4.24 are the same as those used to fit the data in Figure 4.23 concerning response time, amplitude, standard deviation, and Weber fraction. These model simulations strongly resemble the multiple timing data of Millenson, Kehoe, and Gormezano (1977) that are summarized in Figure 4.25. Millenson, Kehoe, and Gormezano (1977) presented rabbits in a nictitating membrane paradigm with a tone CS followed by a shock US at two randomly alternating ISI's of 200 and 700 msec. The CS terminated at US onset, and the US had a 50 msec duration. Each row in Figure 4.25 corresponds to a different experimental condition. The experiment summarized in row one used a 200 msec ISI throughout. The experiment in row five used a 700 msec ISI throughout. Compare these relative peak times, amplitudes, and Weber fractions with the model simulation in Figure 4.24.

Experiments summarized in the middle three rows used varying fractions of the two ISI delays during learning trials. In the second row, the ISI equaled 200

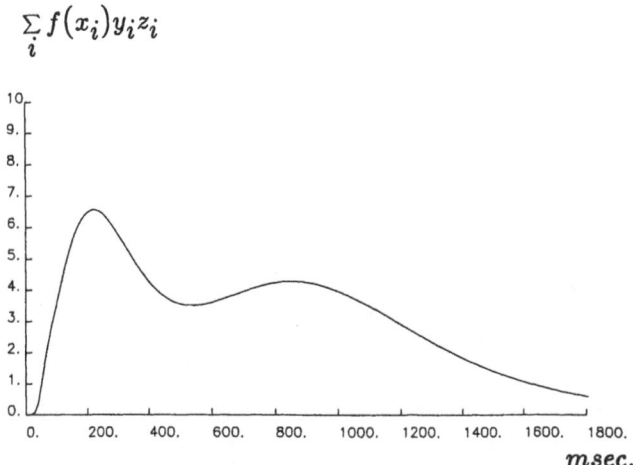

$$\sum_i f(x_i) y_i z_i$$

FIG. 4.24.   Multiple timing peaks due to learning with more than one ISI. The output signal function $R(t)$ is plotted on a test trial after 20 learning trials during which a US of intensity 10 was presented alternately at an ISI of 200 msec and 800 msec.

msec on $^7\!/_8$ of the learning trials and 700 msec on $^1\!/_8$ of the learning trials. In the third row, the ISI equaled each of these values on $^1\!/_2$ of the learning trials. In row four, the ISI equaled 200 msec on $^1\!/_8$ of the trials and 700 msec on $^7\!/_8$ of the trials.

Each column in Figure 4.25 corresponds to a different test condition subsequent to a set of learning trials. During such a test, a CS, but no US, was presented. In column one, the CS duration was 200 msec. In column two, the CS duration was 700 msec. In each panel, a test curve is displayed after three days and after 10 days of prior learning.

The data curves of greatest interest are in row three, column two. These curves are strikingly similar to the model simulation in Figure 4.24. Row three, column one is also of interest, because it shows that termination of a CS of 200 msec duration under these conditions prevents strong preservation of its I_{CS} curve for the additional 500 msec needed to read out a large response at 700 msec.

19.   LOCATING THE TIMING CIRCUIT WITHIN A SELF-ORGANIZING SENSORY-COGNITIVE AND COGNITIVE-REINFORCEMENT ART NETWORK

The timing circuit is hypothesized to form part of the interacting sensory-cognitive and cognitive-reinforcement circuits summarized in Figures 4.1 and 4.2 As in Figure 4.1a, a sensory-cognitive ART circuit is broken up into an attentional subsystem and an orienting subsystem. The attentional subsystem learns ever

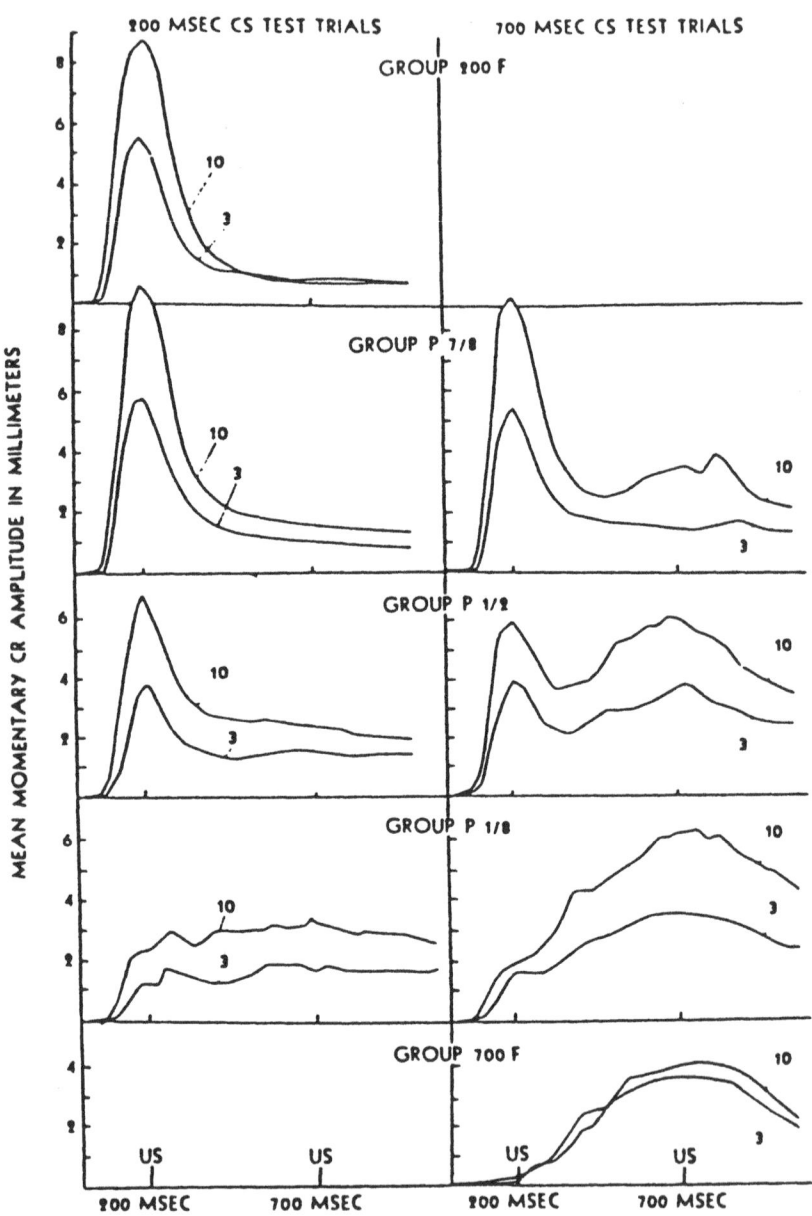

FIG. 4.25. Conditioning data from the nictitating membrane learning paradigm of Millenson, Kehoe, and Gormenzano (1977). Data shown after learning trials 3 and 10 using a tone CS of duration 200 msec and 700 msec, ISI's of 200 msec and 700 msec, and a shock US of 50 msec duration. See text for details. Reprinted with permission.

114

more precise internal representations of and responses to events as they become more familiar. The attentional subsystem also learns the top-down expectations that help to stabilize memory of the learned bottom-up codes of familiar events. The orienting subsystem resets the internal representation that is active in short term memory (STM) in the attentional subsystem when an unfamiliar or unexpected event occurs. The orienting subsystem also energizes the orienting response, including the orienting movements triggered by novel events that enable such events to be more efficiently processed.

The orienting subsystem is activated when a sufficiently large mismatch occurs within the attentional subsystem between bottom-up sensory input signals and learned top-down expectations. In Figure 4.1a, the learned top-down expectations are read-out form level F_2 to level F_1, and matching of expectations with bottom-up input patterns occurs at level F_1. When a mismatch occurs, the orienting subsystem A is activated and causes an STM reset wave to be delivered to level F_2. This STM reset wave resets the sensory representations of cues that are currently being stored in STM at F_2.

As noted in Section 13, one function of the timing circuit is to prevent spurious resets of active internal representations in response to mismatches due to *expected nonoccurrences* of sensory events. In addition, the timing circuit should not prevent registration of bottom-up input patterns and their matching with active top-down expectations. Thus the timing circuit does not interfere with processing within the attentional subsystem.

Instead, we hypothesize that the timing circuit inhibits read-out of the STM reset wave from the orienting subsystem A (Figure 4.1a). Thus when the timing circuit is active, both STM reset within the attentional subsystem and the orienting response are inhibited. When the timing circuit is inactive, an *unexpected nonoccurrence* of an event is able to trigger the STM reset and orienting response needed to cope with the unexpected event. How this may be accomplished will now be described.

20.   ADAPTIVE TIMING AS SPECTRAL CONDITIONED REINFORCER LEARNING

The Spectral Timing circuit is assumed to be a variant of the conditioned reinforcer learning that occurs from sensory representations to opponent drive representations built up from gated dipoles. The main new idea is that the population of neurons in the on-channel of the drive representation is broken up into subpopulations whose membrane properties enable them to respond to inputs at different rates α_i, as in equation (16). In other words, a sloppy parametric specification of cell reaction rates generates an adaptive timing mechanism *if* such sloppiness occurs at the proper processing stage of a gated dipole circuit. We summarize this fact by saying that adaptive timing is a type of *spectral conditioned reinforcer learning*.

21.   TIMED INHIBITION OF THE ORIENTING
SUBSYSTEM BY DRIVE REPRESENTATIONS

It remains to describe how a timing circuit embedded within the on-channels of gated dipole drive representations can achieve the functional properties described in Section 13. These properties follow if we assume, in addition, that the drive representations D inhibit the orienting subsystem A, as in Figure 4.26. In Figure 4.1a, level F_1 also inhibits A. Thus several processing levels within the attentional subsystem are assumed to inhibit the orienting subsystem. The hypothesis of competition from D to A representations was first made wtihin the context of ART-type models in Grossberg (1975; reprinted in Grossberg, 1982, pp. 284–286). Given $D \rightarrow A$ inhibition, spectral conditioned reinforcer learning generates the desired adaptive timing properties as follows:

After CS-US conditioning at a fixed ISI, presentation of the CS activates its sensory representation S_{CS}, which activates its conditioned drive representation D with a response curve $R(t)$ of the form depicted in Figure 4.20. Each of these response curves $R(t)$ begins to grow right after I_{CS} read-out; remains positive throughout an interval whose total width covaries with the ISI, due to the approximate constancy of the Weber fraction W (Figure 4.23); and peaks at the ISI. Inhibition of A by D thus prevents STM reset by expected nonoccurrences of the US throughout a time interval that is centered at the expected delay of the US whose width covaries with this delay.

22.   TIMED ACTIVATION OF THE HIPPOCAMPUS
AND THE CONTINGENT NEGATIVE VARIATION

Because it is activated by the drive representations D, positive feedback from D to S along the $D \rightarrow S$ incentive motivational pathways is also timed to provide peak motivational support for release of a conditioned response (Figure 4.2) at the expected delay of the US.

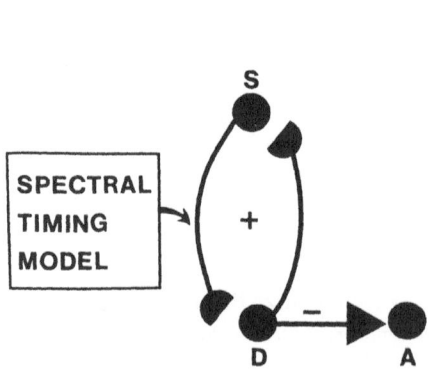

FIG. 4.26. Inhibition of the orienting subsystem A by the output from a drive representation D. The Spectral Timing Model is assumed to be part of the network whereby conditioning of a sensory representation S to a drive representation D endows S with conditioned reinforcer properties. As S reads-out spectrally timed conditioned signals to D, D inhibits output signals from A and thereby prevents expected nonoccurrences of the US from resetting STM and triggering orienting responses.

SPECTRAL TIMING MODEL

S

+

D   A

In Grossberg (1975, Section VII, and 1978, Section 16; reprinted in 1982), such $D \rightarrow S$ feedback was first interpreted to be a formal analog of the contingent negative variation, or CNV, event-related potential. The CNV had earlier been experimentally shown to be sensitive to an animal's expectancy, decision (Walter, 1964), motivation (Cant & Bickford, 1967; Irwin, Rebert, McAdam, & Knott, 1966), preparatory set (Low, Borda, Frost, & Kellaway, 1966), and arousal (McAdam, 1969). It is also a conditionable wave whose timing tends to match the ISI. Until the present work, development of our conditioning theory, as summarized in Grossberg (1987, chap. 1, sections 23 and 25, and chap. 2, sections 30, 43, 53, 57, and 60), suggested how the CNV is conditioned and how it is related to expectancy, decision, motivation, preparatory set, and arousal. The theory had not, however, heretofore explained how the learning process enables CNV timing to mimic the ISI. The present extension of the theory provides an explanation through the hypothesis of spectral conditioned reinforcer learning. The interpretation of drive representations in terms of hypothalamo-hippocampal interactions (Grossberg, 1971, 1982, 1987) provides an anatomical marker for directly testing the existence of spectral activation.

The hypotheses that drive representations include hippocampus and that the hippocampus is involved in conditioned timing have also received support from neurophysiological experiments (Berger & Thompson, 1978; Delacour & Houcine, 1980; Hoechler & Thompson, 1980; Rawlins, 1985; Rawlins, Feldon, & Gray, 1982; Solomon, 1979, 1980; Solomon, van der Schaff, Thompson, & Weisz, 1986); also see Section 4.4 .

23. A SOLUTION OF THE TIMING PARADOX

The Spectral Timing Model considers only the type of timing that enables an organism to time and differentially respond to an expected nonoccurrence, an expected occurrence, and an unexpected nonoccurrence of a sensory event subsequent to a prior sensory event or action. In so doing, the model provides a solution to a Timing Paradox that becomes apparent upon closer inspection of this type of timing problem.

On the one hand, in response to *any* fixed choice of conditionable ISI, it is desired that the learned *optimal* response delay approximate the ISI. Thus the model must be capable of an accurate discrimination of individual temporal delays. On the other hand, it is also desired that spurious orienting responses be inhibited in response to expected nonoccurrences that may occur *throughout* the ISI interval subsequent to a CS onset. Thus the inhibitory signal must be temporally distributed throughout the ISI interval.

The Spectral Timing Model reconciles the two requirements of accurate optimal temporal delay and temporally distributed activation via the Weber law property (see Figures 4.20 and 4.21). According to this property, the standard deviation of the model response scales with its peak time. Consequently the model begins to immediately generate an output signal that may be used to inhibit the orienting subsystem, even though its peak output is accurately located at the ISI.

24.   TIMING WITH AND WITHOUT ENDOGENOUS OSCILLATORS: CIRCADIAN RHYTHMS

Church (see chap. 9 in this volume by Church & Broadbent) has suggested a timing model to explain timing data collected from instrumental conditioning, rather than Pavlovian conditioning, experiments. The Church timing model is similar to the Spectral Timing Model in some respects, but differs in others.

For example, both models use a type of working memory. In our model, this function is accomplished by the stored representation of the CS, which controls the signal $I_{CS}(t)$, and the activation spectrum that responds to $I_{CS}(t)$ in equation (16). In contrast to the Spectral Timing Model model, the Church model assumes that timing is controlled by endogenously active oscillators whose periods are set in a logarithmic progression: 1,2,4,8, ... This logarithmic assumption is used to explain the Weber Law property (Figures 4.20 and 4.21). Such a logarithmic progression of periods is assumed to include oscillators fluctuating on a fast time scale as well as much slower oscillators, such as those controlling circadian rhythms.

Church's assumption that endogenous oscillators control adaptive timing of the sort considered here will require for its elaboration a number of mechanistic innovations. These include: how phase-resetting of the oscillators is caused by reinforcing events, how endogenous cycling of the oscillators is prevented from interfering with inappropriate behaviors, and how the logarithmic progression of periods is generated and maintained. None of these problems needs to be solved in the Spectral Timing Model.

This is true because the Spectral Timing Model does not use endogeneously active oscillators. On the other hand, a specialized gated dipole circuit has been shown capable of generating clock-like endogenous oscillations on a circadian time scale, and has been used as the core of a model capable of quantitatively simulating a large data base concerning the control of circadian rhythms by the suprachiasmatic nuclei of the mammalian hypothalamus (Carpenter & Grossberg, 1983, 1984, 1985). It has also been shown that each term in this circadian gated dipole model is computationally homologous to mechanisms in the gated dipole model for a drive representation, which are also assumed to involve hypothalamic circuitry (Carpenter & Grossberg, 1985).

Our results thus support the spirit of Church's analysis, if not its modeling details, by analysing how specialized gated dipole circuitry can be used both for exogenously reactive timing and endogenously active timing on multiple time scales.

ACKNOWLEDGMENTS

Supported in part by the Air Force Office of Scientific Research (AFOSR F49620-86-C-0037 and AFOSR F49620-87-C-0018) and the National Science Foundation (NSF IRI-84-17756 and NSF IRI-87-16960). The author thanks Carol Y. Jefferson and Cynthia Suchta for their assistance in the preparation of the manuscript and illustrations.

REFERENCES

Barto, A. G., Sutton, R. S., & Anderson, C. W. (1983). Neuron-like adaptive elements that can solve difficult learning control problems. *IEEE Transactions, SMC-13*, 834–846.

Berger, T. W., & Thompson, R. F. (1978). Neuronal plasticity in the limbic system during classical conditioning of the rabbit nictitating membrane response, I: The hippocampus. *Brain Research, 145*, 323–346.

Bower, G. H. (1981). Mood and memory. *American Psychologist, 36*, 129–148.

Bower, G. H., Gilligan, S. G., & Monteiro, K. P. (1981). Selectivity of learning caused by adaptive states. *Journal of Experimental Psychology: General, 110*, 451–473.

Cant, B. R., & Bickford, R. G. (1967). The effect of motivation on the contingent negative variation (CNV). *Electroencephalography and Clinical Neurophysiology, 23*, 594.

Carpenter, G. A., & Grossberg, S. (1983). A neural theory of circadian rhythms: The gated pacemaker. *Biological Cybernetics, 48*, 35–59.

Carpenter, G. A., & Grossberg, S. (1984). A neural theory of circadian rhythms: Aschoff's rule in diurnal and nocturnal mammals. *American Journal of Physiology, 247*, R1067–R1082.

Carpenter, G. A., & Grossberg, S. (1985). A neural theory of circadian rhythms: Split rhythms, after-effects, and motivational interactions. *Journal of Theoretical Biology, 113*, 163–223.

Carpenter, G. A., & Grossberg, S. (1987a). A massively parallel architecture for a self-organizing neural pattern recognition machine. *Computer Vision, Graphics, and Image Processing, 37*, 54–115.

Carpenter, G. A., & Grossberg, S. (1987b). ART 2: Self-organization of stable category recognition codes for analog input patterns. *Applied Optics, 26*, 4919–4930.

Carpenter, G. A., & Grossberg, S. (1988). The ART of adaptive pattern recognition by a self-organizing neural network. *Computer, 21*, 77–88.

Carpenter, G. A., & Grossberg, S. (1989). Neural dynamics of category learning and recognition: Structural invariants, reinforcement, and evoked potentials. In M. L. Commons, R. J. Herrnstein, S. M. Kosslyn, & D. B. Mumford (Eds.), *Computational and clinical approaches to pattern recognition and concept formation*. Hillsdale, NJ: Lawrence Erlbaum Associates.

Delacour, J., & Houcine, O. (1980). Conditioning to time: Evidence for a role of hippocampus from unit recording. *Neuroscience, 23*, 87–94.

Ellias, S., & Grossberg, S. (1975). Pattern formation, contrast control, and oscillations in the short term memory of shunting on-center off-surround networks. *Biological Cybernetics, 20*, 69–98.

Grossberg, S. (1968). Some physiological and biochemical consequences of psychological postulates. *Proceedings of the National Academy of Sciences, 60*, 758–765.

Grossberg, S. (1969a). On learning and energy-entropy dependence in recurrent and nonrecurrent signed networks. *Journal of Statistical Physics, 1*, 319–350.

Grossberg, S. (1969b). On the production and release of chemical transmitters and related topics in cellular control. *Journal of Theoretical Biology, 22*, 325–364.

Grossberg, S. (1971). On the dynamics of operant conditioning. *Journal of Theoretical Biology, 33,* 225–255.

Grossberg, S. (1972a). A neural theory of punishment and avoidance, I: Qualitative theory. *Mathematical Biosciences, 15,* 39–67.

Grossberg, S. (1972b). A neural theory of punishment and avoidance, II: Quantitative theory. *Mathematical Biosciences, 15,* 253–285.

Grossberg, S. (1975). A neural model of attention, reinforcement, and discrimination learning. *International Review of Neurobiology, 18,* 263–327.

Grossberg, S. (1976a). Adaptive pattern classification and universal recoding, I: Parallel development and coding of neural feature detectors. *Biological Cybernetics, 23,* 121–134.

Grossberg, S. (1976b). Adaptive pattern classification and universal recoding, II: Feedback, expectation, olfaction, and illusions. *Biological Cybernetics, 23,* 187–202.

Grossberg, S. (1978). A theory of human memory: Self-organization and performance of sensory-motor codes, maps, and plans. In R. Rosen, & F. Snell (Eds.), *Progress in theoretical biology* (Vol. 5). New York: Academic Press.

Grossberg, S. (1982). *Studies of mind and brain: Neural principles of learning, perception, development, cognition, and motor control.* Boston: Reidel Press.

Grossberg, S. (1984). Neuroethology and theoretical neurobiology. *Behavioral and Brain Sciences, 7,* 388–390.

Grossberg, S. (Ed.) (1987). *The adaptive brain, Volumes I and II.* Amsterdam: Elsevier/North-Holland.

Grossberg, S. (Ed.). (1988). *Neural networks and natural intelligence.* Cambridge, MA: MIT Press.

Grossberg, S., & Kuperstein, M. (1986). *Neural dynamics of adaptive sensory-motor control: Ballistic eye movements.* Amsterdam: North-Holland.

Grossberg, S., & Kuperstein, M. (1989). *Neural dynamics of sensory-motor control, Expanded edition.* Elmsford, NY: Pergamon Press.

Grossberg, S., & Levine, D. S. (1975). Some developmental and attentional biases in the contrast enhancement and short term memory of recurrent neural networks. *Journal of Theoretical Biology, 45,* 341–380.

Grossberg, S., & Levine, D. S. (1987). Neural dynamics of attentionally-modulated Pavlovian conditioning: Blocking, inter-stimulus interval, and secondary reinforcement. *Applied Optics, 26,* 5015–5030.

Grossberg, S., & Schmajuk, N. A. (1987). Neural dynamics of attentionally-modulated Pavlovian conditioning: Conditioned reinforcement, inhibition, and opponent processing. *Psychobiology, 15,* 195–240.

Grossberg, S., & Schmajuk, N. A. (1989). Neural dynamics of adaptive timing and temporal discrimination during associative learning. *Neural Networks, 2,* 79–102.

Hawkins, R. D., Abrams, T. W., Carew, T. J., & Kandel, E. R. (1983). A cellular mechanism of classical conditioning in *Aplysia:* Activity dependent amplification of presynaptic facilitation. *Science, 219,* 400–405.

Hawkins, R. D., & Kandel, E. R. (1984). Is there a cell biological alphabet for simple forms of learning? *Psychological Review, 9,* 375–391.

Hebb, D. O. (1949). *The organization of behavior.* New York: Wiley.

Henneman, E. (1957). Relation between size of neurons and their susceptibility to discharge. *Science, 26,* 1345–1347.

Henneman, E. (1985). The size-principle: A deterministic output emerges from a set of probabilistic connections. *Journal of Experimental Biology, 115,* 105–112.

Hoechler, F. K., & Thompson, R. F. (1980). Effect of the interstimulus (CS-UCS) interval on hippocampal unit activity during classical conditioning of the nictitating membrane response of the rabbit (*Oryctolagus cuniculus*). *Journal of Comparative and Physiological Psychology, 94,* 201–215.

Irwin, D. A., Rebert, C. S., McAdam, D. W., & Knott, J. R. (1966). Slow potential change (CNV) in the human EEG as a function of motivational variables. *Electroencephalography and Clinical Neurophysiology, 21,* 412–413.

Jones, R., & Kreck, M. J. (1978). Visual evoked response as a function of grating spatial frequency. *Investigative Ophthalmology and Visual Science, 17,* 652–659.

Kamin, L. J. (1969). Predictability, surprise, attention, and conditioning. In B. A. Campbell & R. M. Church, (Eds.), *Punishment and aversive behavior.* New York: Appleton-Century-Crofts.

Killeen, P. R., & Weiss, N. A. (1987). Optimal timing and the Weber function. *Psychological Review, 94,* 455–468.

Klopf, A. H. (1988). A neuronal model of classical conditioning. *Psychobiology, 16,* 85–125.

Levy, W. B., Brassel, S. E., & Moore, S. D. (1983). Partial quantification of the associative synaptic learning rule of the dentate gyrus. *Neuroscience, 8,* 799–808.

Levy, W. B., & Desmond, N. L. (1985). The rules of elemental synaptic plasticity. In W. B. Levy, J. Anderson, & S. Lehmkuhle (Eds.), *Synaptic modification, neuron selectivity, and nervous system organization* (pp. 105–121). Hillsdale, NJ: Lawrence Erlbaum Associates.

Low, M. D., Borda, R. P., Frost, J. D., & Kellaway, P. (1966). Surface negative slow potential shift associated with conditioning in man. *Neurology, 16,* 711–782.

McAdam, D. W. (1969). Increases in CNS excitability during negative cortical slow potentials in man. *Electroencephalography and Clinical Neurophysiology, 26,* 216–219.

Millenson, J. R., Kehoe, E. J., & Gormezano, I. (1977). Classical conditioning of the rabbit's nictitating membrane response under fixed and mixed CS-US intervals. *Learning and Motivation, 8,* 351–366.

Moore, J. W., Desmond, J. E., Berthier, N. E., Blazis, D. E. J., Sutton, R. S., & Barto, A. G. (1986). Simulation of the classically conditioned nictitating membrane response by a neuron-like adaptive element: Response topography, neuronal firing and interstimulus intervals. *Behavioral Brain Research, 21,* 143–154.

Musselwhite, M. J., & Jeffreys, D. A. (1985). The influence of spatial frequency on the reaction times and evoked potentials recorded to grating pattern stimuli. *Vision Research, 25,* 1545–1555.

Optican, L. M., & Robinson, D. A. (1980). Cerebellar dependent adaptive control of primate saccadic system. *Journal of Neurophysiology, 44,* 1058–1076.

Parker, D. M., & Salzen, E. A. (1977a). Latency changes in the human visual evoked response to sinusoidal gratings. *Vision Research, 17,* 1201–1204.

Parker, D. M., & Salzen, E. A. (1977b). The spatial selectivity of early and late waves within the human visual evoked response. *Perception, 6,* 85–95.

Parker, D. M., Salzen, E. A., & Lishman, J. R. (1982a). Visual evoked responses elicited by the onset and offset of sinusoidal gratings: Latency, waveform, and topographic characteristics. *Investigative Ophthalmology and Visual Sciences, 22,* 675–680.

Parker, D. M., Salzen, E. A., & Lishman, J. R. (1982b). The early waves of the visual evoked potential to sinusoidal gratings: Responses to quadrant stimulation as a function of spatial frequency. *Electroencephalography and Clinical Neurophysiology, 53,* 427–435.

Plant, G. T., Zimmern, R. L., & Durden, K. (1983). Transient visually evoked potentials to the pattern reversal and onset of sinusoidal gratings. *Electroencephalography and Clinical Neurophysiology, 56,* 147–158.

Rauschecker, J. P., & Singer, W. (1979). Changes in the circuitry of the kitten's visual cortex are gated by postsynaptic activity. *Nature, 280,* 58–60.

Rawlins, J. N. P. (1985). Associations across time: The hippocampus as a temporary memory store. *The Behavioral and Brain Sciences, 8,* 479–496.

Rawlins, J. N. P., Feldon, J., & Gray, J. A. (1982). Behavioral effects of hippocampectomy depend on inter-event intervals. *Society for Neuroscience Abstracts, 8,* 22.

Ron, S., & Robinson, D. A. (1973). Eye movements evoked by cerebellar stimulation in the alert monkey. *Journal of Neurophysiology, 36,* 1004–1021.

Schneiderman, N., & Gormenzano, I. (1964). Conditioning of the nictitating membrane response of the rabbits as a function of the CS-US interval. *Journal of Comparative and Physiological Psychology, 57,* 188–195.

Singer, W. (1983). Neuronal activity as a shaping factor in the self-organization of neuron assemblies. In E. Basar, H. Flohr, H. Haken, & A. J. Mandell (Eds.), *Synergetics of the brain.* New York: Springer-Verlag.

Skrandies, W. (1984). Scalp potential fields evoked by grating stimuli: Effects of spatial frequency and orientation. *Electroencephalography and Clinical Neurophysiology, 58,* 325–332.

Smith, M. C. (1968). CS-US interval and US intensity in classical conditioning of the rabbit's nictitating membrane response. *Journal of Comparative and Physiological Psychology, 3,* 679–687.

Smith, M. D., Coleman, S. R., & Gormezano, I. (1969). Classical conditioning of the rabbit's nictitating membrane response at backward, simultaneous, and forward CS-US intervals. *Journal of Comparative and Physiological Psychology, 69,* 226–231.

Solomon, P. R. (1979). Temporal versus spatial information processing views of hippocampal functions. *Psychological Bulletin, 86,* 1271–1279.

Solomon, P. R. (1980). A time and a place for everything? Temporal processing views of hippocampal function with special reference to attention. *Physiological Psychology, 8,* 254–261.

Solomon, P. R., van der Schaaf, E. R., Thompson, R. F., & Weisz, D. J. (1986). Hippocampus and trace conditioning of the rabbit's classically conditioned nictitating membrane response. *Behavioral Neuroscience, 100,* 729–744.

Sutton, R. S., & Barto, A. G. (1981) Towards a modern theory of adaptive networks: Expectation and prediction. *Psychological Review, 88,* 135–170.

Sutton, R. S., & Barto, A. G. (1989). Time-derivative models of Pavlovian reinforcement. To appear in J. W. Moore, & M. Gabriel, (Eds.) *Learning and computational neuroscience.* Cambridge, MA: MIT Press.

Thompson, R. F., Barchas, J. D., Clark, G. A., Donegan, N., Kettner, R. E., Lavond, D. G., Madden, J., Mauk, M. D., & McCormick, D. A. (1984). Neuronal substrates of associative learning in the mammalian brain. In D. L. Alkon & J. Farley (Eds.) *Primary neural substrates of learning and behavioral change.* New York: Cambridge University Press.

Vassilev, A., Manahilov, V., & Mitov, D. (1983). Spatial frequency and pattern onset-offset response. *Vision Research, 23,* 1417–1422.

Vassilev, A., & Strashimirov, D. (1979). On the latency of human visually evoked response to sinusoidal gratings. *Vision Research, 19,* 843–846.

Walter, W. G. (1964). Slow potential waves in the human brain associated with expectancy, attention, and decision. *Arch. Psychiat. Nervenkr., 206,* 309–322.

Walters, E. T., & Byrne, J. H. (1983). Associative conditioning of single sensory neurons suggests a cellular mechanism for learning. *Science, 219,* 405–408.

Williamson, S. J., Kaufman, I., & Brenner, D. (1978). Latency of the neuromagnetic response of the human visual cortex. *Vision Research, 18,* 107–110.

5 Simulations of Conditioned Perseveration and Novelty Preference From Frontal Lobe Damage

Daniel S. Levine
University of Texas at Arlington

Paul S. Prueitt
Georgetown University

ABSTRACT

Neural networks are presented that simulate certain behavioral effects of frontal lobe damage. On some cognitive tasks, frontal lesions cause perseveration of formerly rewarding choices of action. On other tasks, frontal lesions cause approach to objects just because they are novel. Both effects can be explained by weakening of signals between sensory and reinforcement loci. The networks that reproduce these data incorporate neural design principles developed for other purposes by Grossberg and his co-workers. These design principles include adaptive resonance between two layers of sensory processing; attentional gating of synapses between layers; and competition among gated dipoles containing on-units and off-units.

REINFORCEMENT LEARNING: WHEN IT OCCURS AND WHEN IT DOESN'T

One of the major issues in conditioning is selective attention. How does a person or animal select out, from the multitude of stimuli it encounters, the ones that will be predictive of important positive or negative consequences in the future, and learn to act on such stimuli appropriately? The solutions to this problem in a neural network context have broad implications not only for the psychologist but also for the engineer who seeks to build goal direction or planning into intelligent devices. A collection of articles in progress (Levine & Leven, in press) focuses on the interconnected issues of motivation, emotion, and goal direction.

Other neural network modelers (Hinton, 1987; Werbos, 1988) have posed this

123

issue as *reinforcement learning:* how does an organism learn to control the environment so as to maximize positive reinforcement and minimize negative reinforcement? But it is hard to believe that *all* decisions by humans or other mammals are based on reinforcement learning. What about gamblers, or drug addicts, or people trapped in bad love affairs that they lack the courage to leave? What about rats who get into a vicious circle when they learn a response to turn off electric shock and then get shocked for making that very response? At times, a more potentially rewarding alternative is also more risky, and a known alternative yields a reward that is lesser but certain. At other times, an activity is engaged in not because it promises to be rewarding but only because it is novel. So "wanting" in the sense of maximizing expected net positive reinforcement is not the sole reason for doing things.

Grossberg and Levine (1987) use blocking data from classical conditioning as a prototype of reinforcement learning. But Levine and Prueitt (1989) use cognitive data on frontally damaged primates as a prototype of *deviation* from reinforcement learning. The deviating data are results of brain damage but, we believe, represent exaggerations of effects that occur in apparently normal people and animals—effects whereby habit or novelty is sometimes a stronger guide to behavior than is current or anticipated reinforcement.

Figure 5.1 schematizes the well known blocking paradigm (Kamin, 1968, 1969). First, a neutral stimulus (CS_1), such as a tone, is presented several times, followed at a given time interval by an unconditioned stimulus (US) such as an electric shock. This paring occurs until a conditioned response (CR) is established to the CS_1. Next, several trials are given, in which CS_1 and another neutral stimulus (CS_2), such as a light, are presented simultaneously, followed at the same time interval by the US. Finally, the CS_2 is presented alone but not reinforced. On these recall trials, no CR occurs in response to the CS_2.

In order to simulate blocking in a neural network, Grossberg and Levine (1987, p. 5016) broke up the selective attention problem into four subproblems, as follows: (a) "How does the pairing of CS_1 with US in the first phase of the blocking experiment endow the CS_1 cue with properties of a conditioned, or secondary, reinforcer? (b) How do the reinforcing properties of a cue, whether

1. CS_1 —— US

    CS_1 ——→ CR

2. $CS_1 + CS_2$ —— US

    CS_2 ↛ CR

FIG. 5.1. Typical blocking paradigm; see text for details. (From Grossberg & Levine, 1987; reprinted with permission of the Optical Society of America.)

primary (US) or secondary (CS_1), shift the focus of attention toward its own processing? (c) How does the limited capacity of attentional resources arise, so that a shift of attention toward one set of cues . . . can prevent other cues . . . from being attended? (d) How does withdrawal of attention from a cue prevent that cue from entering into new conditioned relationships?"

The prefrontal cortex is a connecting link between limbic-hypothalamic reward loci and neocortical sensory processing loci. Hence, this brain region plays a large role in solutions to the four subproblems listed above, particularly the first two. Damage to the prefrontal cortex lessens the influence of reinforcement on responses. Figure 5.2 shows the schematic conditioning network from Grossberg and Levine (1987). Other articles (for example, Grossberg, 1975) have postulated analogies between parts of this circuit and brain regions. Although detailed anatomical ascriptions are beyond the scope of this article, we suggest that frontal damage correlates with considerable weakening, but not complete severing, of the feedback connections between sensory and drive loci in Figure 5.2.

The weakening of reinforcement criteria with frontal damage means that behavioral decisions will often be made using other criteria. For example, frontal

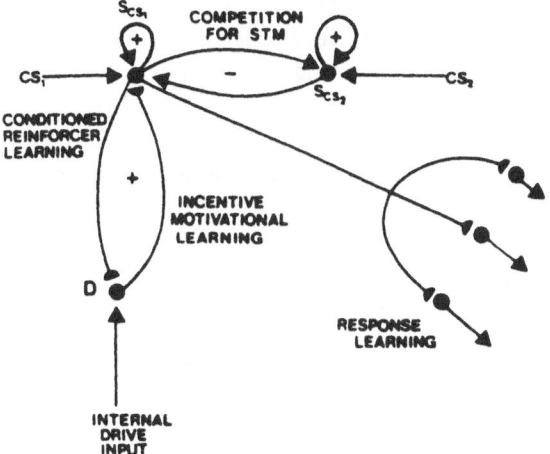

FIG. 5.2.   Schematic conditioning circuit. Conditioned stimuli (CS_i) activate sensory representations (Scs) that compete among themselves for limited capacity short term memory activation and storage. The activated Scs elicit conditioned signals to drive representations and motor command representations. Learning from an Scs to a drive representation D is called conditioned reinforcer learning. Learning from D to Scs is called incentive motivational learning. Signals from D to Scs are elicited when the combination (additive or multiplicative?) of external sensory and internal drive inputs is sufficiently large. Frontal damage reduces the strength of these feedback connections, particularly the incentive motivation connections. (From Grossberg & Levine, 1987; reprinted with permission of the Optical Society of America.)

lesions often lead to perseveration in formerly rewarding behavior after it ceases to be rewarding (Konorski & Lawicka, 1964; Milner, 1963, 1964). In addition, frontal damage leads to deficits in responding to stimuli after a delay (Jacobsen, 1935; Spaet & Harlow, 1943; Stamm, 1964). These delay deficits were at first thought to reflect impairment of short term memory but were later found instead to illustrate forms of perseveration.

Yet if environmental changes involve the introduction of novel objects, frontally lesioned monkeys change their behavior to approach the novel objects more readily than do normal monkeys (Pribram, 1961). How can the same brain lesions lead to increased perseveration if there are changes in reinforcement contingencies, but to decreased perseveration if there are changes in the stimuli presented?

One possible explanation is that the primary locus of the lesions within the prefrontal cortex is different for the two sets of experiments. In the experiment of Milner (1964), the perseverating patients were damaged in the dorsolateral frontal area. In the experiment of Pribram (1961), the novelty-preferring monkeys were lesioned in the orbital (or ventral) frontal area. These two areas have functional differences, for which a complete theory is lacking. In general, the dorsal area has reciprocal connections with the secondary sensory cortices, whereas the orbital area has reciprocal connections (some via the mediodorsal thalamus) with the hypothalamus and limbic system. Hence (Fuster, 1980, p. 74):

> Lesion studies indicate that the cortex of the dorsal and lateral prefrontal surface is primarily involved in cognitive aspects of behavior. The rest of the prefrontal cortex, medial and ventral, appears to be mostly involved in affective and motivational functions and in the inhibitory control of both external influences and internal tendencies that interfere with purposive behavior and provoke inappropriate motor acts.

But Levine and Prueitt (1989) simulated both the novelty and perseveration data using a simplified neural network model that ignored differences between subregions. They simply modeled frontal damage as the weakening of certain connections in their network. The effects of this weakening were shown to mirror aspects of a "frontal lobe syndrome" whose existence is widely accepted, a syndrome that includes both the novelty and perseveration effects. The cognitive effects of dorsal damage and the motivational effects of orbital damage were both treated as part of a general defect in the guidance of behavior by motivational influences.

Nauta (1971) emphasized that the frontal cortex integrates sensory information from the neocortex with visceral information from the hypothalamus and limbic system. The lack of such integration, resulting from frontal damage, leads to a variable set of effects that Nauta (1971, p. 182) has called "interoceptive agnosia".

These effects may include distractibility (see also Grueninger & Pribram, 1969, and Wilkins et al., 1987), lack of foresight, and situationally inappropriate behavior. As influence of reward or punishment on behavior is lessened, non-affective influences on behavioral decisions, including *both* entrenched habits and attraction to novelty, are disinhibited. Which of those two is stronger depends on the task involved, as will be discussed in the next two sections.

CONDITIONED PERSEVERATION AND BEHAVIORAL STEREOTYPY

Milner (1963, 1964) compared patients with different brain lesions and normal subjects on the Wisconsin card sorting test. In this test, a subject is given a sequence of 128 cards, each displaying a number, color, and shape, as shown in Figure 5.3. The subject has to match the card shown to one of four template cards. The experimenter then says whether the match is right or wrong, not giving a reason. After ten correct matches in a row to color, the experimenter (without warning) switches the criterion to shape. Then if ten correct matches are made in a row to shape, the criterion shifts to number, then back to color, and so on. Milner found that most patients with damage to the dorsolateral frontal cortex learned the color criterion as quickly as normals, but never switched to shape after having learned color. Normals and patients with other brain lesions, by contrast, achieved on the average four or five criteria (color, shape, number, color, shape) over 128 trials.

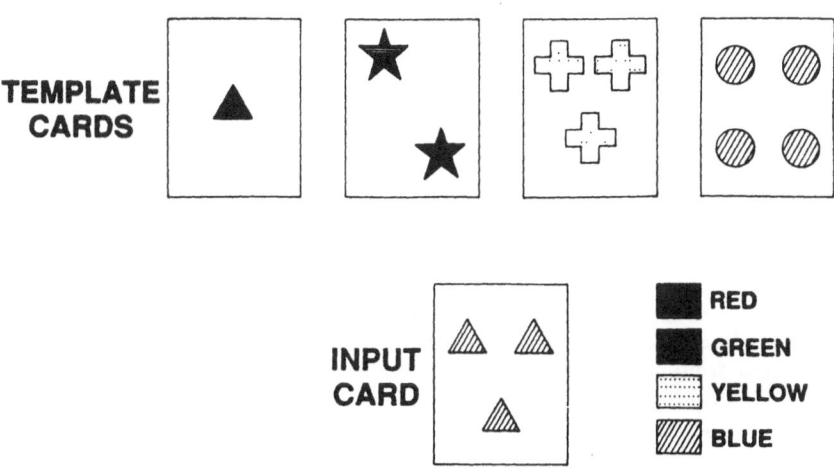

FIG. 5.3. Cards used in the Wisconsin card sorting test; see text for details. (From Levine & Prueitt, 1989; reprinted with permission of Pergamon Press.)

Milner's results were simulated, first in Leven and Levine (1987), using the neural network of Figure 5.4. This network builds on the adaptive resonance theory network of Carpenter and Grossberg (1987a) (ART 1), which is designed to classify input patterns. The ART 1 network includes a field of nodes coding input features and another field of nodes coding categories. A given pattern,

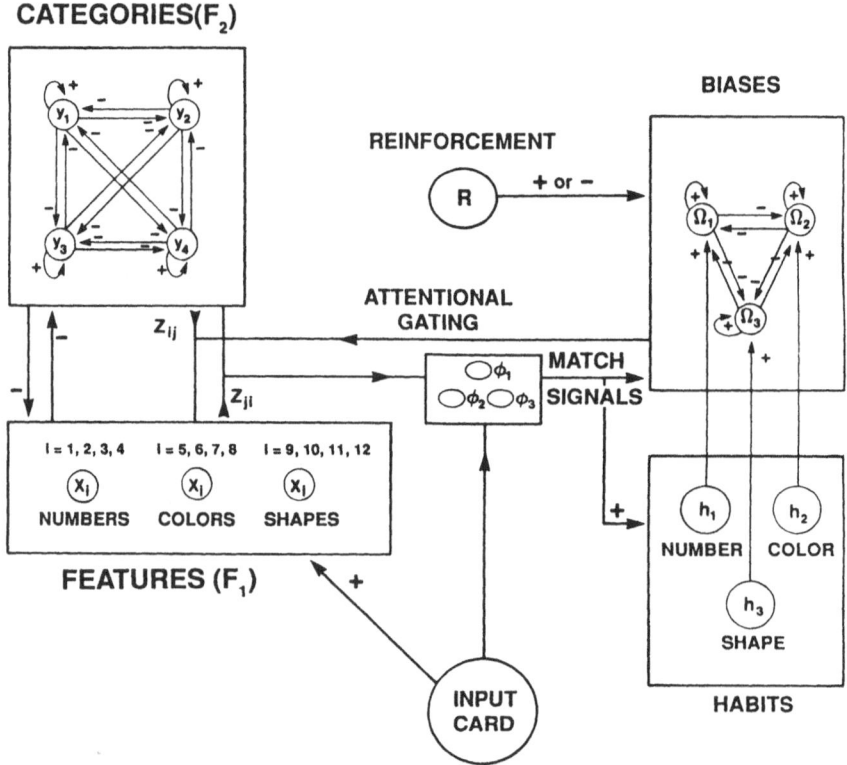

FIG. 5.4. Neural network, based on adaptive resonance theory, used in the simulation of Milner's card sorting data. Frontal lobe damage is modeled by a sharp reduction in gain of the positive and negative signals from the reinforcement node R to the bias nodes Ω_i (i=1 for number, 2 for color, 3 for shape). The bias nodes in turn gate signals from feature to category nodes. Each bias node also receives a positive signal from the corresponding habit node, which encodes past decisions (regardless of correctness) based on the given matching criterion. The match signal generators ϕ_i send positive signals to the corresponding habit nodes, and positive or negative signals to the bias nodes depending on the sign of reinforcement. Inhibition from F_1 to F_2 allows F_2 to reset to new inputs. Inhibition from F_2 to F_1 prevents top-down signals from being seen as inputs. (Modified from Leven & Levine (1987), copyright 1987 I.E.E.E., with permission).

treated as a vector of feature node activities, differentially excites category nodes, which competitively inhibit each other. We added to ART 1 a mechanism for attentional biases within the feature field (Grossberg & Levine, 1975).

In the network of Figure 5.4, nodes in the field F_1 code individual features (4 numbers, 4 colors, 4 shapes). Nodes in the field F_2 code template cards, each establishing the category of cards "similar" to it. F_1 divides into three "subfields" (number, color, and shape). To each subfield corresponds a "habit node" and a "bias node," separate from F_1 and F_2. Habit nodes detect how often classifications have been made, rightly or wrongly, on the basis of the given criterion. Bias nodes are affected by both habit node activities and reinforcement signals (the experimenter's "Right" or "Wrong").

Synaptic strengths z_{ij} and z_{ji} between F_1 and F_2 are large when node X_i represents a feature present in card y_j. Attentional gating from bias nodes selectively enhances F_1-to-F_2 signals; for example, if color bias is high and shape bias is low, the "one red triangle" node at F_2 is more excited by the "red" node at F_1 than by the "triangle" node. When an input card is presented, the template card whose activity y_j is largest in response to the input is chosen as the one matched. If the card chosen and the input card share a feature (color, shape, or number) a "match signal" is sent to the habit and bias nodes corresponding to that feature. This signal either excites or inhibits the bias node, depending on the sign of reinforcement. (Equations for this network, not shown here, are listed in Levine & Prueitt, 1989.)

Figure 5.5 shows results of our simulations. The parameter α that was varied measures gain of signals (positive or negative) from the reinforcement node to the bias nodes. With high α, the network acted like a normal subject in the card

	CRITERION	TRIAL
$\alpha = 4$	Color	13
("Normal")	Shape	40
	Number	82
	Color (again)	96
	Shape (again)	115
$\alpha = 1.5$	Color	13
("Frontally	Thereafter, classified by color	
Damaged")	for all remaining trials	

FIG. 5.5. Results of simulations on the network of Figure 5.4. Trial number listed is the first one that the network achieved ten correct matches in a row based on the given criterion. (From Levine & Prueitt, 1989; reprinted with permission of Pergamon Press.)

sorting test. It reached ten correct responses in a row five times during 128 trials. With low α, the network acted like one of Milner's dorsal frontal patients. It learned the color criterion on trial 13, as rapidly as did the "normal" network, but classified on the basis of color for all remaining trials.

Separation of habit and reinforcement loci is compatible with macaque monkey data showing that memories of motor responses and memories of the reinforcement values of events are encoded in interacting, but separate, neural systems (Mishkin et al., 1984; Mishkin & Appenzeller, 1987). The architecture involving positive feedback from habit nodes is intended to reproduce perseverative behavior in general, of which the Milner data comprise but one example.

Figure 5.6 shows a possible alternative network for simulating the Milner data. In this network, input patterns at F_1 are compared with prototype patterns stored at the synaptic weights from F_2 to F_1, as in Carpenter and Grossberg (1987a, 1987b). In the Carpenter-Grossberg network, a vigilance parameter is used to determine the degree of match that is recognized by the network. This vigilance parameter can be interpreted as the gain of signals from a preprocessing layer to a node that shuts off reset at the F_2 level. The bias nodes, instead of gating F_1-to-F_2 signals as in Figure 5.4, selectively modulate the contributions of different parts of F_1 to the match measurement.

The role of bias node signals in Figure 5.6 is similar to one that has been suggested for norepinephrine synapses from the midbrain locus coeruleus (Foote & Morrison, 1987; Hestenes, in press). This network has not been simulated, but comparative capabilities of the networks of Figures 5.4 and 5.6 are currently under investigation.

The behavior of frontal patients on the card sorting test can be thought of as *conditioned* perseveration; although the perseverative tendency itself is hard-wired, *what* response is perseverated depends on learning. This interpretation is supported by a variety of data from classical conditioning experiments whereby frontal lobe damage tends to replace reinforcement-dependent behavior with fairly stereotyped response-dependent behavior.

For example, Pribram (1961) tested normal and frontally lesioned rhesus monkeys on a lever pressing task with a fixed interval of food reinforcement. It is well-known (Ferster & Skinner, 1957) that in such fixed-interval tasks, the response rate will tend to increase exponentially over the time interval between reinforcements, leading to a "scalloped" response curve. With frontal damage, Pribram found, the scalloped curve was replaced by a nearly linear curve. This signified that the amount of lever pressing by the lesioned animal was nearly uniform over that time interval. Stamm (1963) found that frontally damaged monkeys exhibited less variability of interresponse rates than normal monkeys when subjected to operant conditioning on a differential reinforcement of low rates (DRL) schedule. Finally, Crow and McWilliams (1979), studying a water

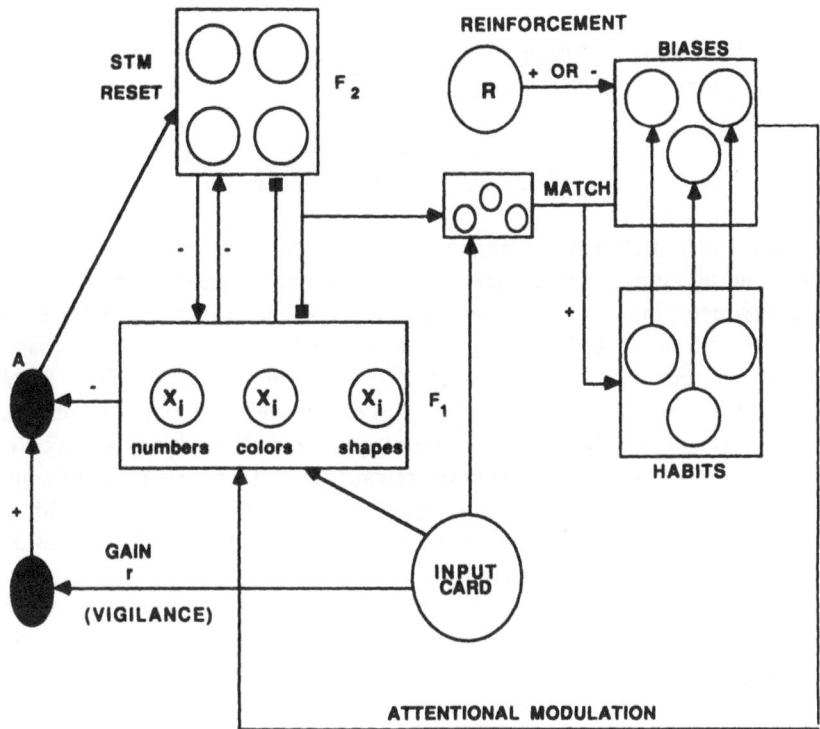

FIG. 5.6.   An alternative to the network of Figure 5.3 for explaining the card sorting data. If [X] is the total activity at F_1 and [I] the total activity at the input (card) level, the reset node A is inhibited if and only if [X] > r[I], which is the criterion for sufficient match between the input and the active top-down category prototype. As in Carpenter and Grossberg (1987a), when A is activated, reset occurs at F_2, allowing a new category to be tested. When A is inhibited, the input is classified as being in the active category. Bias nodes, instead of modulating bottom-up synapses as in Figure 5.3, selectively modulate the contributions of different features to overall F_1 activity.

ingestion paradigm in rats, found that frontally damaged animals maintained a more constant response rate and extinguished more slowly than did normals.

Some of this literature on frontal lobes and conditioning has been tied to the notion of *variability generation* (Crow, 1985). That article lists other experimental ministrations besides frontal damage that can reduce variability of behavior and thereby increase perseveration; among these treatments are amphetamine, ethanol, and some kinds of stress. Also, variability can be reduced in such mental illnesses as schizophrenia (Pishkin & Williams, 1976) and obsessive-compulsive disorder (Rapoport, 1989). Effective conditioning in complex environments depends on having a wide range of behavioral responses from which to choose those that are most rewarding or least punishing. The frontal lobes appear to be important in ensuring such a range.

NOVELTY PREFERENCE

As for novelty preference, Pribram (1961) compared normal and frontally lesioned rhesus monkeys in a scene with several junk objects. In the first step of the experiment, the monkey is presented with one cue, a junk object placed over one of twelve holes drilled in a wooden board. The experimenter has placed a peanut under this cue. After a certain fixed number of trials, in which the monkey first lifts the cue and is thus rewarded, a second (novel) junk object is introduced while the board is hidden from the monkey's view. This object is placed over another of the twelve holes. Again the reward (peanut) is placed under the novel cue. This same process is repeated until all the holes are covered with junk objects. Each time, the peanut remains under the same (novel) cue until the animal finds the peanut a certain number of times.

Figure 5.7 shows the results of Pribram's experiment. The number of repetitive errors (liftings of the familiar object) before the monkey first selects the novel object is shown for both normal and frontally damaged ("frontal") subjects. In general, frontal animals made fewer errors than normals, being more attracted to

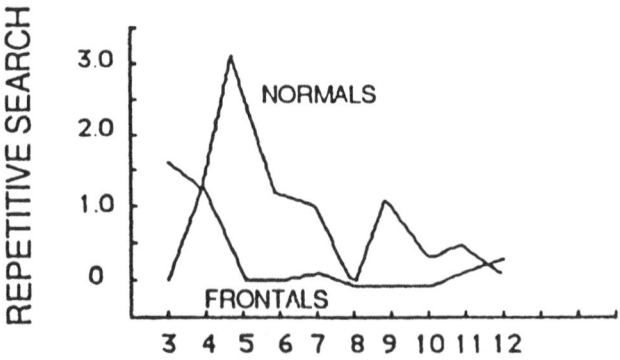

FIG. 5.7.   Experimental data on frontally damaged versus normal rhesus monkeys in a scene with up to 12 junk objects. A reward is placed each time under the novel object. The graph shows how many times, on the average, the monkey approaches the previously rewarded object before reaching a criterion number of consecutive responses to the novel one. (From Pribram, 1961; reprinted with permission of Academic Press.)

the novel object. The greater attraction to novelty on the part of frontals was most pronounced when the number of cues was between five and nine. The performance of frontal animals was worse if there were very few cues, and the performance of normals improved with very many cues.

We return at the end of this section to a qualitative explanation of the effects at the extreme ends of the graph in Figure 5.7. The main part of this section is devoted to discussing a network simulated by Levine and Prueitt (1989), which explains the effect described by the middle range of that graph.

Hence, our first task is to explain attraction to novelty in any animals (brain damaged or normal). To do so, let us review briefly the neural network notion of *gated dipole* (Figure 5.8). Gated dipoles are devices for comparing current values of stimulus or reinforcement variables with recent past values of the same variables. In this way, reactions to novel or unexpected events are more enhanced than reactions to familiar or expected events, other things being equal. The comparison operation in such networks uses chemical transmitters that are depressed with repeated activation, as will be explained later.

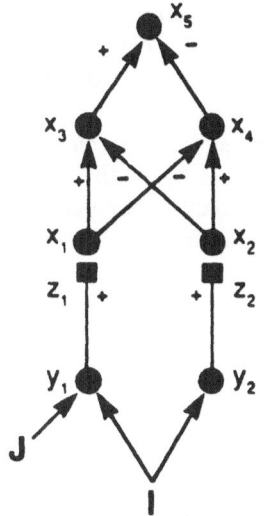

$$y_i = \text{FIRST INPUT STAGE}$$

$$z_i = \text{DEPLETABLE TRANSMITTER}$$

$$x_i = \text{COMPETITIVE STAGE,}$$
$$i = 1, 2$$

FIG. 5.8.   A schematic gated dipole with two competing channels (representing, for example, "fear and relief" if the input J is electric shock, or "on and off" if J is a given sensory stimulus), and its system equations. I is a nonspecific arousal input to both channels. Synapses with a depletable transmitter are indicated by filled in squares. While J is on, the greater input to the left channel overcomes that channel's greater transmitter depletion. Hence $x_3 > x_4$, and x_5 is excited. After J is turned off, the greater depletion of the left channel combined with equal channel inputs I causes rebound activity in the right channel. Hence, transiently, $x_4 > x_3$, and x_5 is inhibited. (Adapted from Levine & Prueitt, 1989, with permission of Pergamon Press.)

Gated dipoles were first introduced by Grossberg (1972a, 1972b) to answer the following question about reinforcement. Suppose an animal receiving steady electric shock presses a lever that turns off the shock. Later, in the same context, the animal's tendency to press the lever is increased. How can a motor response associated with the *absence* of negative reinforcement (shock) become itself positively reinforcing?

Figure 5.8 shows a schematic gated dipole. The synapses marked with squares have a chemical transmitter that tends to be depleted with activity. The input J represents shock, for example. The input I is nonspecific arousal to both channels y_1-to-x_1-to-x_3 and y_2-to-x_2-to-x_4. While shock is on, the left channel receives more input than the right channel, hence the transmitter is more depleted at z_1 than at z_2. But the greater input overcomes the more depleted transmitter, so the left channel activity x_1 exceeds the right channel activity x_2. This leads, by feedforward competition between channels, to net positive activity from the left channel output node x_3. But for a short time after the shock ends, both channels receive equal inputs I but the right channel is less depleted of transmitter than the left channel. Hence, the right channel activity x_2 now exceeds x_1, until the depleted transmitter recovers. Again, competition leads to net positive activity from the right channel output node x_4. Whichever channel has greater activity either excites or inhibits x_5, thereby enhancing or suppressing a particular motor or cognitive response.

Equations for a single gated dipole are listed in the Appendix. Characteristic output of one gated dipole is graphed in Figure 5.9. As that figure indicates, while the input J is on, x_1 is larger than x_2 so that x_3 is activated. After J is shut off, x_2 transiently exceeds x_1, because of less depletion of transmitter in the right channel. Hence x_4 is activated for a period of time.

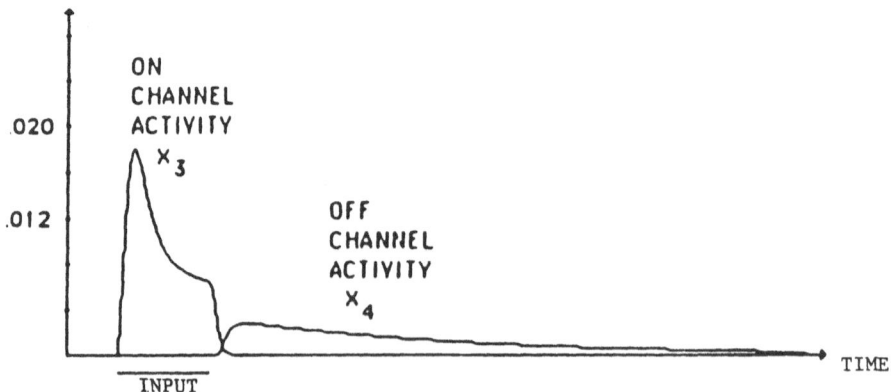

FIG. 5.9.   Typical time courses of the channel outputs x_3 and x_4 in a single gated dipole as in Figure 5.7. (From Levine & Prueitt, 1989; reprinted with permission of Pergamon Press.)

More recently (Grossberg, 1980; Levine & Prueitt, 1989), gated dipoles have been generalized from the reinforcement domain to the sensory domain. In the sensory domain, a gated dipole consists of "on cells" and "off cells" responding to the presence or absence of a particular sensory stimulus. On cells and off cells for different stimuli are joined into a *dipole field*. Transient rebounds in a dipole field were used by Grossberg (1980) to a model such visual phenomena as color-dependent tilt after-effects.

Figure 5.10 shows a dipole field used to model Pribram's data. Two dipole channel pairs are shown, one corresponding to an old cue and one to a novel cue. The nodes $x_{1,5}$ and $x_{2,5}$ of the dipoles represent tendencies to approach the given cues. Inhibitory links between these nodes, and a node $x_{3,5}$ representing the output of an additional dipole corresponding to some other cue in the environment, denote competition between attractions to different cues. (For simplicity, the rest of the dipole whose output is $x_{3,5}$ is not shown.) The cue with largest $x_{1,5}$ at a given time is approached.

The network of Figure 5.10 incorporates two competing rules. There is a net positive output at $x_{i,5}$ for each of the cues that have been presented. But for the novel cue, the on channel is less depleted than for the old cue, so that, in the absence of reward, the net positive output is greater for the novel cue than for the old cue. Hence competition among $x_{i,5}$ nodes favors those corresponding to novel cues, all else equal. But also, the reward node is excited when the monkey finds the peanut. Each $x_{i,5}$ connects with the reward node via modifiable synapses. Hence competition among $x_{i,5}$ nodes favors those with strong links to the reward node, all else equal. The latter effect is reminiscent of blocking.

Levine and Prueitt (1989) simulated the network of Figure 5.10, whose equations are shown in the Appendix. Representative graphs of their simulations are shown in Figure 5.11. The critical variable is again the gain α of signals from the reward locus to sensory loci. In Figure 5.11a, α is high, as in a normal monkey. With both cues present, $x_{1,5}$ has greater activity than $x_{2,5}$, so the previously rewarded cue is chosen over the novel cue. In Figure 5.11b, α is low, as in a frontally damaged monkey. In this case, $x_{2,5}$ has greater activity and $x_{1,5}$, so the novel cue is chosen. The choice of the novel object occurs because "on" channel transmitter is less depleted for the novel object than for the familiar object. Reward signals, which counteracted the novelty preference when α was high, are now too weak to have that effect.

Levine and Prueitt (1989) extended the model defined by Figure 5.10 to dipole fields with larger numbers of dipoles up to 12, to represent the cues in Pribram's experiment. Their simulations, not reproduced here, showed that the qualitative result of greater novelty preference on the part of "frontally damaged" networks extends to an arbitrary number of cues. But recall that the extreme ends of the graph of Pribram's data (Figure 5.8) show that in actual monkeys, performance is sensitive to how many cues are present. With small numbers of cues, the error rate of frontally damaged animals decreases, and with large numbers of cues, the

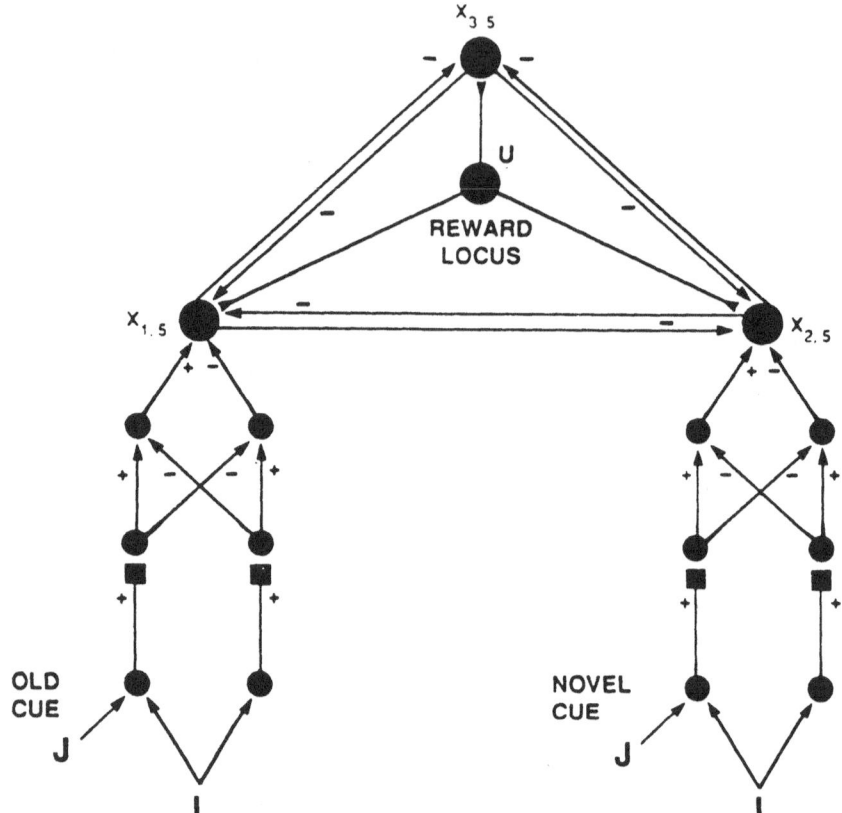

FIG. 5.10.   A dipole field, with two of the gated dipoles shown, used
to model Pribram's data. One dipole corresponds to an old cue and one
to a novel cue. The dipole outputs $x_{i.5}$ represent competing tendencies
to approach given cues. Competition among $x_{i.5}$ nodes is biased by
(a) modifiable connections with u, giving an advantage to previously
rewarded cues, and (b) transmitter depletion at the "square" synapses
within the dipoles, giving an advantage to novel cues. Which of (a) or
(b) is stronger is determined by the gain of reward signals to the $x_{i.5}$'s;
frontal damage is assumed to lower that gain. (Adapted from Levine &
Prueitt, 1989; reprinted with permission of Pergamon Press.)

error rate of normal animals decreases. Hence, the network of Figure 5.10 must
be extended to account for these additional effects.

The better performance of normal animals with large numbers of cues might
be explained by the formation of a higher-order rule that novelty per se is
rewarding (see Levine & Prueitt, 1989 for discussion). A simplified mechanism
that can account for some of this qualitative behavior, however, is shown in
Figure 5.12. The network of Figure 5.10 is extended so that the gain coefficient

α is no longer constant but instead is the dynamically changing activity of an additional node. This node is in turn inhibited by all of the sensory nodes $x_{1.5}$. In the network combining Figure 5.10 with Figure 5.12, the more cues are present, other things being equal, the less should be the influence of reward on the competition between cues because of high "background arousal." Hence, as in frontally lesioned animals, the influence of novelty is disinhibited.

The worse performance of frontal animals with small numbers of cues in Pribram's experiment might be explained by faulty segmentation of the perceptual environment. In the early stages of the experiment, the monkey may not sufficiently isolate the junk object cues mentally from the rest of the scene. Such segmentation also is likely to include higher cortical functions; hence frontally damaged monkeys could be slower than normal monkeys to separate the junk objects from the scene, a precondition for discrimination *between* these objects.

Alternatively, this effect might be explained by involvement of the habit system utilized in Figure 5.4 but not included in Figure 5.10. With few cues, positive feedback from habit nodes could tend to reinforce attraction to previously rewarded cues, which would perseverate for longer in frontally damaged than in normal animals. With an increase in the number of cues, the habits of approach to several previously rewarded cues would compete with each other so that no single habit predominates. This in turn would disinhibit the approach to novel cues that is also enhanced by frontal damage.

DISCUSSION

Integration of motivational with cognitive information, as simulated herein, covers a large number but not all of the functions of the prefrontal cortex. The cognitive-motivational linkage subsystems discussed here appear to be part of a larger function of forming strategies for goal directed behavior (see Fuster, 1980, and Stuss & Benson, 1986 for summaries). Hence, these subsystems need to be coordinated with other subsystems that link past events or actions across time (Fuster, 1980, 1985) and anticipate future events or actions (Gevins et al., 1987; Ingvar, 1985).

Levine (1986) proposed some network hypotheses for how the motivational and the timing functions of the frontal lobes could be coordinated. It was suggested there that different, interacting subsystems of the frontal cortex supply inhibition of the reticular activating (nonspecific arousal) system; inhibition of inhibitory cells in secondary sensory cortical areas, enabling greater attentional focus on a smaller number of cues; and selective activation of areas coding longer rather than shorter sequences of stimuli or movements, so that events can be linked across time. A schematic network architecture for the latter function is shown in Figure 5.13.

The process of Figure 5.13 is based on the notion that goal directed behavior

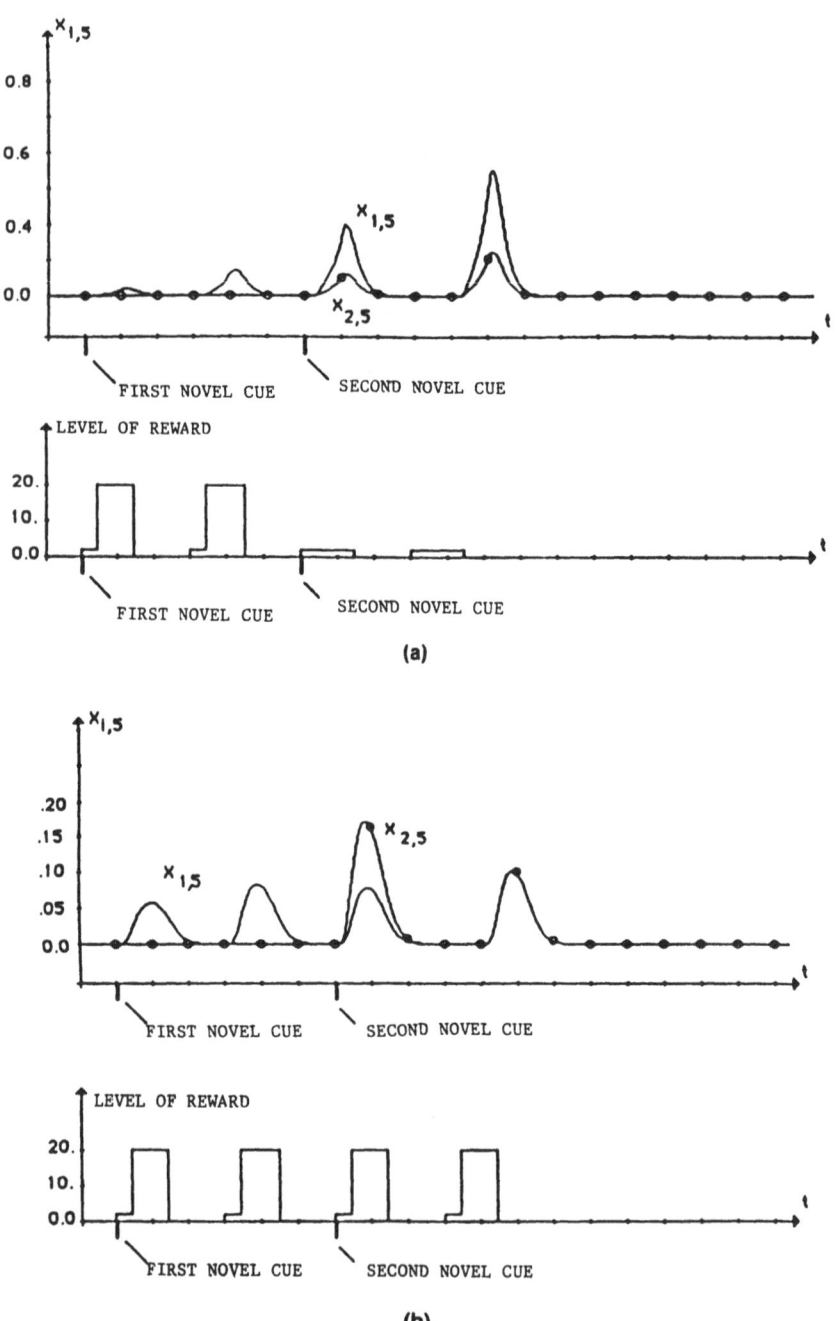

FIG. 5.11. Results of simulations of the dipole field network of Figure 5.10. $x_{1,5}$ denotes approach to the junk object cue that is first presented, and $x_{2,5}$ is approach to the second cue presented. Dark vertical tick marks show times of first presentation of novel cues. The bottom graph (labeled "level of reward") denotes the value at the present time of the

138

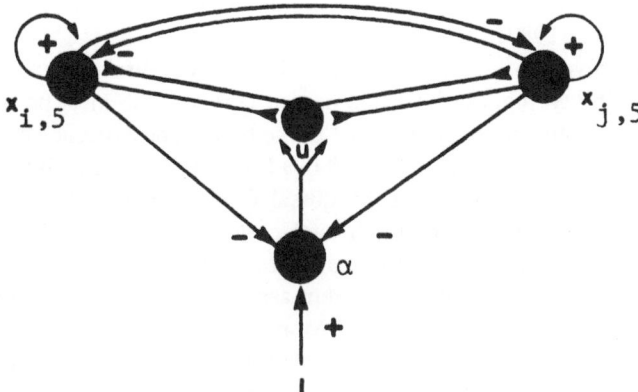

FIG. 5.12. Architecture in which coupling coefficient between reward and cue approach nodes is itself a dynamically changing node activity. The node ﹀, which modulates synapses between each $x_{i,5}$ and the reward node u, is in turn inhibited by each $x_{i,5}$ and excited by a tonic arousal input I. (Adapted from Levine & Prueitt, 1989, with permission of Pergamon Press.)

depends on the encoding by cortical cells of *chunks* or sequences of either stimuli or movements. The chunking idea was discussed in Grossberg (1978) and further developed in the masking field architecture of Cohen and Grossberg (1987), an example of which is shown in Figure 5.14. (The masking field idea may also help to provide more detailed explanations for the cognitive regrouping and perceptual segmentation processes that we have used to explain the data at the extreme ends of the graph in Pribram, 1961).

The masking field architecture incorporates a rule described by Grossberg (1978, p. 325) as follows: "*Self-Similar Coding Rule:* Other things being equal, higher order chunks have greater STM (short term memory) activity and longer duration than lower order chunks." This means that chunk representations compete in an on-center off-surround network, with a competitive advantage to the "higher order" ones that encode longer sequences. Hence, the response to a given stimulus is biased in favor of actions that are potentially appropriate to the whole sequence of preceding stimuli as well as the immediate one. Masking fields have

reward node activity u. (a) In the "normal monkey" network, a = 2.0. When both cues are present, $x_{1,5} > x_{2,5}$, so the previously rewarded cue is chosen (and not rewarded). (b) In the "frontal monkey" network, a = 0.1. When both cues ar present, $x_{2,5} > x_{1,5}$, so the novel cue is chosen (and rewarded). (From Levine & Prueitt, 1989; reprinted with permission of Pergamon Press.)

particularly been applied to the simulation of context-dependent speech parsing (Cohen et al., 1987).

The involvement of the prefrontal cortex in goal direction and in suppression of distractibility suggests that this part of cortex is the most probable locus for masking field anatomies if any exist in the brain. The existence of such architectures in the frontal lobes is supported in a fragmentary manner by single cell data. Fuster et al. (1982), working the dorsal frontal cortex, and Rosenkilde et al. (1981), working in the ventral frontal cortex, recorded from cells during the performance of rewarded discrimination tasks with delays. They found different groups of cells responding to different aspects of the task (visual stimuli, delay periods, rewards, and movements). Moreover, they found a columnar organization, reminiscent of that of primary sensory cortices, whereby cells in nearby locations tend to have similar firing properties during the task.

But the masking field only answers part of the timing question. How do such higher order and lower order chunks arise in the first place? The answer involves extensions of previously discussed network notions of category formation. Many commonly used neural network models of categorization (Carpenter & Grossberg, 1987a, 1987b; Edelman, 1987; Rumelhart et al., 1986; Rumelhart & Zipser, 1985) involve a layer that codes individual input features and another layer that codes *spatial patterns* of feature node activities. In order to model prefrontal functions, we must extend this categorization notion to *space-time patterns.*

Indeed, the difference between space-time patterns and time-independent spatial patterns is one of the two most cogent functional distinctions between the prefrontal cortex and other multimodal association areas like the temporal and parietal cortices. (The other distinction is the prefrontal area's greater ability to

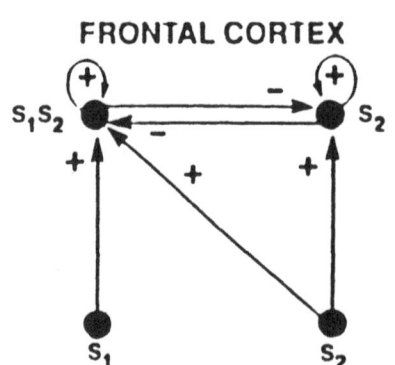

FRONTAL CORTEX

$S_1 S_2$ S_2

S_1 S_2

SECONDARY SENSORY CORTEX

FIG. 5.13. An on center off surround network at the frontal cortex could have nodes encoding sequences of one or more sensory events. Here two of these events, called S_1 and S_2, are multiply represented. A form of self-similarity is assumed: those frontal nodes encoding longer sequences, which receive more inputs from the sensory level, also have stronger self-excitation (shown in figure by a darker arrow at the $S_1 S_2$ node than at the S_2 node). This leads to bias toward longer sequences and hence increased attention to plans (Modified from Levine, 1986, with permission of Lawrence Erlbaum Associates.)

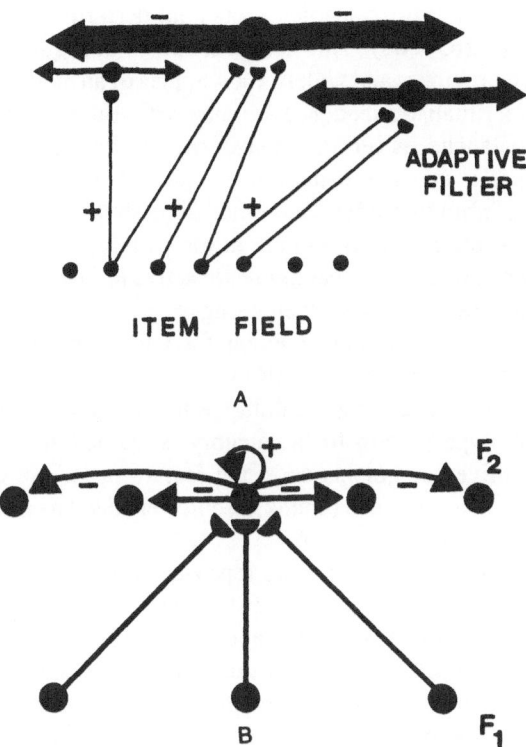

FIG. 5.14. Masking field interactions; F_1 and F_2 are as in the adaptive resonance networks, but F_1 nodes are interpreted as coding items in sequences and F_2 as coding sequences. (a) Connections from F_1 to F_2 grow randomly along positionally sensitive gradients. The nodes in the masking field F_2 grow so that larger item groupings, up to some optimal size, can activate nodes with broader and stronger inhibitory interactions. Thus the F_1-to-F_2 connections and the within-F_2 interactions both exhibit self-similarity. (b) The interactions within the masking field F_2 include positive feedback from a node to itself and negative feedback from a node to its neighbors. Long term memory traces at the ends of F_1-to-F_2 pathways adaptively tune the filter defined by these pathways to amplify the F_2 reaction to item groupings that have previously succeeded in activating their target F_2 nodes. (From Cohen & Grossberg, 1987; reprinted with permission of The Optical Society of America.)

process motivational inputs.) Grossberg and Kuperstein (1986), in their theory of saccadic eye movements, have located *target position maps* in the parietal cortex and *target position sequences* in the frontal eye field. (Although the frontal eye field lies just outside the boundaries of the prefrontal cortex, it is similar enough to the prefrontal cortex in its structure and connections that it can be studied as a microcosm of the region (Fuster, personal communication).)

One possible architecture for learned sequential performance is discussed in Grossberg (1978, pp. 265–266). This network, shown in Figure 5.15, was built on Grossberg's own previous, and widely known, idea of an *avalanche*, a network for performance of a ritualistic sequence of motor acts. But the network of Figure 5.15, unlike the avalanche, is sensitive to external feedback; therefore, the goal directed actions that it encodes can be interrupted if a significant contextual change occurs. The inhibition after each input stage shuts off the representation of a sensory stimulus after it has been present for a while, or of a motor act after it has performed. The nodes $v_{i,1}$ in that figure are active in succession, but external events can alter the exact timing of their firings.

Figure 5.16 shows one possible method for combining the categorization capability of the adaptive resonance (ART) network (Carpenter & Grossberg, 1987a, 1987b) with the sequencing capability of the network in Figure 5.15. Just as in ART the prototype pattern for a category is learned at a set of synaptic weights from the active category node to the feature nodes, so do the sensory events or motor acts in a sequence (chunk) become encoded as a set of top-down weights in the network of Figure 5.16.

All of these functional subsystems are reproduced in neural networks using, in different combinations, such familiar organizing principles as associative learning, competition, opponent processing, and interlevel resonant feedback (see Levine, 1989, and Hestenes, in press, for theoretical discussions of these principles). Suitable concatenation of these principles, in addition to illuminating the complex interactions involved in mammalian conditioning (see Grossberg, chap. 4 in this volume), can also yield insights into key brain regions implicated in

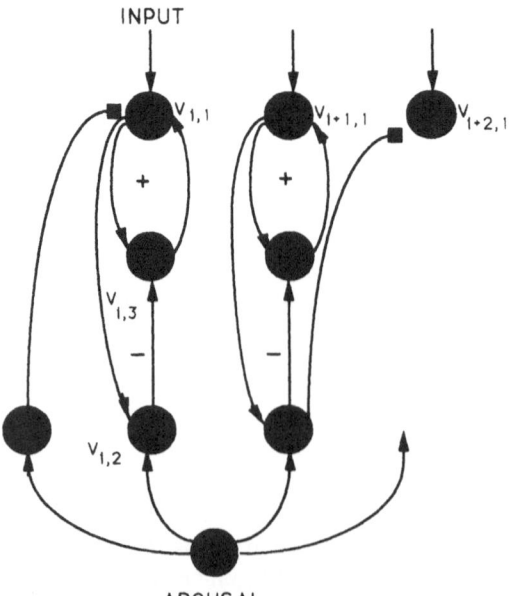

FIG. 5.15. Minimal network for learned sequential performance with modulation by arousal. The sensory or motor representations $v_{i,1}$ are activated in succession. Each $v_{i,1}$ has a corresponding $v_{i,3}$ to keep it reverberating in short-term memory as long as needed, and a corresponding $v_{i,2}$ (influenced by an arousal source) to shut off its reverberation. The $v_{i,2}$ also activates the next stage $v_{i+1,1}$ of the sequence. (Modified from Grossberg, 1978 with permission of Academic Press.)

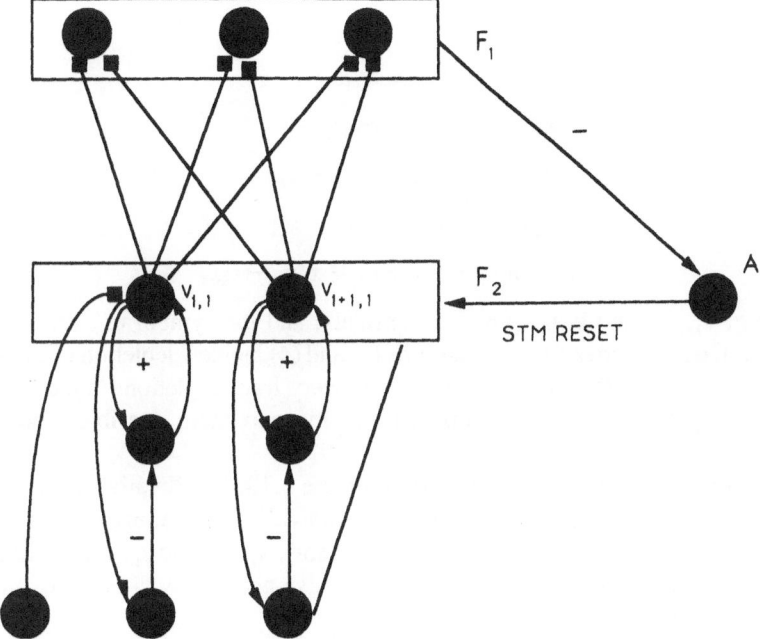

FIG. 5.16. Minimal synthesis of part of the sequential performance
network of Figure 5.15 with the adaptive resonance network of Carpenter and Grossberg (1987a). Each $v_{i,1}$ is a node of the category level F_2.
The nodes at F_2 learn categories of activity vectors at the input level F_1.
During the learning process, if there is a mismatch between the input
pattern at F_1 and the pattern of synaptic weights from F_2 to F_1, the node
A causes short-term memory reset, leading to the testing of a new
category at F_2.

conditioning (see also the model of hippocampal function in Schmajuk & DiCarlo,
chap. 6 in this volume). The combination of all the architectures shown here into
a system model of the prefrontal cortex and its interconnections is a subject of
ongoing work.

APPENDIX: EQUATIONS FOR THE DIPOLE FIELD
USED TO SIMULATE NOVELTY PREFERENCE

The gated dipole network defined by Figure 5.8 obeys the following equations:

$$dy_1/dt = -gy_1 + I + J \tag{1}$$

$$dy_2/dt = -gy_2 + I \tag{2}$$

$$dz_1/dt = a_1 (.5 - z_1) - a_2y_1z_1 \tag{3}$$

$$dz_2/dt = a_1(.5 - z_2) - a_2y_2z_2 \tag{4}$$

$$dx_1/dt = -gx_1 + by_1z_1 \tag{5}$$

$$dx_2/dt = -gx_2 + by_2z_2 \tag{6}$$

$$dx_3/dt = -gx_3 + b[x_1 - x_2]^+ \tag{7}$$

$$dx_4/dt = -gx_4 + b[x_2 - x_1]^+ \tag{8}$$

$$dx_5/dt = -gx_5 + (1 - x_5)x_3 - x_5x_4 \tag{9}$$

where g, a_1, a_2, and b are positive constants, and the symbol u^+, for any real number u, denotes max $(u,0)$. Equations (3) and (4) include depletion of transmitter (in the terms of the form $a_2y_iz_i$), and recovery from depletion (in the terms of the form $a_1(.5 - z_i)$, with .5 taken to be the maximum possible amount of transmitter).

The equations defining the network of Figure 5.10, for all variables except $x_{i,5}$ and the reward node u, are (1)–(9) for each individual dipole, except that for the i th cue and for each index k (up to 4 as relevant), x_k, y_k, and z_k, are replaced by $x_{i,k}$, $y_{i,k}$, and $z_{i,k}$. To connect $x_{i,5}$ in a competitive network with other $x_{j,5}$ nodes, we replace (9) by

$$dx_{i,5}/dt = -gx_{i,5} + (1 - x_{i,5})(\alpha uz_{i,5} + x_{i,5}) \tag{9'}$$
$$-cx_{i,5}(x_{i,4} + \sum_{j \neq i} x_{j,5})$$

In (9'), the coupling factor α between reward and sensory loci is assumed to be large in normal animals and small in frontally damaged animals.

Equation (9') illustrates three interacting sets of influences on the value of $x_{i,5}$, the tendency to approach a given object. These influences are: (a) exponential decay of activity back to a baseline level, with decay rate g; (b) shunting excitation (see Grossberg, 1973) proportional to the difference from its maximum 1; (c) shunting inhibition proportional to $x_{i,5}$. Shunting excitation is from the reward locus u, via a synapse with strength $z_{i,5}$, and from the output $x_{i,5}$ of the "on" channel for a given sensory cue. Shunting inhibition is from competing $x_{j,8}$, j not equal to i, and from the output $x_{i,4}$ of the "off" channel for the given cue.

The synaptic strength $z_{i,5}$ between the reward node and $x_{i,5}$ for a given cue obeys the equation

$$dz_{i,5}/dt = -f_1z_{i,5} + f_2ux_{i,5} \tag{10}$$

with f_1 and f_2 positive constants. Reward node activity obeys an equation of the form

$$du/dt = gu + r \tag{11}$$

where r represents the actual reward input.

In the network extended by the extra node described in Figure 5.12, it is assumed that the coefficient α of (9′) is no longer constant but is instead a dynamically changing node activity. This node is inhibited by all of the sensory nodes $x_{i,5}$, resulting in a new equation of the form

$$d\alpha/dt = -f(\sum_{i=1}^{n} x_{i,5}) + (k-\alpha)I$$

where f is a monotone increasing function, I a tonic arousal input, and k the maximum activity of the node whose activity is α.

REFERENCES

Carpenter, G. A., & Grossberg, S. (1987a). A massively parallel architecture for a self-organizing neural pattern recognition machine. *Computer Vision, Graphics, and Image Processing, 37,* 54–115.

Carpenter, G. A., & Grossberg, S. (1987b). ART 2: self-organization of stable category recognition codes for analog input patterns. *Applied Optics, 26,* 4919–4930.

Cohen, M. A., & Grossberg, S. (1987). A massively-parallel architecture for learning, recognizing, and predicting multiple groupings of patterned data, *Applied Optics, 26,* 1866–1891.

Cohen, M. A., Grossberg, S., & Stork, D. (1987). Recent developments in a neural model of real time speech analysis and synthesis. *Proceedings of the first international conference on neural networks* (Vol. IV, pp. 443–454). San Diego: IEEE/ICNN.

Crow, L. T. (1985). More on variability as a behavioral concept. *The Psychological Record, 35,* 293–300.

Crow, L. T., & Mc Williams, L. S. (1979). Relative stereotypy of water ingestive behavior induced by frontal cortical lesions. *Neuropsychologia, 17,* 393–400.

Edelman, G. (1987). *Neural Darwinism.* New York: Basic.

Ferster, C. B., & Skinner, B. F. (1957). *Schedules of reinforcement.* New York: Appleton-Century-Crofts.

Foote, S. W., & Morrison, J. H. (1987). Extrathalamic cortical modulation. *Annual review of Neuroscience, 10,* 67–95.

Fuster, J. (1980). *The prefrontal cortex.* New York: Raven. (Reprinted 1989).

Fuster, J. M. (1985). The prefrontal cortex: mediator of cross-temporal contingencies. *Human Neurobiology, 4,* 169–179.

Fuster, J. M., Bauer, R. H., & Jervey, J. P. (1982). Cellular discharge in the dorsolateral prefrontal cortex of the monkey during cognitive tasks. *Experimental Neurology, 77,* 679–694.

Gevins, A. S., Morgan, N. H., Bressler, S. L., Cutillo, B. A., White, R. M., Illes, J., Greer, D. S., Doyle, J. C., & Zeitlin, G. M. (1987). Human neuroelectric patterns predict performance accuracy. *Science, 235,* 580–584.

Grossberg, S. (1972a). A neural theory of punishment and avoidance. I. Qualitative theory. *Mathematical Biosciences, 15,* 39–67.

Grossberg, S. (1972b). A neural theory of punishment and avoidance. II. Quantitative theory. *Mathematical Biosciences, 15,* 253–285.

Grossberg, S. (1973). Contour enhancement, short term memory, and constancies in reverberating neural networks. *Studies in Applied Mathematics, 52,* 213–257.

Grossberg, S. (1975). A neural model of attention, reinforcement, and discrimination learning. *International Review of Neurobiology, 18,* 263–327.

Grossberg, S. (1978). A theory of human memory: self-organization and performance of sensory motor codes, maps, and plans. In R. Rosen & F. Snell (Eds.), *Progress in theoretical biology* (Vol. 5, pp. 233–374).

Grossberg, S. (1980). How does a brain build a cognitive code? *Psychological Review, 87,* 1–51.

Grossberg, S., & Kuperstein, M. (1986). *Neural dynamics of adaptive sensory-motor control: Ballistic eye movements.* Amsterdam: Elsevier/North-Holland.

Grossberg, S., & Levine, D. S. (1975). Some developmental and attentional biases in the contrast enhancement and short-term memory of recurrent neural networks. *Journal of Theoretical Biology, 53,* 341–380.

Grossberg, S., & Levine, D. S. (1987). Neural dynamics of attentionally modulated Pavlovian conditioning: blocking, interstimulus interval, and secondary reinforcement. *Applied Optics, 26,* 5015–5030.

Grueninger, W. E., & Pribram, K. H. (1969). Effects of spatial and non-spatial distractors on performance latency of monkeys with frontal lesions. *Journal of Comparative and Physiological Psychology, 68,* 203–209.

Hestenes, D. (in press). A neural network theory of manic-depressive illness. In D. S. Levine & S. J. Leven (Eds.), *Motivation, emotion, and goal direction in neural networks.* Hillsdale, NJ: Lawrence Erlbaum Associates.

Hinton, G. (1987). Connectionist learning procedures. Technical Report No. CMU-CS-87-115. Pittsburgh: Carnegie-Mellon University, Computer Science Department.

Ingvar, D. (1985). Memory of the future: an essay on the temporal organization of conscious awareness. *Human Neurobiology, 4,* 124–136.

Jacobsen, C. F. (1935). Functions of the frontal association area in primates. *Archives of Neurology and Psychiatry 33,* 558–569.

Kamin, L. J. (1968). Attention-like processes in classical conditioning. In M. R. Jones (Ed.), *Miami symposium on the prediction of behavior: aversive stimulation.* Miami, FL: University of Miami Press.

Kamin, L. R. M. J. (1969). Predictability, surprise, attention, and conditioning. In B. A. Campbell & R. M. Church, (Eds.), *Punishment and aversive behavior* (pp. 279–296). New York: Appleton-Century-Crofts.

Konorski, J., & Lawicka, W. (1964). Analysis of errors of prefrontal animals on the delayed response test. In J. Warren & K. Akert (Eds.), *The frontal granular cortex and behavior* (pp. 313–334). New York: McGraw-Hill.

Leven, S. J., & Levine, D. S. (1987). Effects of reinforcement on knowledge retrieval and evaluation. *Proceedings of the first international conference on neural networks* (Vol. II, pp. 269–279). San Diego. IEEE/ICNN.

Levine, D. S. (1986). A neural network theory of frontal lobe function. In C. Clifton (Ed.), *Proceedings of the eighth annual conference of the Cognitive Science Society* (pp. 716–727). Hillsdale, NJ: Lawrence Erlbaum Associates.

Levine, D. S. (1989). Neural network principles for theoretical psychology. *Behavior Research Methods, Instruments, and computers, 21,* 213–224.

Levine, D. S., & Leven, S. J. (Eds.). (in press). *Motivation, emotion, and goal direction in neural networks.* Hillsdale, NJ: Lawrence Erlbaum Associates.

Levine, D. S., & Prueitt, P. S. (1989) Modeling some effects of frontal lobe damage: novelty and perseveration. *Neural Networks, 2,* 103–116.

Milner, B. (1963). Effects of different brain lesions on card sorting. *Archives of Neurology, 9,* 90–100.

Milner, B. (1964). Some effects of frontal lobectomy in man. In J. Warren & K. Akert (Eds.), *The frontal granular cortex and behavior* (pp. 313–334). New York: McGraw-Hill.

Mishkin, M., & Appenzeller, T. (1987, June). The anatomy of memory. *Scientific American,* 80–89.

Mishkin, M., Malamut, B., & Bachevalier, J. (1984). Memories and habits: two neural systems. In G. Lynch, J. McGaugh, & N. Weinberger (Eds.), *Neurobiology of learning and memory* (pp. 65–77). New York: Guilford.

Nauta, W. J. H. (1971). The problem of the frontal lobe: a reinterpretation. *Journal of Psychiatric Research, 8,* 167–187.

Pishkin, V., & Williams, W. V. (1976). Cognitive rigidity in information processing of undifferentiated schizophrenics. *Journal of Clinical Psychology, 33,* 625–630.

Pribram, K. H. (1961). A further experimental analysis of the behavioral deficit that follows injury to the primate frontal cortex. *Experimental Neurology, 3,* 432–466.

Rapoport, J. L. (1989, March). The biology of obsessions and compulsions. *Scientific American,* 83–89.

Rosenkilde, C. E., Bauer, R. H., & Fuster, J. M. (1981). Single cell activity in ventral prefrontal cortex of behaving monkeys. *Brain Research, 209,* 375–394.

Rumelhart, D. E., Hinton, G. E., & Williams, R. J. (1986). Learning internal representations by back propagation. In D. E. Rumelhart, J. L. McClelland, an the PDP Research Group (Eds.), *Parallel distributed processing* (Vol. 1, pp. 365–422). Cambridge, MA: MIT Press.

Rumelhart, D. E., & Zipser, D. (1985). Feature discovery by competitive learning. *Cognitive Science, 9,* 75–112.

Spaet, T., & Harlow, H. F. (1943). Problem solution by monkeys following bilateral removal of the prefrontal areas. II. Delayed reaction problems involving use of the matching-to-sample method. *Journal of Experimental Psychology, 32,* 424–434.

Stamm, J. S. (1963). Function of prefrontal cortex in timing behavior of monkeys. *Experimental Neurology, 7,* 87–97.

Stamm, J. S. (1964). Retardation and facilitation in learning by stimulation of frontal cortex in monkeys. In J. Warren & K. Akert (Eds.), *The frontal granular cortex and behavior* (pp. 102–125). New York: McGraw-Hill.

Stuss, D. T., & Benson, D. F. (1986). *The frontal lobes.* New York: Raven.

Werbos, P. J. (1988). Backpropagation: past and future. *Proceedings of the IEEE international conference on neural networks, 1988* (Vol. I, pp. 343–353). San Diego: IEEE.

Wilkins, A. J., Shallice, T., & McCarthy, R. (1987). Frontal lesions and sustained attention. *Neuropsychologia, 25,* 359–365.

Neural Dynamics of Hippocampal Modulation of Classical Conditioning

Nestor A. Schmajuk
James J. DiCarlo
Department of Psychology
Northwestern University

ABSTRACT

This chapter describes hippocampal participation in classical conditioning in terms of Grossberg's (1975) attentional theory. According to this theory, pairing of a conditioned stimulus (CS) with an unconditioned stimulus (US) causes both an association of the sensory representation of the CS with the US (conditioned reinforcement learning) and an association of the drive representation of the US with the sensory representation of CS (incentive motivation learning). Sensory representations compete among themselves for a limited capacity short-term memory (STM) activation that is reflected in a long-term memory (LTM) storage.

We propose that the hippocampus controls self-excitation and competition among sensory representations and stores incentive motivation associations, thereby regulating the contents of a limited capacity short-term memory. Based on this hypothesis, the model predicts that hippocampal lesions impair phenomena that depend on the competition for short-term memory, such as blocking and overshadowing. The model also predicts that hippocampal long-term potentiation or kindling facilitate the acquisition of classical discrimination by increasing the stored values of incentive motivation associations. In addition, the model describes hippocampal neural activity as proportional to the strength of CS-US associations. Predictions generated by the neural network under the present hypothesis regarding hippocampal function are contrasted with experimental data.

Attentional theories of hippocampal function (Douglas, 1972; Douglas & Pribram, 1966; Grastyan, Lissak, Madarasz, & Donhoffer, 1959; Kimble, 1968; Moore, 1979; Solomon & Moore, 1975) emphasize that the hippocampus is involved in the modulation of the level of processing assigned to environmental stimuli. An important problem with most of these theories is that they do not specify the nature of interactions between attention and associative learning, and consequently, none of the theories can provide unequivocal predictions of the effects of hippocampal manipulations on classical conditioning. Typical hippocampal manipulations include hippocampal lesions (HL), hippocampal induction of long-term potentiation (LTP), hippocampal kindling, and hippocampal neural recording. Because the actual meaning of attention in classical conditioning depends on the particular model in which it is defined, a precise understanding of how attentional variables affect conditioning requires a computational model in which the variables are incorporated. Computational models should be able to describe normal conditioning and, with changes in their attentional variables, they should describe the consequences of different hippocampal manipulations.

Computational models, however, are not enough. Although computational models may offer accurate descriptions of behavior of normal animals and animals with hippocampal manipulations, a computational model providing trial-to-trial predictions cannot describe *how* information is processed in the brain. Such insight is only provided by computational models that describe behavior in real time. Real time models can describe simultaneously animal behavior and the variables controlling behavior. Behavioral descriptions can be compared to behavioral results, and the dynamics of hypothetically intervening variables can be compared to neural activity in different regions of the brain.

At least three real time attentional models are capable of describing classical conditioning in normal animals. Grossberg (1975) presented a real time neural model explaining attentional-associative interactions during conditioning, Moore and Stickney (1980) presented a real time version of Mackintosh's (1975) attentional model, and Schmajuk and Moore (1985) proposed a real time version of Pearce and Hall's (1980) attentional model. Because the attentional mechanisms in these models can be related to hippocampal function, modifications in attentional variables potentially allow the description of hippocampal manipulations.

Schmajuk and Moore (1985, 1989) studied the effects of various hippocampal manipulations on the classically conditioned nictitating membrane (NM) response in a rendering of the Moore and Stickney (M-S) model, called the M-S-S model. The M-S-S model incorporates an attentional rule that "tunes in" relevant CSs and "tunes out" irrelevant CSs. When "tuned in," a CS increases the rate at which it changes its associations with other CSs and the US. When "tuned out," a CS decreases the rate at which it changes its associations with other CSs and the US.

Moore and Stickney (1980) proposed that hippocampal lesions prevent poor predictors from being "tuned out." Under the "tuning out" hypothesis, the M-S-S model correctly describes the experimental effects of hippocampal lesions on delay conditioning, conditioning under optimal interstimulus interval (ISI), conditioned inhibition, extinction, latent inhibition, blocking, and mutual overshadowing. The model, however, is inconsistent with experimental findings describing the effects of hippocampal lesions on trace conditioning with shock as the US under short and long ISIs, trace conditioning with air puff as the US under long ISIs, discrimination reversal and sensory preconditioning. Schmajuk (1986) suggested that long-term potentiation facilitates the "tuning in" of good predictors. Under the "tuning in" hypothesis the M-S-S model was not able to describe the effects of long-term potentiation on acquisition of classical discrimination. Under the assumption that hippocampal neuronal activity is proportional to the magnitude of CS-US associations, the M-S-S model describes hippocampal neuronal activity during acquisition and extinction of classical conditioning.

Schmajuk (1986, 1989; Schmajuk & Moore, 1988) introduced a real time version of Pearce and Hall's (1980) attentional model, namely the S-P-H model. Schmajuk (1984) proposed the "aggregate prediction" hypothesis of hippocampal function. When a CS is followed by a US, an association is formed. This association can be regarded as the prediction of the US by the CS. The "aggregate prediction" hypothesis suggests that the hippocampus *computes* the sum of individual predictions of environmental events used to control associative learning. Hippocampal lesions imply impairments in the computation of the aggregate prediction. Under the "aggregate prediction" hypothesis, the S-P-H model correctly describes the effect of hippocampal lesions on delay conditioning, conditioning with short, optimal, and long ISI with a shock as US, conditioning with long ISI and air puff as the US, extinction, latent inhibition, generalization, blocking, overshadowing, discrimination reversal, and sensory preconditioning. However, under the aggregate prediction hypothesis, the S-P-H model has difficulty describing the effect of hippocampal lesions on conditioned inhibition and mutual overshadowing. The "aggregate prediction" hypothesis assumes that long-term potentiation induction increases the integration of multiple predictions into the aggregate prediction by way of increasing CS-CS associations. Under the aggregate prediction hypothesis, the S-P-H model has difficulty describing the effects of long-term potentiation on discrimination acquisition. The "aggregate prediction" hypothesis assumes that neural activity in the hippocampus is proportional to the instantaneous value of the aggregate prediction. Under the aggregate prediction hypothesis, the S-P-H model correctly describes neural activity in hippocampus during acquisition but not during extinction of delay conditioning.

Although both the M-S-S and the S-P-H are *computational* models that describe the effects of hippocampal manipulations in many classical conditioning paradigms, neither one has been implemented as a *neural network*. This limitation is particularly important for the design of a model, in which structure and function are closely related, that is,, a model *functionally isomorphic* with hippocampal processes and *structurally isomorphic* with the hippocampal circuitry. In contrast to both the M-S-S and the S-P-H model, Grossberg (1975; Grossberg & Levine, 1987) proposed an *attentional neural network* that describes classical conditioning. In Grossberg's model, as in the M-S-S and S-P-H models, when circuits that modulate attention are related to hippocampal function, modifications in these circuits allow the description of the functional effects of different hippocampal manipulations. Behavioral descriptions of classical conditioning provided by the model can be compared with behavioral results and the dynamics of the model variables can be compared with neural activity in different regions of the brain. In addition, the neural architecture of the model might provide clues regarding the functional anatomy of hippocampal circuitry.

Therefore, the present study contrasts experimental results regarding hippocampal manipulations in classical conditioning with computer simulations using Grossberg's (1975) attentional model under the hypothesis that hippocampal manipulations affect attentional mechanisms. Relevant data include the effect of HL, LTP, and kindling on classical conditioning paradigms and hippocampal neural activity recording during classical conditioning.

GROSSBERG'S (1975) ATTENTIONAL NETWORK

Grossberg (1975) proposed a neural model of attention, reinforcement, and discrimination learning, that describes how animals pay attention to and discriminate among certain cues while ignoring others. According to this theory, pairing of a CS with a US causes both an association of the sensory representation of the CS with the US (conditioned reinforcement learning) and an association of the drive representation of the US with the sensory representation of CS (incentive motivation learning). Sensory representations are stored in a short-term memory (STM) in order to bridge the temporal gap between the presentation of the CS and the presentation of the US. Sensory representations compete among themselves for a limited capacity STM activation. Associations of the sensory representation of the CS with the US and associations of the drive representation with the sensory representation of the CS are stored in long-term memory (LTM).

A piece of Grossberg's (1975) network is shown in Figure 6.1. CS_i activates sensory representation node X_{i1}. The activity of node X_{i1} represents the STM of CS_i. The US unconditionally activates neural populations of the drive representation (Y). Simultaneous activation of the drive representation and of sensory representations, causes X_{i1} to become associated with the output of the drive representation Y. This association is stored in LTM by increasing the synaptic weight V_i. After X_{i1} becomes associated with the drive representation, Y, it becomes a secondary reinforcer for other CSs.

Simultaneous activation of the drive representation and of sensory representations causes the output of the drive representation Y to become associated with X_{i1}. This association is stored in LTM by increasing the synaptic weight Z_i. Conditioning of the Y-X_{i2} pathway increases sensory representation X_{i1} by incen-

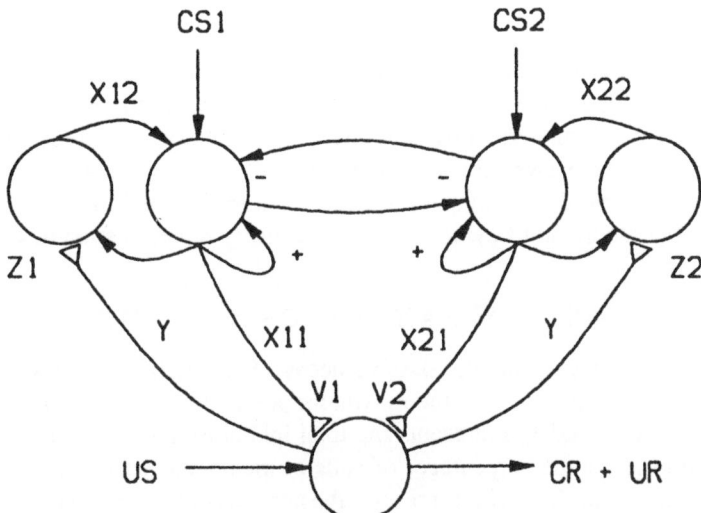

FIG. 6.1. Diagram of Grossberg's (1975; Grossberg & Levine, 1987) attentional network for classical conditioning. Conditioned stimuli, CS$_1$ and CS$_2$, activate sensory representations, X$_{11}$ and X$_{21}$, which compete among themselves for a limited capacity short-term memory. Sensory representations, X$_{11}$ and X$_{21}$, become associated with the US by changing synaptic weights V$_1$ and V$_2$ (conditioned reinforcement learning). Output Y becomes associated with X$_{11}$ and X$_{21}$ by changing synaptic weights Z$_1$ and Z$_2$ (incentive motivation learning). Arrows indicate non-modifiable synapses. Triangles indicate plastic synapses.

tive motivation. A sensory cue with large V_i and Z_i can augment the STM activity of its sensory representation. STM is sustained by a positive feedback loop.

Sensory representations that excite the drive representation node, and receive back incentive motivational signals, compete among themselves for a limited STM capacity. This limited STM capacity is implemented through an on-center off-surround architecture. In such architecture, an input CS_i excites STM activity X_{i1} and inhibits all other STM activities X_{j1}.

In blocking, an animal is first conditioned to CS_1, and this training is followed by conditioning to a compound consisting of CS_1 and a second stimulus CS_2. This procedure results in a weaker conditioning to CS_2. The network applies to blocking as follows. When CS_1 is paired with the US, associations V_1 and Z_1 are created. After CS_1 becomes conditioned, a strong X_{11}-Y-X_{12} feedback and competition between X_{11} and X_{21}, makes X_{21} activity so small that CS_2 does not gain much association with the US when it is presented with CS_1. Therefore, the selective attention assigned to a CS_1 depends on three circuits: (a) the X_{11}-Y-X_{12}-X_{11} incentive motivation loop, (b) the X_{11}-X_{21} competition loop, and (c) the X_{11} self-excitation.

A Formal Description of the Model

Grossberg and Levine (1987) presented a neural network, based on Grossberg (1975), that describes several classical conditioning paradigms, including blocking, acquisition under different interstimulus intervals, and secondary reinforcement. In the present chapter we use a version of the Grossberg and Levine (1987) model that differs in some aspects from the original model.

As in Grossberg and Levine (1987), the sensory representation of CS_i, X_{i1}, is defined by

$$d(X_{i1}) / dt = - A X_{i1} + B (C - X_{i1}) I_{i1} - D X_{i1} J_{i1}, \qquad (1)$$

where $- A X_{i1}$ represents the passive decay of STM, B represents the rate of increase of X_{i1}, constant C is the maximum possible value of X_{i1}, I_{i1} is the total excitatory input, and J_{i1} represents the total inhibitory input. C can be regarded as the number, or the percentage, of cells or membrane active sites that can be excited. Therefore, $(C - X_{i1})$ represents the number, or the percentage, of inactive sites that can be excited, and X_{i1} the number of active sites that can be inhibited.

Grossberg and Levine (1987) assumed that positive feedback signals trigger a process of habituation that steadily attenuates the net size of the feedback signals. For simplicity, the present paper assumes no attenuation, and the total excitatory input, I_{i1}, is given by

$$I_{i1} = CS_i + E X_{i1} + F CS_i X_{i2}, \qquad (2)$$

where CS_i represents the CS_i intensity, and $E X_{i1}$ represents a positive feedback from X_{i1} to itself. $F CS_i X_{i2}$ represents a signal from X_{i2} to X_{i1} that is active only if CS_i is present. This last term, that can be regarded as a presynaptic modulation of X_{i2} by CS_i, avoids the activation of sensory representations by drive representations in the absence of CS_i.

Like Grossberg and Levine (1987), we assume that the total inhibitory input to X_{i1}, J_{i1}, is the sum of the activities of all other nodes, X_{j1}, is given by

$$J_{i1} = \Sigma_{j \neq i} X_{j1}. \qquad (3)$$

Activity in the drive representation node is given by

$$dY / dt = - H Y + G (\Sigma_i X_{i1} V_i + US), \qquad (4)$$

where $- H Y$ represents the passive decay of drive representation activity, $\Sigma_i X_{i1} V_i$ is the sum of sensory representations gated by the corresponding LTM traces, and US is the US input to the drive representation node. $\Sigma_i X_{i1} V_i$ represents the intensity of the CR. Notice that whereas Grossberg and Levine (1987) assumed that the US acts on the sensory input, thereby explaining ISI effects, for simplicity we have assumed that the US acts only on the drive node. We assume that Y is the output of the system, and is proportional to the sum of the conditioned response (CR), $\Sigma_i X_{i1} V_i$, and the unconditioned response (UR).

The signal from Y activates LTM trace Z_i, and determines the activity of X_{i2}

$$d\,(X_{i2})\,/\,dt = -\,L\,X_{i2} + M\,Y\,Z_i. \tag{5}$$

Whereas Grossberg and Levine (1987) assumed that LTM associations of sensory representation X_{i1} with the drive representation, V_i, undergo extinction even when X_{i1} is zero, we assume that changes in V_i are possible only when X_{i1} is active. Also, contrasting with Grossberg and Levine (1987), we assume that V_i has a maximum value. Changes in V_i are given by

$$d\,(V_i)\,/\,dt = -\,N\,V_i\,X_{i1} + P\,(Q - V_i)\,Y\,X_{i1}, \tag{6}$$

where $-\,N\,V_i\,X_{i1}$ is the active decay in V_i when X_{i1} is active, and $P\,(Q - V_i)\,Y\,X_{i1}$ is the increment in V_i when Y and X_{i1} are active together. Q can be regarded as the number, or percentage, of cells or membrane patches that can be modified by learning. Therefore, $(Q - V_i)$ represents the number, or percentage, of unmodified sites that can increase the efficacy of their connection, and V_i the number of already modified sites that can decrease their connectivity.

As in Grossberg and Levine (1987), Equation 6 generates secondary reinforcement. This is so because, by Equation 4, Y is activated either by presentation of the US or of a CS already associated with the US. Therefore, if CS_1 is associated with the US, it generates activity Y that becomes associated with X_{21} by Equation 6.

For simplicity, Grossberg and Levine (1987) assumed that Z_i was identical to V_i. In order to explain savings effect seen in acquisition-extinction series (described later), Z_i should vary at a slower rate than V_i. Therefore, we have computed Z_i with

$$d\,(Z_i)\,/\,dt = -\,R\,Z_i\,Y + S\,(T - Z_i)\,X_{i1}\,Y, \tag{7}$$

where $-\,R\,Z_i\,Y$ is the active decay in Z_i when Y is active, and $S\,(T - Z_i)\,X_{i1}\,Y$ is the increment in Z_i when Y and X_{i2} are active together.

Hippocampal function: The "STM regulation" hypothesis

Grossberg (1975, 1980, Figure 16) suggested how his attentional network could be mapped into the brain. He proposed that sensory representations X_{i1}, the X_{i1} self-excitation loop, the X_{i1}-X_{j1} competition loop, and incentive motivation associations Z_i were located in the neocortex. Self-excitation, competition among drive representations, and sensory-drive associations V_i were located in the hippocampus.

In contrast to Grossberg's (1980) view, and in line with the attentional view of hippocampal function suggested in our previous papers (Schmajuk & Moore, 1988, 1989), we propose that circuits regulating selective attention are part of the hippocampal system. In the Grossberg and Levine (1987) network, the level of processing assigned to CS_i (selective attention), is determined by the magnitude of the sensory representation X_{i1} stored in STM. In the model, STM is modulated by three circuits: (a) the X_{i1}-Y-X_{i2}-X_{i1} incentive motivation loop, (b) the X_{i1}-X_{j1} competition loop, and (c) the X_{i1} self-excitation. Therefore, we assume that these three circuits are part of the hippocampus. We refer to this hypothesis as the "STM regulation" hypothesis of hippocampal function.

According to the "STM regulation" hypothesis, Grossberg and Levine's (1987) neural network is embedded in the brain as represented in Figure 6.2. Sensory-drive associations, V_i, are stored in different brain circuits, such as the cerebellum in the case of classical conditioning of the rabbit's nictitating membrane. Positive and negative feedbacks regulating STM are mediated through the hippocampus. Notice that, although for simplicity Figure 6.2 shows only reciprocal negative feedback between X_{11} and X_{21}, each sensory representation receives negative feedback from the sum of all other X_{ii}s. This sum is computed in the hippocampus. In addition, according to the "STM regulation" hypothesis, the hippocampus stores incentive motivation associations, Z_i, in the form of long-term potentiation (LTP).

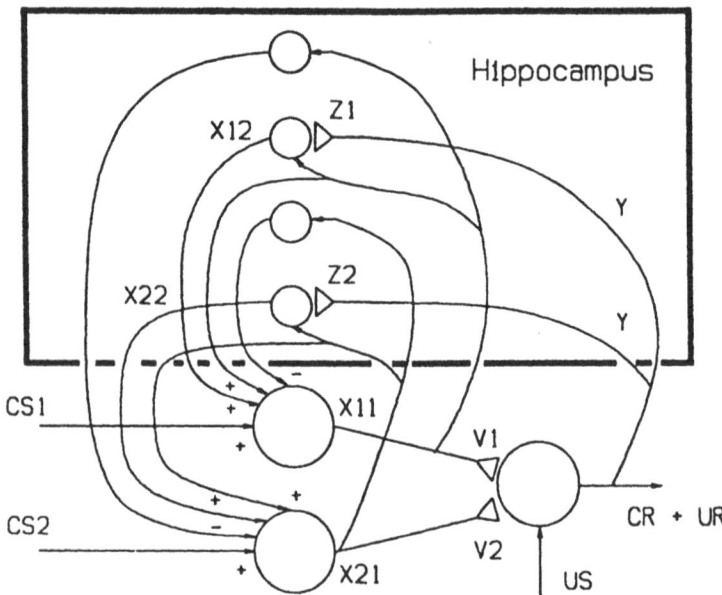

FIG. 6.2.   The "Short-term Memory Regulation" hypothesis: Mapping of Grossberg's (1975) network onto the hippocampus. Circuits regulating short-term memory by storing incentive motivation values (Z_1 and Z_2), mediating self-excitation, and mediating competition among sensory representations are assumed to be part of the hippocampal circuit. Nodes in the STM circuit may be regarded as hippocampal pyramidal cells. Circuits storing CS-US associations (V_1 and V_2) are assumed to be in other areas of the brain, such as the cerebellum.

According to the "STM regulation" hypothesis, hippocampal circuits regulate the content of STM by controlling incentive motivation, self-excitation, and competition of sensory representation X_{i1}. Therefore, hippocampal lesions can be represented by removal of nodes X_{i2}, the self-excitation of nodes X_{i1}, and the reciprocal inhibition among nodes X_{i1}. After HL, sensory representation X_{i1} is given by

$$d(X_{i1}) / dt = - A X_{i1} + B (C - X_{i1}) I_{i1}. \qquad (8)$$

After HL, the total excitatory input, I_{i1}, is given by

$$I_{i1} = CS_i. \qquad (9)$$

After HL, activity in the drive representation node remains unchanged, and is given by

$$dY / dt = - H Y + G (\Sigma_i X_{i1} V_i + US). \qquad (10)$$

After HL, changes in V_i are still given by

$$d (V_i) / dt = - N V_i X_{i1} + P (Q - V_i) Y X_{i1}. \qquad (11)$$

According to the "STM regulation" hypothesis, synapses storing Z_i values are assumed to be part of the hippocampal circuit. In consequence, the effect of inducing LTP or kindling in the hippocampus is equivalent to increasing Z_i.

According to the "STM regulation" hypothesis, nodes X_{i2} are assumed to be part of the hippocampal circuitry. Therefore, hippocampal neural activity is assumed to be proportional to their input Y. Consequently, hippocampal neural activity reflects the real time activity of the drive representation and of the CR.

METHODS

Although the model is a real time model, computer simulations of the model generate values of the relevant variables at discrete time instants. In our simulations we assume that one computer time step is equivalent to 10 msec. Each trial consisted of 100 steps, equivalent to 1 sec. Unless specified, the simulations assumed 200 msec CSs, the last 50 msec of which overlaps the US. CS onset was at 200 msec. Parameters were selected so that simulated asymptotic values of V_i were reached in around 10 acquisition trials. Since asymptotic conditioned NM responding is reached in approximately 200 real trials (Gormezano, Kehoe, & Marshall, 1983), one simulated trial is approximately equivalent to 20 experimental trials.

The right panels of all figures displaying simulation results shows the average values of V_i, Z_i, and X_{i1}. The left panel shows the real time value of the CR, $\Sigma_i X_{i1} V_i$, as a function of trials.

Parameter values were $A = .1, B = .1, C = 1, D = 5, E = .1, F = 30, G =$

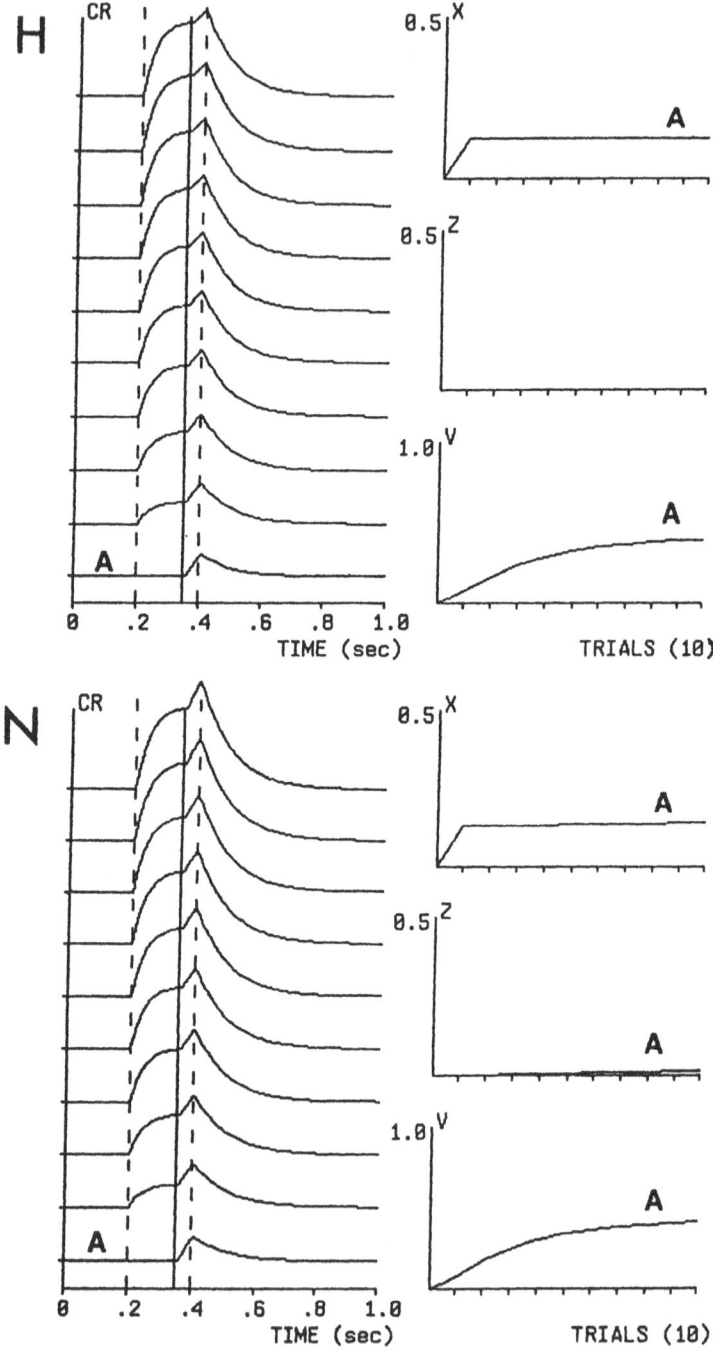

FIG. 6.3. Effect of hippocampal lesions on delay conditioning. H: HL case. N: normal case. Left Panels: Simulated conditioned response (CR) in real time, over 10 reinforced trials. Upper-Right Panels: Average sensory representation (X_i) value on each trial as a function of trials.

1, H = 1, L = 1, M = 1, N = .025, P = .05, Q = 1, R = .00015, S = .0004, T = 1. Initial values of V_i and Z_i were 0. These values were kept constant for all simulations. The effects of LTP and kindling were simulated by assigning to all Z_is an initial value equal to 0.2.

RESULTS

This section presents computer simulations for normal and HL cases in the following paradigms: (a) acquisition of delay conditioning, (b) acquisition of trace conditioning, (c) acquisition/extinction series of delay conditioning, (d) blocking, (e) overshadowing, and (f) discrimination reversal. Also, it presents computer simulations for (g) the effect of LTP induction and kindling on discrimination acquisition, and (h) the effect of kindling on discrimination reversal. Finally, computer simulations of (i) hippocampal neural activity are also presented. This section presents relevant experimental data and contrasts the data with the results of the computer simulations.

Effects of Hippocampal Lesions

Acquisition of Delay Conditioning

Experimental Data.   Schmaltz and Theios (1972) found faster than normal acquisition in HL rabbits but Port, Mikhail, and Patterson (1985), Solomon (1977), and Solomon and Moore (1975) found no difference in the rate of acquisition between normal and HL rabbits trained in delay acquisition. In summary, delay acquisition rates sometimes become accelerated, but often remain unaffected by HL.

Computer Simulations.   Figure 6.3 shows simulations of 10 acquisition trials of delay conditioning for normal and HL cases. Acquisition was simulated with a 200-msec CS, a 50-msec US, and a 150-msec ISI. In agreement with experimental data, the model shows similar rates of acquisition for normal and HL cases.

Acquisition of Trace Conditioning

Experimental Data.   Port, Mikhail, and Patterson (1985) found that acquisition rates become accelerated in HL rabbits trained in trace conditioning. James, Hardiman, and Yeo (1987) and Port, Romano, Steinmetz, Mikhail, and Patterson (1986) reported similar rates of acquisition of trace conditioning in normal and

---

Middle-Right Panels: Average incentive motivation association (Z_i) on each trial as a function of trials. Lower-Right Panels: Average drive association (V_i) on each trial as a function of trials.

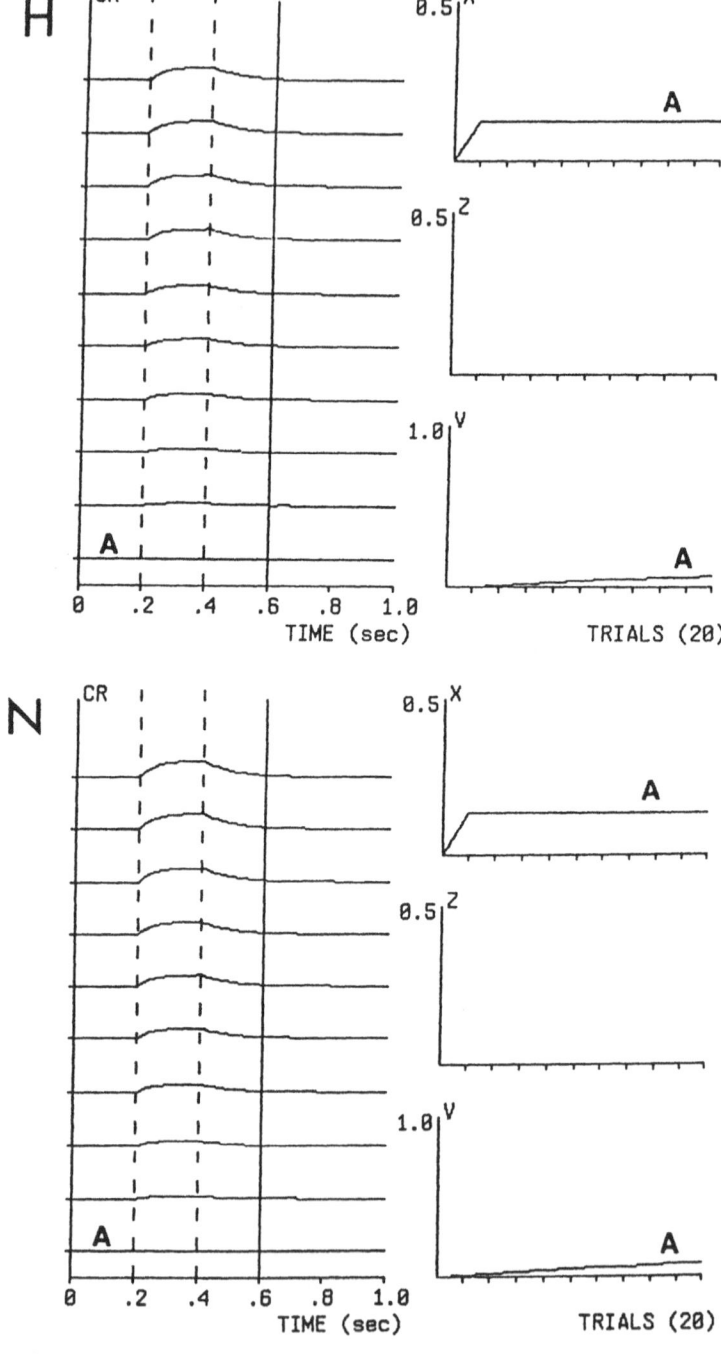

FIG. 6.4.   Effect of hippocampal lesions on trace conditioning. H: HL case. N: normal case. Left Panels: Simulated conditioned response (CR) in real time, over 10 reinforced trials. Upper-Right Panels: Average sensory representation (X_i) value on each trial as a function of trials.

HL rabbits. However, Moyer, Deyo and Disterhoft (1988) and Solomon, Vander Schaaf, Thompson, and Weisz (1986) reported that trace conditioning is partially or completely impaired in HL rabbits.

Computer Simulations.   Figure 6.4 shows simulations of acquisition of trace conditioning. Acquisition of trace conditioning was simulated with a 200-msec CS, a 50-msec US, and a 400-msec ISI. In agreement with experimental results (James et al., 1987; Port et al., 1986) both normal and HL cases show similar levels of conditioning after 10 simulated trials.

Extinction

Experimental Data.   Two studies describe the effect of HL on extinction. After initial acquisition, extinction of conditioned NM response in rabbits appeared to be unaffected by HL (Berger & Orr, 1983; Schmaltz & Theios, 1972). Schmaltz and Theios (1972) report that following alternating acquisition-extinction-reacquisition sessions, normal rabbits decreased the number of trials to reach criterion, whereas HL rabbits increased the number of trials to criterion.

Computer Simulations.   Figure 6.5 shows simulations of 10 acquisition and 20 extinction trials. In agreement with experimental data, both acquisition and extinction are essentially unaffected after HL.

Figure 6.6 shows simulations of three alternating acquisition-extinction sessions, each one with 10 acquisition and 20 extinction trials. In the normal case X_{11} receives an increasing amount of incentive motivation over trials. As a result, after the first session both acquisition and extinction proceed faster in the normal but not in the HL case. As mentioned before, incentive motivation associations are assumed to be stored in the form of LTP. These results are in partial agreement with the acquisition-extinction series of the Schmaltz and Theios (1972) study showing that normal animals decreased the number of trials to extinction criterion. However, in conflict with Schmaltz and Theios' (1972) results, simulations do not show an increase in the number of trials to extinction criterion in HL animals.

Blocking and Overshadowing

Experimental Data.   Solomon (1977) found that HL disrupted blocking of the rabbit NM response. Rickert, Bennett, Lane, and French (1978) also found impairment in a blocking paradigm in rats. Rickert, Lorden, Dawson, Smyly, and Callahan (1979) and Schmajuk, Spear, and Isaacson (1983) found that

---

Middle-Right Panels: Average incentive motivation association (Z_i) on each trial as a function of trials. Lower-Right Panels: Average drive association (V_i) on each trial as a function of trials.

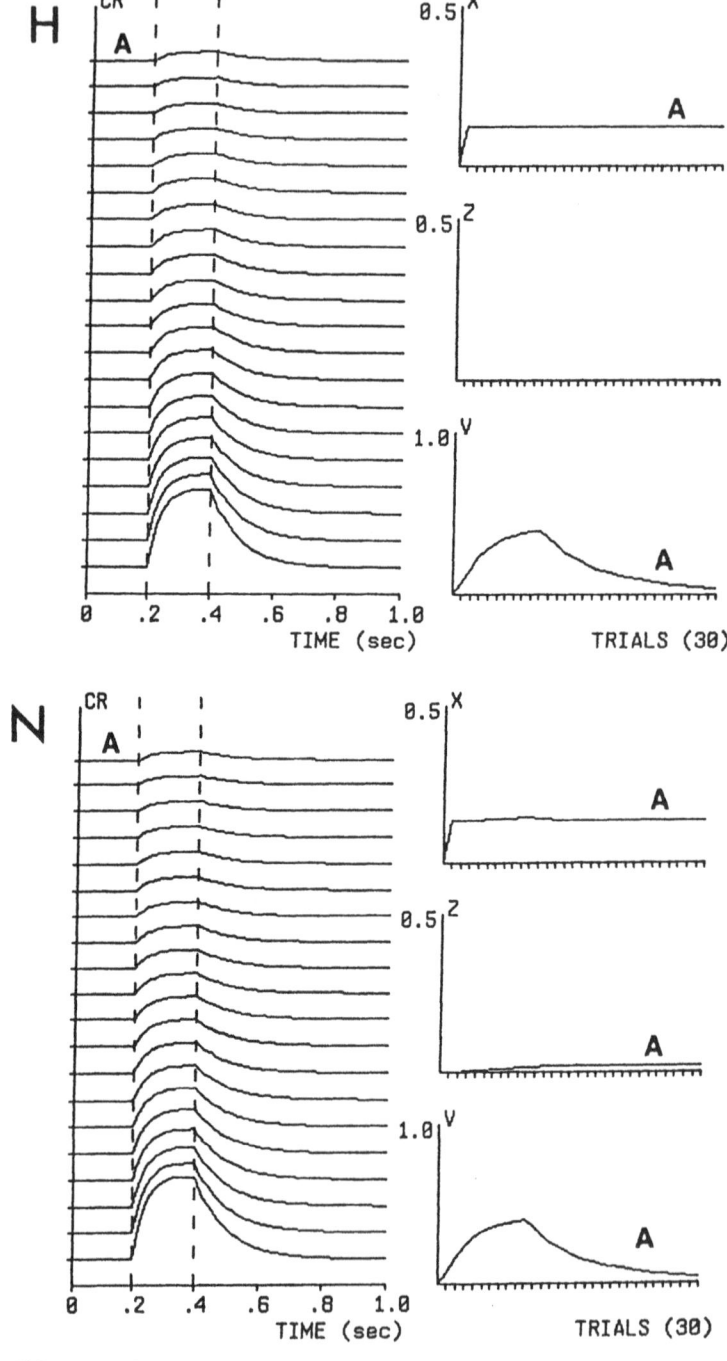

FIG. 6.5. Effect of hippocampal lesions on extinction. H: HL case. N: normal case. Left Panels: Simulated conditioned response (CR) in real time, over 20 nonreinforced extinction trials after 10 reinforced trials.

overshadowing is disrupted in rats with HL. In contrast to these findings, Garrud, Rawlins, Mackintosh, Goodal, Cotton, and Feldon (1984) found that neither blocking nor overshadowing was affected by HL, and Solomon (1977) reported no deficit in overshadowing.

Computer Simulations.   Figure 6.7 shows simulations of a blocking paradigm. The paradigm consisted of 20 trials, in which CS_1 was paired with the US, followed by 20 trials, in which CS_1 and CS_2 were paired with the US. Figure 6.7 shows that the model simulated blocking in the normal case because the CR for the CS_2 was smaller in the experimental condition than in the control condition, in which two CSs were paired together with the US during 20 trials (See Figure 6.8).

In the normal case, incentive motivation enhances the value of sensory representation X_{11} of the blocker, thereby inhibiting sensory representation X_{21}, of the blocked CS. Consistent with Rickert et al. (1978) and Solomon (1977), simulations show that HL virtually eliminated blocking. Blocking is absent in the HL case because X_{11} does not inhibit X_{21} and therefore X_{21} is able to accrue the same V_2 that it would have accrued in the absence of X_{11}.

Simulations of an overshadowing paradigm were also carried out (Figure 6.8). The paradigm consisted of 20 reinforced presentations of CS_1 and CS_2 together. In the normal case, CRs elicited by CS_1 were smaller than CRs generated by CS_1 when it had been reinforced alone, that is, the model yielded overshadowing in the normal case. HL did affect overshadowing because both CSs had larger values of X_{11}. They therefore accumulated V_i at the same rate. These results are in agreement with Rickert et al.,'s (1979) and Schmajuk et al.'s (1983) data showing that overshadowing is impaired by HL.

Discrimination Reversal

Experimental Data.   The effect of HL on discrimination reversal has been studied in the NM and eyelid preparations in the rabbit. Buchanan and Powell (1982) examined the effect of HL on acquisition and reversal of eyeblink discrimination in rabbits. HL slightly impairs acquisition of discrimination and severely disrupts its reversal by showing an increased responding to CS−. Berger and Orr (1983) contrasted HL and control rabbits in two tone differential conditioning and reversal of the rabbit NM response. Although HL does not affect initial differential conditioning, these animals are incapable of suppressing CRs to the original CS+

Upper-Right Panels: Average sensory representation (X_i) value on each trial as a function of trials. Middle-Right Panels: Average incentive motivation association (Z_i) on each trial as a function of trials. Lower-Right Panels: Average drive association (V_i) on each trial as a function of trials.

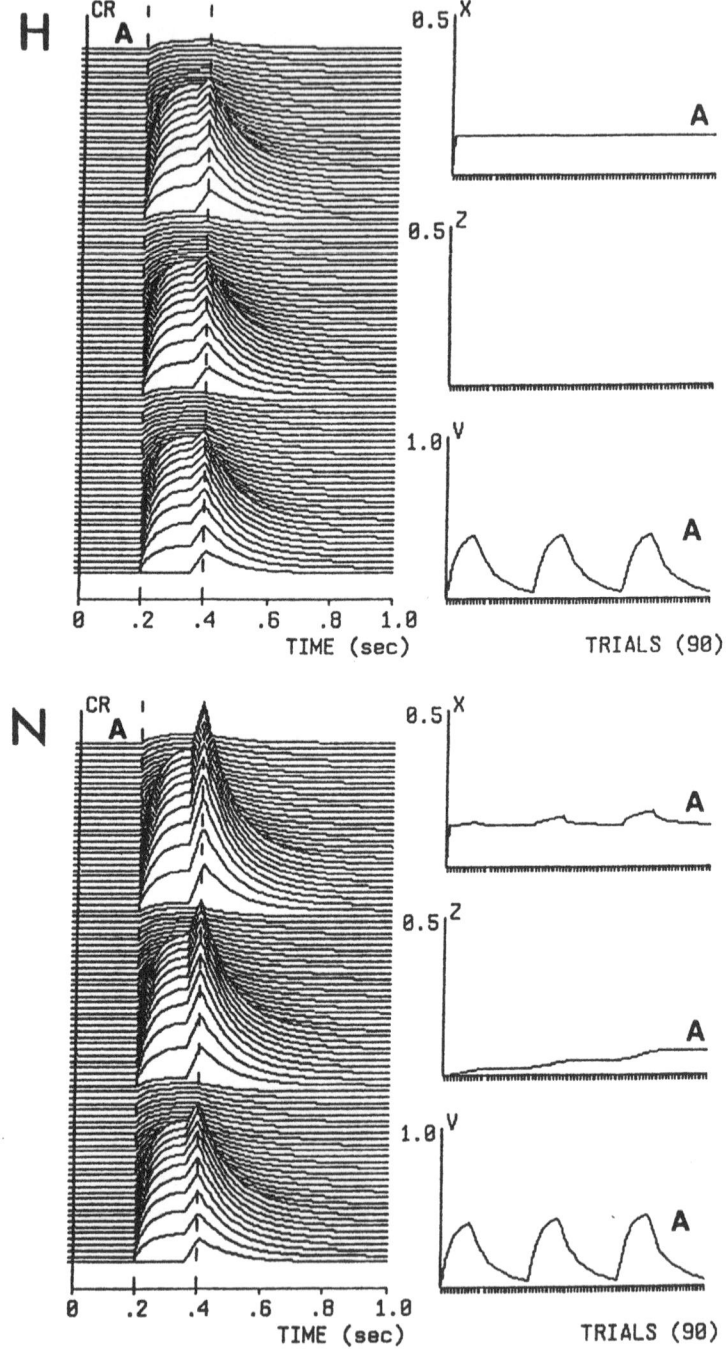

FIG. 6.6.   Effect of hippocampal lesions on acquisition-extinction series. H: HL case. N: normal case. Left Panels: Simulated conditioned response (CR) in real time, over three series of 10 reinforced trials and

after it assumes the role of CS−. This is true even following extended training. Similar results were recently reported by Weikart and Berger (1986) in a tone light discrimination reversal learning paradigm, suggesting that deficits in two tone reversal learning after HL are not due to increased within modality generalization to the tone CS serving as CS+ and CS−. Port, Romano, and Patterson (1986) found that HL impairs the reversal learning of a stimulus duration discrimination paradigm. In addition, Berger, Weikart, Bassett, and Orr (1986) found that lesions of the retrosplenial cortex, which connects the hippocampus to the cerebellar region, produce deficits in reversal learning of the rabbit nictitating membrane response.

Computer Simulations.   Figure 6.9 shows simulations of a discrimination reversal paradigm. In the differential conditioning phase, 10 reinforced trials with CS_1 were alternated with 10 nonreinforced trials with a second CS_2. During reversal, the original nonreinforced CS_2 was reinforced for 10 trials; these trials were alternated with 10 trials in which CS_1, the reinforced CS in the first phase, was presented without the US.

Figure 6.9 shows similar discrimination reversal in normal and HL cases. This result is at odds with results obtained by Berger and Orr (1983), Berger et al. (1986), Buchanan and Powell (1982), Port et al. (1986), and Weikart and Orr (1986).

Effect of Hippocampal LTP or Kindling

Discrimination Acquisition

Experimental data.   Berger (1984) found that entorhinal cortex stimulation that produced long-term potentiation (LTP) increased the rate of acquisition of a two tone classical discrimination of the rabbit NM response.

Computer Simulations.   Figure 6.10 shows simulations of acquisition of classical discrimination, in which 10 reinforced trials with CS_1 alternated with 10 nonreinforced trials with CS_2. Simulations assumed that Z_1 and Z_2 were increased by induction of LTP in the hippocampus. Simulations show that discrimination acquisition proceeded faster in the treated group than in the control group because of the larger incentive motivation. These results are in agreement with experimental data obtained by Berger (1984) and Robinson, Port, and Berger (1989).

---

20 nonreinforced trials. Upper-Right Panels: Average sensory representation (X_i) value on each trial as a function of trials. Middle-Right Panels: Average incentive motivation association (Z_i) on each trial as a function of trials. Lower-Right Panels: Average drive association (V_i) on each trial as a function of trials.

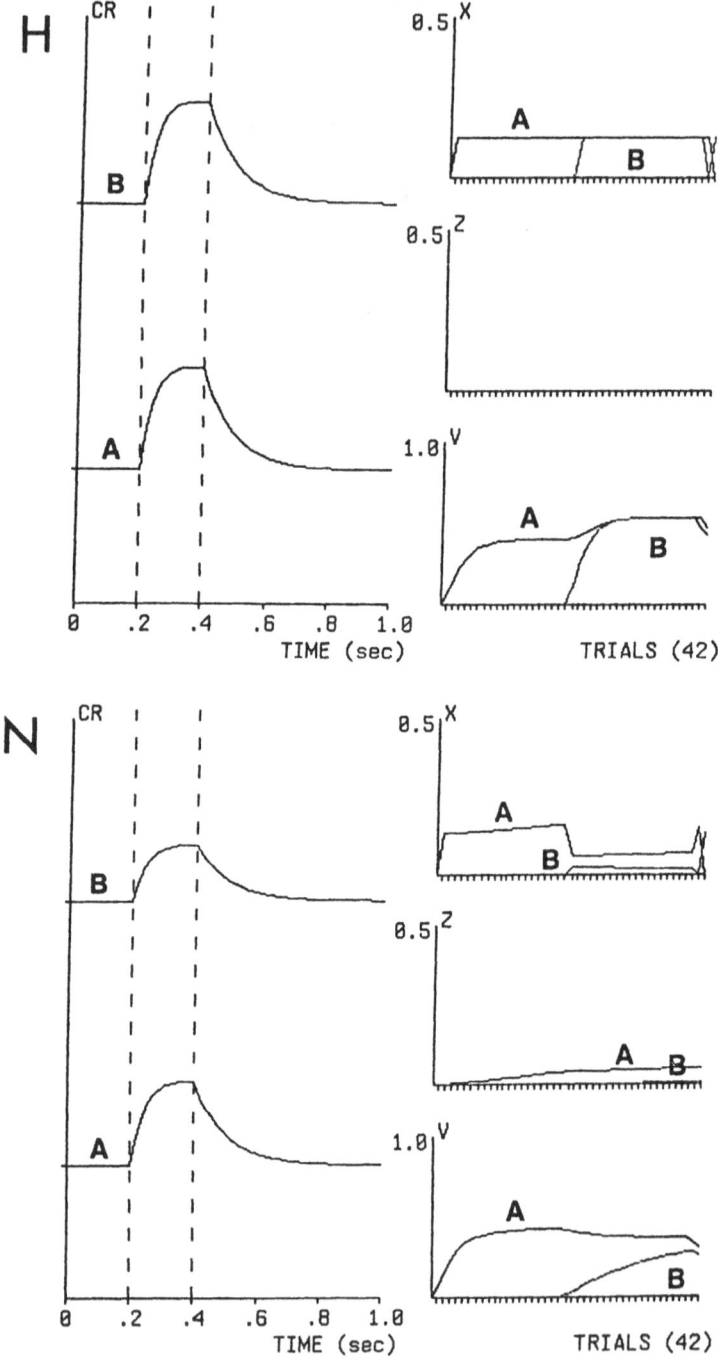

FIG. 6.7. Effect of hippocampal lesions on blocking. H: HL case. N: normal case. Left Panels: Simulated conditioned response (CR) in real time for CS(A) and CS(B) on the 41st and 42nd trials respectively. Upper-

Discrimination Reversal

Experimental Data.   Robinson et al. (1989) showed that kindling of the hippocampal perforant path-dentate projection (a procedure that, among other effects, induces LTP) facilitates discrimination acquisition but impairs discrimination reversal of the rabbit NM response.

Computer Simulations.   Figure 6.11 shows simulations of discrimination reversal, in which 10 reinforced trials with CS_2 alternated with 10 nonreinforced trials with CS_1. During reversal, the original nonreinforced CS_2 was reinforced for 10 trials; these trials alternated with 10 trials in which CS_1, the reinforced CS in the first phase, was presented without the US. As in the case of LTP induction, simulations assumed that Z_1 and Z_2 were increased by induction of kindling in the hippocampus. Simulations show that discrimination reversal proceeded at the same rate in both treated and control groups. These results are in disagreement with experimental data obtained by Robinson et al. (1989).

Hippocampal Neuronal Activity

Acquisition of Delay Conditioning

Experimental Data.   In the rabbit nictitating membrane preparation, hippocampal activity during classical conditioning is positively correlated with the topography of the CR (Berger, Alger & Thompson, 1976; Berger & Thompson, 1978a, 1978b). More specifically, Berger, Rinaldi, Weisz, and Thompson (1983) found that pyramidal cells were characterized by an increase in frequency of firing over conditioning trials, and by a within-trial pattern of discharge that models the CR. Lesions of the dentate and interpositus cerebellar nuclei ipsilateral to the trained eye caused abolition of both the CR and the conditioned increases in hippocampal CA1 neural activity evoked by the CS (Clark, McCormick, Lavond, & Thompson, 1984).

Computer simulations.   Figure 6.12 shows simulations of single unit recordings of pyramidal cells during classical conditioning: unit activity increases over conditioning trials with a within-trial pattern that models the CR. This is in agreement with experimental results.

Right Panels: Average sensory representation (X_i) value on each trial as a function of trials. Middle-Right Panels: Average incentive motivation association (Z_i) on each trial as a function of trials. Lower-Right Panels: Average drive association (V_i) on each trial as a function of trials.

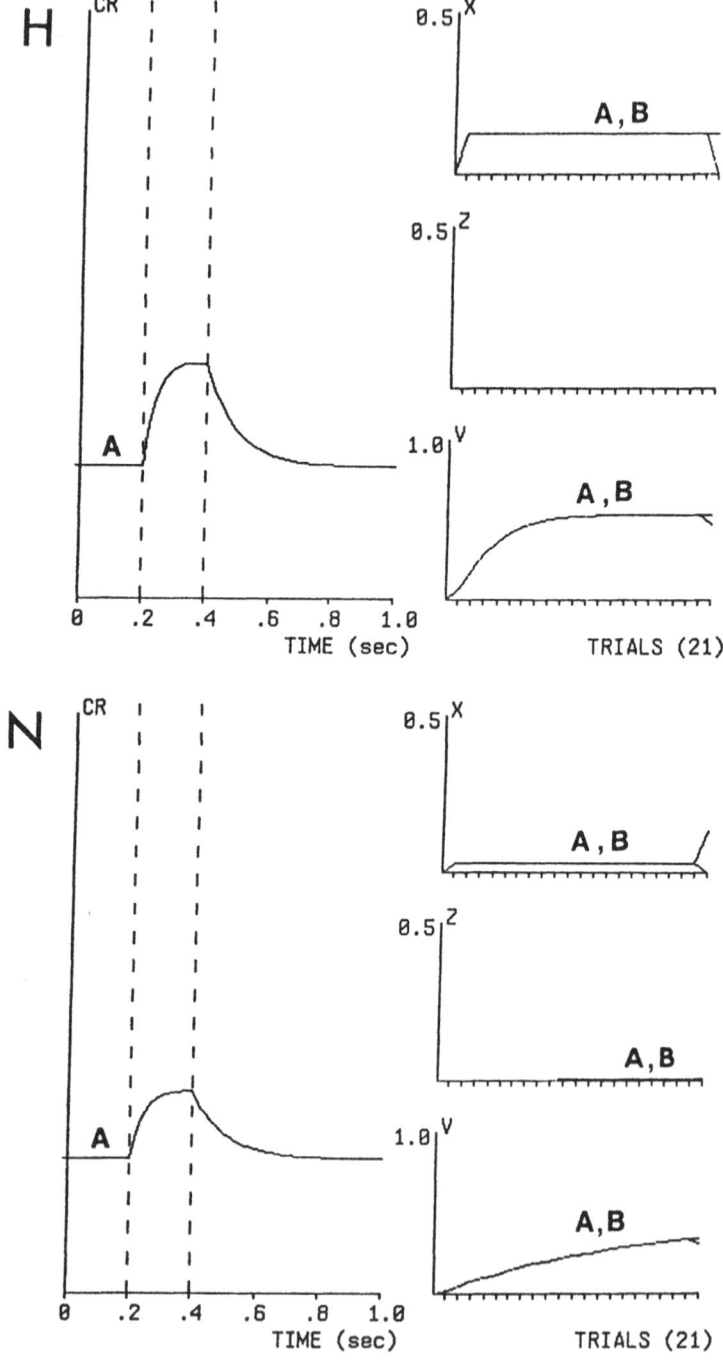

FIG. 6.8. Effect of hippocampal lesions on overshadowing. H: HL case. N: normal case. Left Panels: Simulated conditioned response (CR) in real time for CS(A) on the 21st trial. Upper-Right Panels: Average sen-

DISCUSSION

The present chapter introduces, in the context of Grossberg's (1975) attentional model, the "STM regulation" hypothesis regarding hippocampal function. The "STM regulation" hypothesis suggests that the hippocampus controls self-excitation and competition among sensory representations and stores incentive motivation associations, thereby regulating the contents of a limited capacity STM. Under the "STM regulation" hypothesis, Grossberg's (1975) model is able to describe several aspects of hippocampal modulation of classical conditioning. Our computer simulation results offer insights into how HL affects classical conditioning and the storage of memory in other regions of the brain, what type of memory is stored in the hippocampus in the form of LTP, and the functional meaning of hippocampal neural activity.

Under the "STM regulation" hypothesis, the model is able to describe the effects of HL on delay conditioning, trace conditioning, extinction, acquisition-extinction series, blocking, and overshadowing. In addition, under the assumption that incentive motivation increases with LTP or kindling, the model is able to describe the LTP effects on discrimination acquisition. Also, under the "STM regulation" assumption, the model describes hippocampal neural activity during acquisition of classical conditioning. Under the "STM regulation" hypothesis, however, the model has difficulty simulating the effect of HL in mutual overshadowing and discrimination reversal, and the effect of hippocampal kindling in discrimination reversal. Table 6.1 summarizes the results of the simulation experiments for HL, LTP, and recording simulations.

In addition to the previously mentioned paradigms, the model predicts that secondary reinforcement is facilitated by HL because the CS that acts as conditioned reinforcer does not block the reinforced CS. This prediction awaits experimental testing.

When compared with the M-S-S and the S-P-H models, Grossberg's (1975) model under the STM hypothesis provides a similar number of correct predictions for the paradigms shown in Table 6.1. In contrast to the M-S-S model, under the STM hypothesis Grossberg's model incorrectly predicts the outcome of a mutual overshadowing experiment in HL animals. In contrast to the S-P-H model, under the STM hypothesis Grossberg's model incorrectly predicts the outcome of discrimination reversal in HL animals. In all other paradigms, all three models correctly predict the effect of HL on classical conditioning. However, under the "STM regulation" hypothesis, Grossberg's model is the only one to provide a

---

sory representation (X$_i$) value on each trial as a function of trials. Middle-Right Panels: Average incentive motivation association (Z$_i$) on each trial as a function of trials. Lower-Right Panels: Average drive association (V$_i$) on each trial as a function of trials.

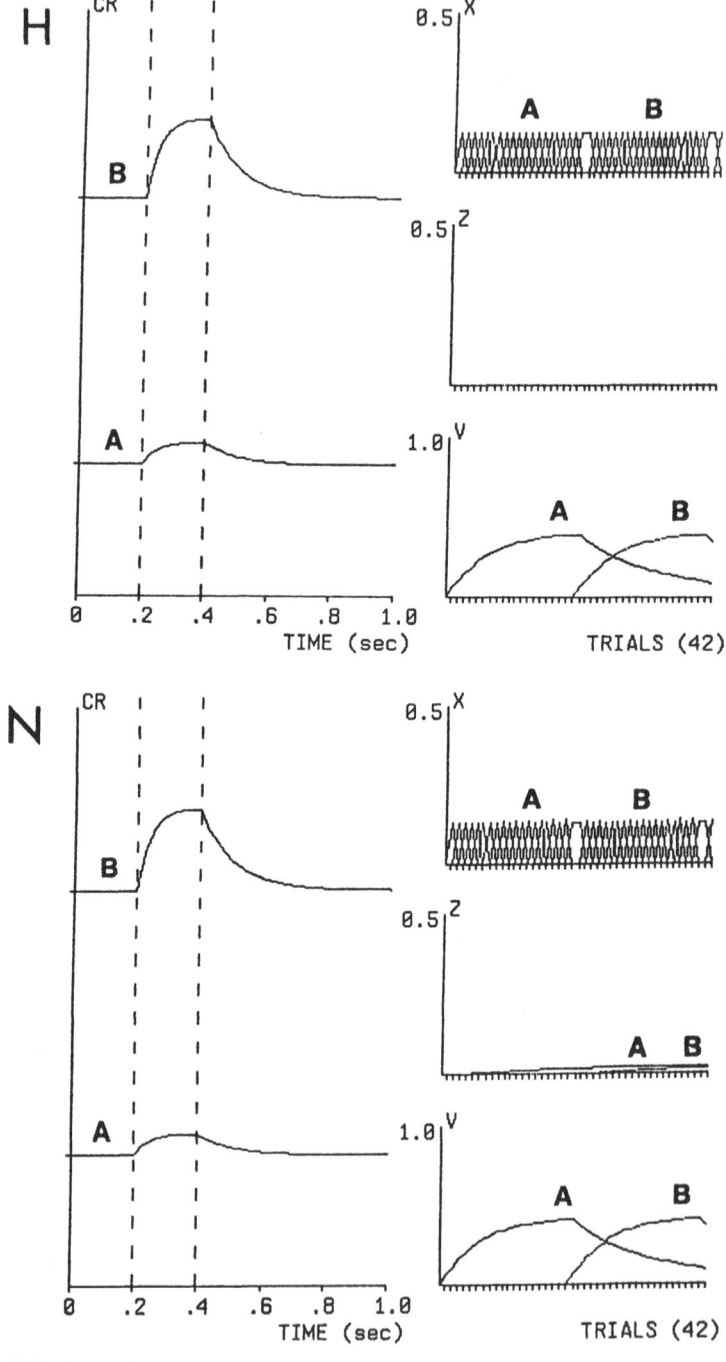

FIG. 6.9. Effect of hippocampal lesions on discrimination reversal. H: HL case. N: normal case. Left Panels: Simulated conditioned response (CR) in real time on CS(A) and CS(B) trials after discrimination reversal.

TABLE 6.1.
Simulations of Grossberg's (1975) Model Under the "STM
Regulation" Hypothesis Regarding Hippocampal Function,
Compared With Experimental Results in Classical Conditioning

Paradigm	Observed Effect	Simulated Effect
HIPPOCAMPAL LESIONS		
Delay Conditioning	+ , 0	0
Trace conditioning	+ , 0 , −	0
Extinction	0	0
Acquisition-extinction		
Acquisition	−	−
Extinction	—	0 *
Blocking	0 , −	−
Overshadowing	0 , −	−
Discrimination Reversal	−	0 *
Secondary Reinforcement	?	+
LONG-TERM POTENTIATION AND KINDLING		
Discrimination acquisition	+	+
Discrimination reversal	−	0 *
HIPPOCAMPAL NEURAL ACTIVITY		
Acquisition	increases models CR	increases models CR
Extinction	decreases	decreases

Note. − = deficit; + = facilitation; 0 = no effect; ? = no available data; * = the model fails to describe accurately the experimental result.

correct description of the effect of LTP induction on discrimination acquisition. It should also be noted that although both the M-S-S and the S-P-H models describe more classical conditioning paradigms than Grossberg's (1975) model, the latter can be integrated to other neural networks that substantially extend its domain of application (see Grossberg, 1982; Grossberg & Schmajuk, 1987).

A Neural Account of Hippocampal Function

The clear advantage of Grossberg's model over the M-S-S and S-P-H models is that it is described as neural architecture. In consequence, our assumption labelled "STM regulation" hypothesis has been defined by specifying nodes and connections in Grossberg's neural network that are part of the hippocampal circuitry. In

Upper-Right Panels: Average sensory representation (X_i) value on each trial as function of trials. Middle-Right Panels: Average incentive motivation association (Z_i) on each trial as a function of trials. Lower-Right Panels: Average drive association (V_i) on each trial as a function of trials.

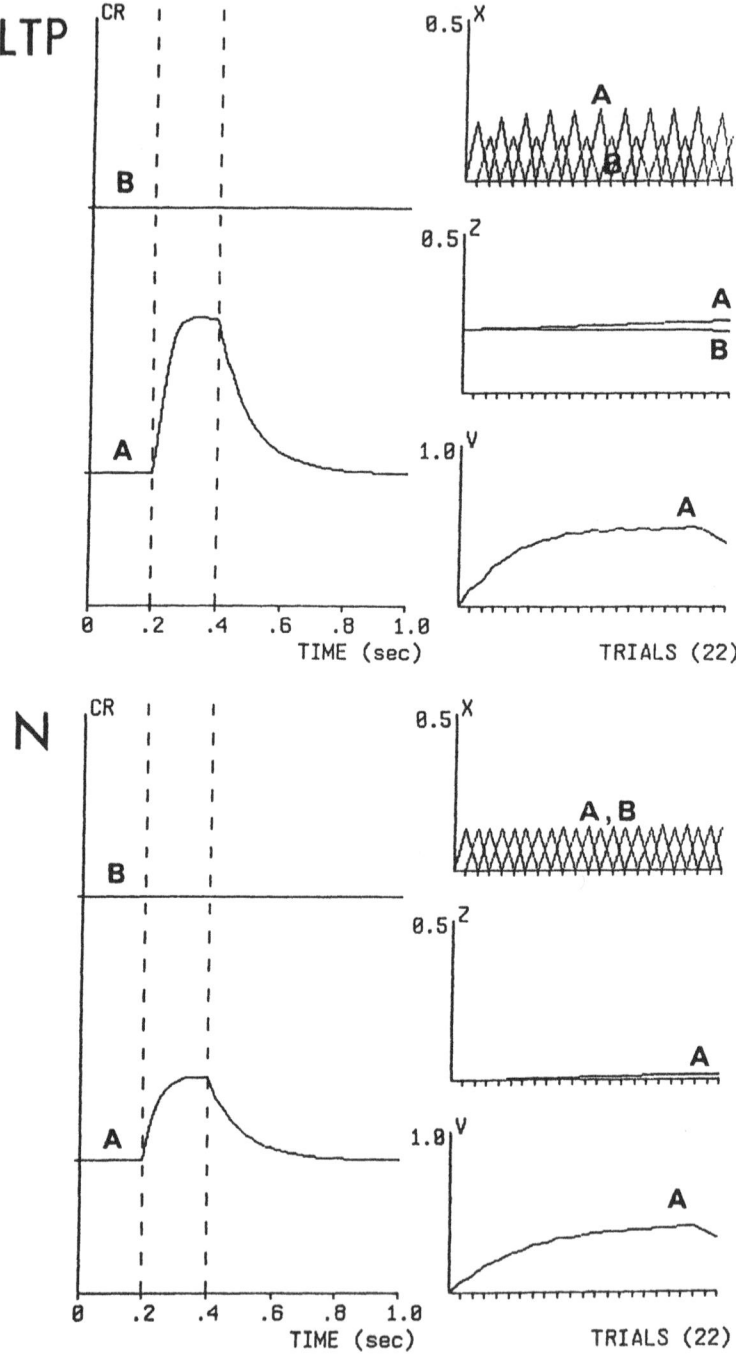

FIG. 6.10 Effect of LTP or kindling on discrimination acquisition. LTP:
LTP case. N: normal case. Left Panels: Simulated conditioned response
(CR) in real time on CS(A) and CS(B) trials after discrimination acquisi-

this section we extend the mapping of the neural network to other brain regions. Because the classically conditioned nictitating membrane response preparation in the rabbit has been extensively studied, our descriptions refer to this preparation.

Experimental evidence from the NM response preparation (Desmond & Moore, 1982; Lincoln, McCormick, & Thompson, 1982; McCormick, Guyer, & Thompson, 1982; McCormick, Lavond, & Thompson, 1983; McCormick & Thompson, 1984; Woodruf-Pak, Lavond, & Thompson, 1986; Yeo, Hardiman, & Glickstein, 1985a, 1985b) suggests that the association of sensory representations X_i and the US would be mediated by plastic changes at the interpositus nucleus of the cerebellum and/or at the Purkinje cells of the hemispheric portion of cerebellar lobule VI. Sensory representations X_i reach the interpositus nucleus and the cerebellar cortex via mossy fibers from the pontine nuclei and the US reaches the interpositus nucleus and the cerebellar cortex via climbing fibers from the inferior olive. Drive representation activity, Y, originates in the cerebellar lobule VI and/or the interpositus nuclei, is relayed to the contralateral red nucleus, and reaches the contralateral accessory abducens nuclei where the NM response is controlled.

As mentioned earlier, Berger and Thompson (1978a, 1978b) found that pyramidal cells in dorsal hippocampus increased their frequency of firing over conditioning trials with a pattern that correlates with the amplitude-time course of the rabbit NM response. As previously shown, hippocampal activity can be adequately represented by the drive representation activity, Y, which also controls the conditioned response. Clark et al. (1984) found that this hippocampal neuronal activity disappears after cerebellar ablation, supporting the idea that V_i information is stored in cerebellum, and not in the hippocampus. Information about Y might be conveyed to the hippocampus through cerebellar-limbic system pathways, such as those reported by Harper and Heath (1973; Heath, 1973; Heath & Harper, 1974).

Berger, Clark, and Thompson (1980) found that the activity correlated with the CR was present also in the entorhinal cortex, but was amplified over trials in CA1 and CA3 hippocampal regions. Berger et al. (1980) suggested that this amplification might take place at the perforant path-dentate synapse or the mossy fiber-CA3 synapse. According to the STM regulation hypothesis, this amplification effect is equivalent to the multiplication of the drive representation activity Y by the incentive motivation Z_i value in Grossberg's model.

Finally, the hippocampus might modulate classical conditioning of the nictitat-

tion for LTP and control groups. Upper-Right Panels: Average sensory representation (X_i) value on each trial as a function of trials. Middle-Right Panels: Average incentive motivation association (Z_i) on each trial as a function of trials. Lower-Right Panels: Average drive association (V_i) on each trial as a function of trials.

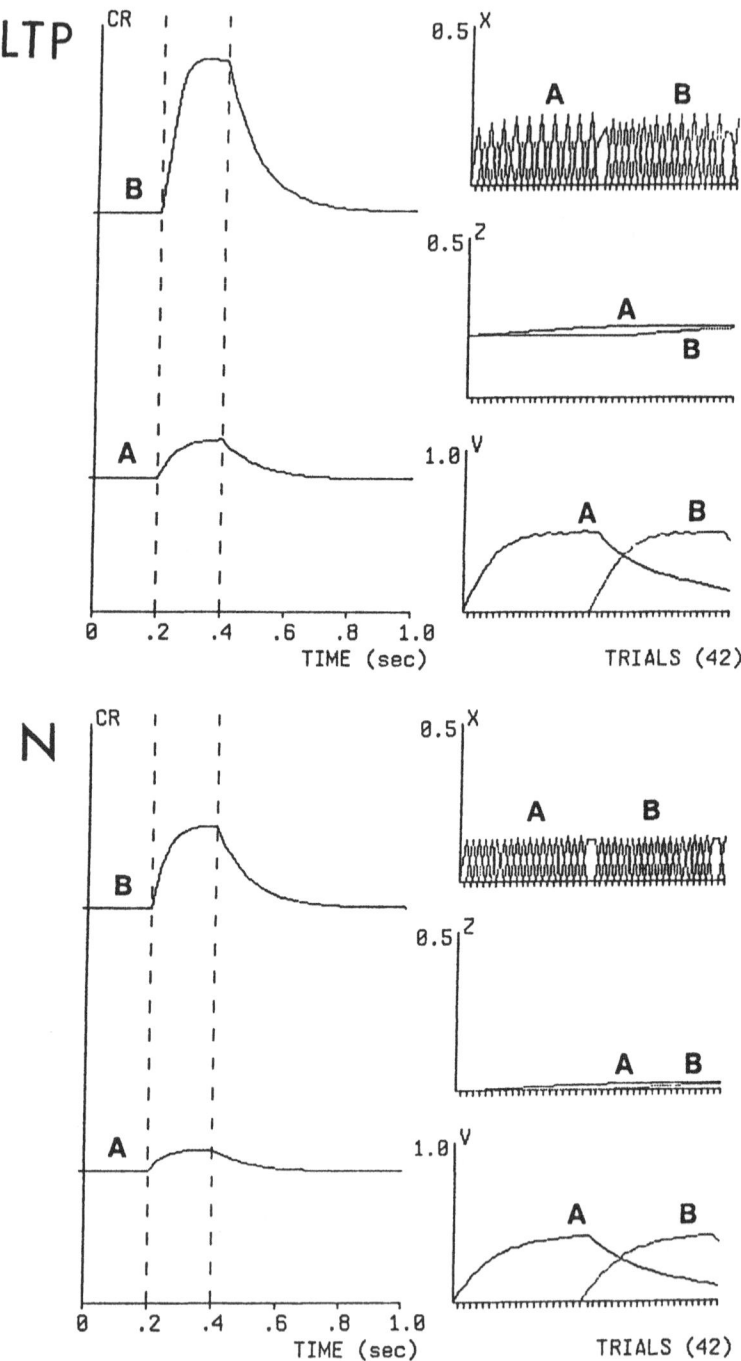

FIG. 6.11. Effect of kindling on discrimination reversal. LTP: kindling case. N: normal case. Left Panels: Simulated conditioned response (CR) in real time on CS(A) and CS(B) trials after discrimination reversal.

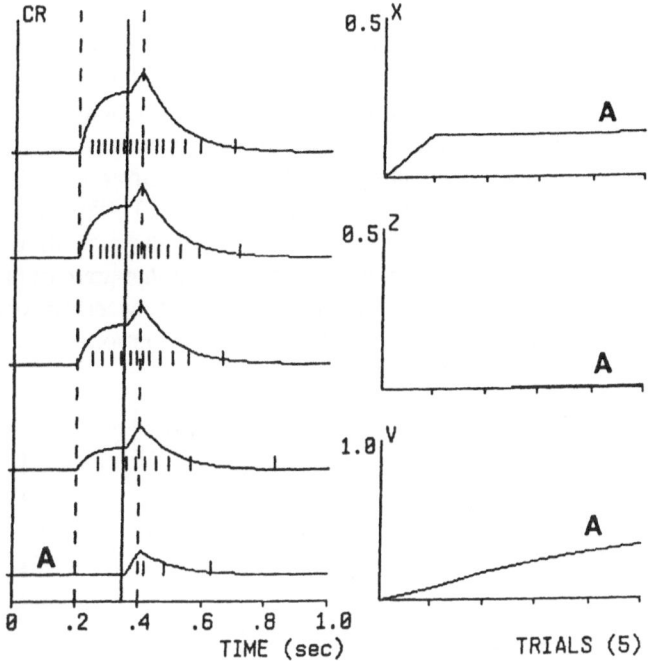

FIG. 6.12. Hippocampal neural activity during delay conditioning. Left Panels: Simulated conditioned response (CR) and neural activity in real time, over 5 reinforced trials. Upper-Right Panels: Average sensory representation (X_i) value on each trial as a function of trials. Middle-Right Panels: Average incentive motivation association (Z_i) on each trial as a function of trials. Lower-Right Panels: Average drive association (V_i) on each trial as a function of trials.

ing membrane through several pathways. A hippocampal-retrosplenial cortex projection via the subiculum reaches the ventral pons (Berger et al., 1986; Berger, Bassett, & Weikart, 1985; Berger, Swanson, Milner, Lynch, & Thompson, 1980; Semple-Rowland, Bassett, & Berger, 1981). A cingulo-pontine projection has been described by Weisendanger and Weisendanger (1982) and confirmed by Wyss and Sripanidkulchai (1984). By modulating the pontine nucleus and thereby its mossy fiber projections to the cerebellar cortex and interpositus nucleus, hippocampal-cerebellar projections would modulate learning processes in the

Upper-Right Panels: Average sensory representation (X_i) value on each trial as a function of trials. Middle-Right Panels: Average incentive motivation association (Z_i) on each trial as a function of trials. Lower-Right Panels: Average drive association (V_i) on each trial as a function of trials.

cerebellum (Berger et al., 1986; Steinmetz, Logan, & Thompson, 1988). Therefore, in agreement with the view of the present paper, experimental evidence
suggests that hippocampal modulation of classical conditioning is exerted on
sensory representations of the CS input to the site of plasticity.

Figure 6.13 shows a schematic diagram of the hippocampal-cerebellar interconnections and their corresponding variables in the Grossberg (1975) model
according to the "STM regulation" hypothesis. Figure 6.13 offers a neural account
of hippocampal function in classical conditioning of the nictitating membrane
response, by specifying nodes and loops in the model that are part of cerebellar
and hippocampal circuits. We propose that associations between sensory representations and the US are stored in cerebellum, thereby controlling the generation

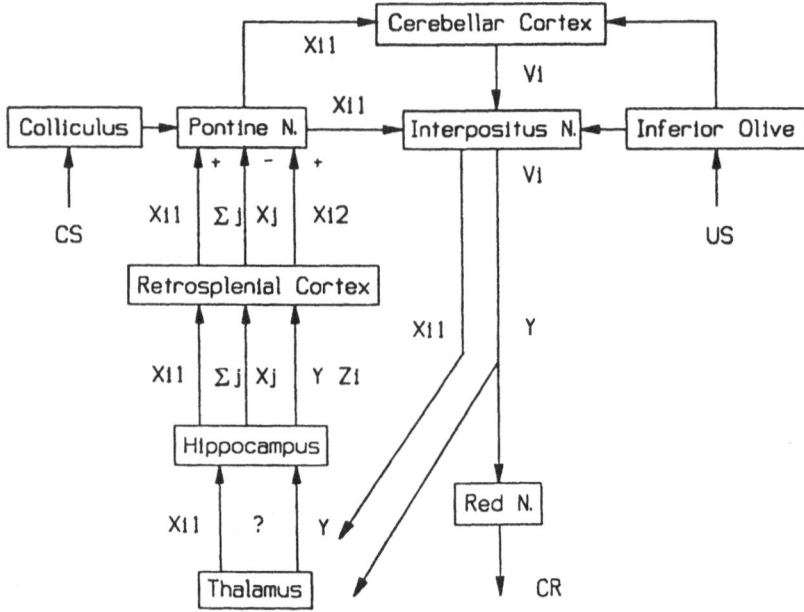

FIG. 6.13. The "Short-term Memory Regulation" hypothesis: Mapping of Grossberg's (1975) network onto a schematic diagram of
hippocampal-cerebellar interconnections. A conditioned stimulus, CS_i,
activates a sensory representation, X_{i1}, in the coliculus, and is relayed
to the pontine nuclei. Sensory representation X_{i1} becomes associated
with the US by changing synaptic weights, V_i, possibly either in the
interpositus nucleus or cerebellar cortex. Output Y becomes associated
with X_{i1} by changing synaptic weights Z_i, in the form of hippocampal
LTP. Sensory representations compete among themselves for a limited
capacity short-term memory. Sensory representations also sustain a
self-excitatory feedback loop. Hippocampal lesions preserve CS_i-US
connections, but eliminate hippocampal modulation of STM of sensory
representations at the pontine nuclei.

of CRs. An output copy of the CR is relayed to the hippocampus, where it is associated with sensory representations (incentive motivation associations) in the form of hippocampal LTP. It is important to recognize that, because the fine structure of the hippocampal circuitry is ignored, only a partial structural isomorphism between the neural network and the hippocampus can be established. Hippocampal outputs controlling self-excitation, incentive motivation, and reciprocal inhibition act on the pontine nuclei, modulating the magnitude of sensory representations. Within the present framework, the effect of hippocampal manipulations is easily described: hippocampal lesions eliminate self-excitation, incentive motivation, and reciprocal inhibition among sensory representations; hippocampal LTP and kindling increase the stored values of incentive motivation; and hippocampal neural activity is correlated with the CR.

CONCLUSION

The present chapter proposes that the hippocampus controls self-excitation and competition among sensory representations and stores incentive motivation associations, thereby regulating the contents of a limited capacity STM in the context of Grossberg's (1975) real time attentional model. Under this hypothesis, called the "STM regulation" hypothesis, the model provides correct descriptions of different hippocampal manipulations on several classical conditioning paradigms.

ACKNOWLEDGMENTS

The authors thank Drs. Stephen Grossberg, John Moore, and John Gabrieli for their comments on this manuscript. This project was supported in part by BRSG S07 RR07028–22 awarded by the Biomedical Research Support Grant Program, Division of Research Resources, National Institutes of Health.

REFERENCES

Berger, T.W. (1984). Long-term potentiation of hippocampal synaptic transmission affects rate of behavioral learning. *Science, 224,* 627–630.
Berger, T.W., Alger, B., & Thompson, R.F. (1976). Neuronal substrate of classical conditioning in the hippocampus. *Science, 192,* 483–485.
Berger, T.W., Bassett, J.L., & Weikart, C. (1985, November). Hippocampal-cerebellar interactions during classical conditioning. Paper presented at the 26th annual meeting of the Psychonomic Society. Boston.
Berger, T.W., Clark, G.A., & Thompson, R.F. (1980). Learning-dependent neuronal responses recorded from limbic system brain structures during classical conditioning. *Physiological Psychology, 8,* 155–167.

Berger, T.W., & Orr, W.B. (1983). Hippocampectomy selectively disrupts discrimination reversal conditioning of the rabbit nictitating membrane response. *Behavioral Brain Research, 8,* 49–68.

Berger, T. W., Rinaldi, P.C., Weisz, D.J., & Thompson, R.F. (1983). Single unit analysis of different hippocampal cell types during classical conditioning of rabbit nictitating membrane response. *Journal of Neurophysiology, 50,* 1197–1219.

Berger, T.W., Swanson, G.W., Milner, T. A., Lynch, G.S., & Thompson, R.F. (1980). Reciprocal anatomical connections between hippocampus and subiculum in the rabbit: Evidence for subicular innervation of regio superior. *Brain Research, 183,* 265–276.

Berger, T.W., & Thompson, R.F. (1978a). Neuronal plasticity in the limbic system during classical conditioning of the rabbit nictitating membrane response. I. The hippocampus. *Brain Research, 145,* 323–346.

Berger, T.W., & Thompson, R.F. (1978b). Neuronal plasticity in the limbic system during classical conditioning of the rabbit nictitating membrane response. II. Septum and mammillary bodies. *Brain Research, 156,* 293–314.

Berger, T.W., Weikart, C. L., Bassett, J. L., & Orr, E.B. (1986). Lesions of the retrosplenial cortex produce deficits in reversal learning of the rabbit nictitating membrane response: Implications for potential interactions between hippocampal and cerebellar brain systems. *Behavioral Neuroscience, 100,* 802–809.

Buchanan, S.L., & Powell, D.A. (1982). Cingulate cortex: Its role in Pavlovian conditioning. *Journal of Comparative and Physiological Psychology, 96,* 755–774.

Clark, G.A., McCormick, D.A., Lavond, D.G., & Thompson, R.F. (1984). Effects of lesions of cerebellar nuclei on conditioned behavioral and hippocampal neuronal responses. *Brain Research, 291,* 125–136.

Desmond, J.E., & Moore, J.W. (1982). Brain stem elements essential for classically conditioned but not unconditioned nictitating membrane response. *Physiology and Behavior, 28,* 1092–1033.

Douglas, R. (1972). Pavlovian conditioning and the brain. In R.A. Boakes & M.S. Halliday (Eds.), *Inhibition and learning.* London: Academic Press.

Douglas, R., & Pribram, K.H. (1966). Learning and limbic lesions. *Neuropsychologia, 4,* 197–220.

Garrud, P., Rawlins, J.N.P., Mackintosh, N.J., Goodal, G., Cotton, M.M., & Feldon, J. (1984). Successful overshadowing and blocking in hippocampectomized rats. *Behavioural Brain Research, 12,* 39–53.

Gormezano, I., Kehoe, E.J., & Marshall, B. S. (1983). Twenty years of classical conditioning research with the rabbit. *Progress in Psychobiology and Physiological Psychology, 10,* 197–275.

Grastyan, E., Lissak, K., Madarasz, I., & Donhoffer, H. (1959). Hippocampal electrical activity during the development of conditioned reflexes. *Electroencephalography and Clinical Neurophysiology, 11,* 409–430.

Grossberg, S. (1975). A neural model of attention, reinforcement, and discrimination learning. *International Review of Neurobiology, 18,* 263–327.

Grossberg, S. (1980). How does a brain build a cognitive code? *Psychological Review, 87,* 1–51.

Grossberg, S. (1982). Processing of expected and unexpected events during conditioning and attention: A psychophysiological theory. *Psychological Review, 89,* 529–572.

Grossberg, S., & Levine, D.S. (1987). Neural dynamics of attentionally modulated Pavlovian conditioning: Blocking, interstimulus interval, and secondary reinforcement. *Applied Optics, 26,* 5015–5030.

Grossberg, S., & Schmajuk, N.A. (1987). Neural dynamics of Pavlovian conditioning: Conditioned reinforcement, habituation, and opponent processing. *Psychobiology, 15,* 195–240.

Harper, J.W., & Heath, R.G. (1973). Anatomic connections of the fastigial nucleus to the rostral forebrain in the cat. *Experimental Neurology, 39,* 285–292.

Heath, R.G. (1973). Fastigial nucleus connections to the septal region in the monkey and cat: A demonstration with evoked potentials of a bilateral pathway. *Biological Psychiatry, 6,* 193–196.

Heath, R.G., & Harper, J.W. (1974). Ascending projections of the cerebellar fastigial nucleus to the hippocampus, amygdala, and other temporal lobes sites: Evoked potential and histological studies in monkeys and cats. *Experimental Neurology, 45*, 268–287.

James, G.O., Hardiman, M.J., & Yeo, C.H. (1987). Hippocampal lesions and trace conditioning in the rabbit. *Behavioral Brain Research, 23*, 109–116.

Kimble, D.P. (1968). Hippocampus and internal inhibition. *Psychological Bulletin, 70*, 285–295.

Lincoln, J.S., McCormick, D.A., & Thompson, R.F. (1982). Ipsilateral cerebellar lesions prevent learning of the classically conditioned nictitating membrane/eyelid response. *Brain Research, 242*, 190–193.

Mackintosh, N.J. (1975). A theory of attention: Variations in the associability of stimuli with reinforcer. *Psychological Review, 82*, 276–298.

McCormick, D.A., Guyer, P.E., & Thompson, R.F. (1982). Superior cerebellar peduncle lesions selectively abolish the ipsilateral classically conditioned nictitating membrane/eyelid response of the rabbit. *Brain Research, 244*, 347–350.

McCormick, D.A., Lavond, D.G., & Thompson, R.F. (1983). Neuronal responses of the rabbit brainstem during performance of the classically conditioned nictitating membrane (NM)/eyelid response. *Brain Research, 271*, 73–88.

McCormick, D.A., & Thompson, R.F. (1984). Cerebellum: Essential involvement in the classically conditioned eyelid response. *Science, 223*, 296–299.

Moore, J.W. (1979). Brain processes and conditioning. In A. Dickinson & R.A. Boakes (Eds.), *Mechanisms of learning and behavior*. Hillsdale, NJ: Lawrence Erlbaum Associates.

Moore, J.W., & Stickney, K.J. (1980). Formation of attentional-associative networks in real time: Role of the hippocampus and implications for conditioning. *Physiological Psychology, 8*, 207–217.

Moyer, J.R., Deyo, R.A., & Disterhoft, J.F. (1988). Effects of hippocampal lesions on acquisition and extinction of trace eye-blink response in rabbits. *Society for Neuroscience Abstracts, 14*.

Pearce, J.M., & Hall, G. (1980). A model for Pavlovian learning: Variations in the effectiveness of conditioned but not of unconditioned stimuli. *Psychological Review, 87*, 532–552.

Port, R.L., Mikhail, A.A., & Patterson, M.M. (1985). Differential effects of hippocampectomy on classically conditioned rabbit nictitating membrane response related to interstimulus interval. *Behavioral Neuroscience, 99*, 200–208.

Port, R.L., Romano, A.G., & Patterson, M.M. (1986). Stimulus duration discrimination in the rabbit: Effects of hippocampectomy on discrimination and reversal learning. *Physiological Psychology, 14*, 124–129.

Port, R.L., Romano, A.G, Steinmetz, J.E., Mikhail, A.A., & Patterson, M.M. (1986). Retention and acquisition of classical trace conditioned responses by rabbits with hippocampal lesions. *Behavioral Neuroscience, 100*, 745–752.

Rickert, E.J, Bennett, T.L., Lane, P., & French, J. (1978). Hippocampectomy and the attenuation of blocking. *Behavioral Biology, 22*, 147–160.

Rickert, E.J., Lorden, J.F., Dawson, R., Smyly, E., & Callahan, M.F. (1979). Stimulus processing and stimulus selection in rats with hippocampal lesions. *Behavioral and Neural Biology, 27*, 454–465.

Robinson, G.B., Port, R.L., & Berger, T.W. (1989). Kindling facilitates acquisition of discriminative responding but disrupts reversal learning of the rabbit nictitating membrane response. *Behavioral Brain Research, 31*, 279–283.

Schmajuk, N.A. (1984). A model of the effects of hippocampal lesions on Pavlovian conditioning. *Abstracts of the 14th Annual Meeting of the Society for Neuroscience, 10*, 124.

Schmajuk, N.A. (1986). *Real-time attentional models for classical conditioning and the hippocampus*. Unpublished PhD Dissertation. University of Massachusetts.

Schmajuk, N.A. (1989). The hippocampus and the control of information storage in the brain. In M.

Arbib & S.I. Amari (Eds.), *Dynamic interactions in neural networks: Models and data*. New York: Springer-Verlag.

Schmajuk, N.A., & Moore, J.W. (1985). Real-time attentional models for classical conditioning and the hippocampus. *Physiological Psychology, 13*, 278–290.

Schmajuk, N.A., & Moore, J.W. (1988). The hippocampus and the classically conditioned nictitating membrane response: A real time attentional-associative model. *Psychobiology, 46*, 20–35.

Schmajuk, N.A., & Moore, J.W. (1989). Effects of hippocampal manipulations on the classically conditioned nictitating membrane response: Simulations by an attentional-associative model. *Behavioral Brain Research, 32*, 173–189.

Schmajuk, N.A., Spear, N.E., & Isaacson, R.L. (1983). Absence of overshadowing in rats with hippocampal lesions. *Physiological Psychology, 11*, 59–62.

Schmaltz, L.W., & Theios, J. (1972). Acquisition and extinction of a classically conditioned response in hippocampectomized rabbits (Oryctolagus cuniculus). *Journal of Comparative and Physiological Psychology, 79*, 328–333.

Semple-Rowland, S.L., Bassett, J.L., & Berger, T.W. (1981). Subicular projections to retrosplenial cortex in the rabbit. *Society for Neurosciences Abstracts, 7*, 886.

Solomon, P.R. (1977). Role of the hippocampus in blocking and conditioned inhibition of rabbit's nictitating membrane response. *Journal of Comparative and Physiological Psychology, 91*, 407–417.

Solomon, P.R., & Moore, J.W. (1975). Latent inhibition and stimulus generalization of the classically conditioned nictitating membrane response in rabbits (Orytolagus cuniculus) following dorsal hippocampal ablation. *Journal of Comparative and Physiological Psychology, 89*, 1192–1203.

Solomon, P.R., Vander Schaaf, E.R., Thompson, R.F., & Weisz, D.J. (1986). Hippocampus and trace conditioning of the rabbit's classically conditioned nictitating membrane response. *Behavioral Neuroscience, 100*, 729–744.

Steinmetz, J.E., Logan, C.G., & Thompson, R.F. (1988). Essential involvement of mossy fibers in projecting the CS to the cerebellum during classical conditioning. In C. Woody, D. Alkon, & J. McGaugh (Eds.), *Cellular Mechanisms of Conditioning and Behavioral Plasticity*. New York: Plenum Press.

Weikart, C., & Berger, T.W. (1986). Hippocampal lesions disrupt classical conditioning of cross-modality reversal learning of the rabbit nictitating membrane response. *Behavioral Brain Research, 22*, 85–90.

Weisendanger, R., & Weisendanger, M. (1982). The corticopontine system in the rat. *Journal of Comparative Neurology, 208*, 227–238.

Woodruf-Pak, D.S., Lavond, D.G., & Thompson, R.F. (1986). Trace conditioning: Abolished by cerebellar nuclear lesions but not lateral cerebellar cortex aspirations. *Brain Research, 348*, 249–260.

Wyss, J.M., & Sripanidkulchai, K. (1984). The topography of the mesencephalic and pontine projections from the cingulate cortex. *Brain Research, 293*, 1–15.

Yeo, C.H., Hardiman, M.J., & Glickstein, M. (1985a). Classical conditioning of the nictitating membrane response of the rabbit: III. Connections of cerebellar lobule HVI. *Experimental Brain Research, 60*, 114–126.

Yeo, C.H., Hardiman, M.J., & Glickstein, M. (1985b). Classical conditioning of the nictitating membrane response of the rabbit: I. Lesions of the cerebellar nuclei. *Experimental Brian Research, 60*, 87–98.

7 Implementing Connectionist Algorithms for Classical Conditioning in the Brain

John W. Moore
University of Massachusetts

ABSTRACT

Simple connectionist models of learning that conform to the Widrow-Hoff rule can be parameterized and extended to describe real-time features of classical conditioning. These features include the dependence of learning on the moment-to-moment status of input to the computational system and on the desired topography of its output. Using the classically conditioned nictitating response (NMR) of the rabbit as a prototypal system, my coworkers and I have devised models that successfully meet these real-time criteria. Two models and neural network architectures are described. The first consists of a single neuron-like processor with learning rules based on the Sutton-Barto model. The second consists of two neuron-like units with input based on a tapped-delay line representation of stimuli. Using anatomical and physiological data, both network models can be aligned with brain stem and cerebellar circuits involved in classical NMR conditioning. These models and their implementation in the brain have testable empirical consequences.

This chapter illustrates how computational models based on abstract neural networks can provide insights into questions such as where in the brain learning occurs and the mechanisms that bring it about. It serves as a tutorial on aligning quantitative learning theory with physiology, thereby establishing a potential conduit for communication between molar and molecular levels of analyses. One might say that such efforts are about bringing abstract models to life in real nervous systems. Specifically, it brings together two lines of research: (a) studies of brain circuits underlying classical conditioning of a simple skeletal response, the rabbit nictitating membrane response (NMR) and (b) developing computational models applicable to real-time conditioning phenomena such as response

181

topography and CS-US interval effects. Our work stands on a foundation provided by the "extended laboratory" (Gabriel, 1988) of those who investigate classical eye blink/NMR conditioning (Gormezano, Prokasy, & Thompson, 1987).

This chapter also illustrates Churchland's (1986) point that a theoretical dimension can be added to the neurosciences through a process of coevolution. Theories, expressed as computational algorithms, can guide decisions about where best to invest experimental resources. Behavioral neuroscientists regard the preparations they investigate as model systems, and therefore they are not indifferent to the broader implications of their work. However, they believe that exploiting these implications requires understanding the model system at multiple levels. At the computational level, this understanding is expressed in terms of algorithms capable of simulating how the system behaves under a range of environmental challenges. Production systems have this capability. More interesting and useful in the long run are algorithms that *emulate* what actually occurs in the brain to cause the system to behave as it does.

I shall recount some of the steps my colleagues and I have taken in evaluating models of classical conditioning and casting them in terms of neural networks that might convincingly be implemented in real brains. By this I refer to discovering *alignments* between models and what is known of the anatomy and physiology underlying a given CR. I say "a given CR" because we are concerned primarily with developing theoretical frameworks useful to neuroscientists. As neuroscientists, our interest in theory cannot stray very far from the nuts and bolts of our experimental work. As neuroscientists, workers in the extended NMR conditioning laboratory seek a deep understanding of classical conditioning. Such an understanding demands rigorous theoretical expression, and thus NMR conditioning has been approached from three directions—behavior, biology, and computation. The overriding objective is to integrate these approaches in much the same way that oculomotor physiologists have done with saccadic eye movement and vestibular oculomotor reflexes (Robinson, 1981, 1989). The oculomotor example is apt because of shared anatomical systems.

Before proceeding, some comment about the title is in order. The title speaks of *implementing* connectionist algorithms[1] in the brain. Others have applied the term *instantiation* to this process. When considering the dictionary definitions of these two terms, I find implementation more apt. According to *Webster's New Collegiate Dictionary*, the infinitive "to instantiate" means "to represent an abstraction by a concrete example" whereas "to implement" means "to provide instruments or means of expression."

I would argue that instantiation is a prerequisite to implementation. To illus-

---

[1] Connectionist algorithms refer to any member of a class of models cast in terms of modifiable connection weights among neuron-like processing elements. The term is not limited to the so-called delta or LMS learning rule or to learning through back-propagated error correction.

trate, my colleagues and I began by selecting one abstract computational theory, the Sutton-Barto (SB) model, which showed promise in describing many of the basic phenomena of classical conditioning, particularly the molar features of classical conditioning of the eye blink/NMR (Barto & Sutton, 1982; Sutton & Barto, 1981). We then sought a specific *instance* of the SB model that could describe in detail real-time features of the conditioned NMR, including response topography. We have referred to the resulting constrained and parameterized version of the model as the Sutton-Barto-Desmond (SBD) model (Moore et al., 1986). By common usage, the SBD model is an *instantiation*—a concrete representation—of the abstract SB model. When we speak of implementing this model (Moore & Blazis, 1989a, 1989b, 1989c), we mean finding brain processes and mechanisms ("instruments") conceivably capable of generating and explaining the measurable consequences of its parameterized and algorithmic ("instantiated") form.

Animal learning theorists are accustomed to evaluating models solely by behavioral criteria. They also prefer models that are parsimonious and elegant. Difficulties arise when there are many serviceable models to account for phenomena at the level of behavior. The instantiation of models into algorithmic and parameterized form adds constraints to models that are useful in their evaluation, but even at this level of exactness the choice of the more valid model can be arbitrary. Animal learning theorists attempt to escape from this dilemma by designing elaborate experiments that test highly refined behavioral predictions. Although not minimizing the importance of this strategy, I suggest that a brain implementation scheme can imply independent experimental assessments of models using criteria at the level of neurobiology as an adjunct to behavioral criteria. It goes without saying that the success of this approach depends on having accurate information about neural substrates of the behavior being modeled. Implementation schemes should be physiologically compelling, not merely plausible. Later on I illustrate this point by showing how implementation schemes my colleagues and I have devised for two models have testable hypotheses in the domain of neurophysiology. The situation is somewhat comparable to neuroanatomists' quest for Renhaw cells, which was stimulated by physiologists' claims that they must exist.

Devising good implementation schemes for instances of mammalian behavior such as NMR conditioning is not easy. It is difficult because of the virtual impossibility of obtaining direct and rigorous proof that mechanisms proposed to account for phenomenology actually exist and function in ways consistent with the model. We must therefore approach the question of mechanisms indirectly through experimental probes—testing behavioral predictions, making lesions or pharmacologic interventions, and recording neural activity. It would be comforting if agreed-upon facts were not open to differing interpretation, but this state of affairs seldom exists. However, it is more important to *identify* agreed-upon experimental evidence than to seek consensus on what the evidence means. What, then, of experimental facts that are in dispute, and which disputes are most crucial to resolve? A good model and implementation scheme should tell us.

NEUROBIOLOGICAL BACKGROUND

Much of the difficulty in addressing questions of loci and mechanisms of learning arises from a reliance on lesion data and other indirect evidence. Although lesion data are vitally important, they have left workers in this field with some puzzles that will not be easily resolved. These unresolved issues are important for implementing neural network models of NMR conditioning in the brain. Before considering them, let us review the facts that I believe are *not* in dispute.[2]

• Telencephalic brain regions and structures are not essential for acquisition, maintenance, or performance of the CR. Telencephalic regions that have been the focus of lesion studies include neocortex, hippocampal formation and other components of Papez circuit, basal ganglia, thalamus, and hypothalamus. At the midbrain level, lesion studies have shown that tectum, tegmental reticular formation, substantia nigra, and periacqueductal grey, to mention a few involved structures, are not essential for conditioning. The only essential midbrain structure identified to date is red nucleus (RN).

• Metencephalic brain regions, the cerebellum and brain stem, are essential for the acquisition, maintenance, and performance of CRs. Within the metencephalon, researchers agree that the following subset of structures are essential for expression of *robust* CRs with normal topography: (a) Cerebellar cortex, specifically the region designated by anatomists as hemispheral lobulus VI (HVI); (b) cerebellar nucleus interpositus, specifically the anterior region (NIA); (c) magnocellular RN, specifically the portion that represents the facial region around the eye; (d) inferior olivary nucleus; specifically the dorsal accessory olive (DAO), which represents the facial regions around the eye and sends climbing fibers to HVI and NIA.

• Lesions that disrupt or eliminate CRs in one eye need not affect acquisition or performance of CRs by the other eye. Nor do such lesions interfere with the normal savings of trials to criterion when the US is no longer applied to the affected eye but is switched to the other eye. Since savings can be demonstrated,

---

[2] Space limitations preclude reference to all of the experimental literature bearing on neurobiological substrates of NMR conditioning. Most of the relevant literature can be gleaned from sources such as Gormezano et al., (1987), Moore (1979), Steinmetz, Lavond, and Thompson (1989), and Yeo (1989).

it is clearly the case that lesions do not impair some general capacity to acquire and store information.

• To a surprising degree, lesions that disrupt or eliminate CRs do not affect the UR, although there may be small modulations of UR amplitude that are understandable in light of known anatomy and physiology.

• The reflex pathways mediating the CR and UR involve motoneurons that innervate the extraocular muscles, especially the retractor bulbi muscles. Most retractor bulbi motoneurons lie in the accessory abducens nucleus (AAN) and are innervated by nearby second-order sensory neurons of spinal trigeminal subnucleus pars oralis (SpoV). Although the brain stem components of these circuits have been well characterized both anatomically and physiologically, the nature and extent of the cerebellum's contribution remains an active area of research. There is strong evidence, outlined later, that the cerebellum makes a substantial *causal* contribution to the generation of CRs.

Workers in the extended laboratory of the conditioned NMR have proposed various tests to determine whether CR-disrupting lesions or pharmacologic interventions involving critical metencephalic structures affect learning or performance. CR-disruption might be due to any number of factors unrelated to learning—these might all be subsumed under the heading "performance factors": (a) motor deficit, (b) sensory deficit, (c) disruption of timing, (d) attentional deficit. Motor deficits can be eliminated as a cause of CR disruption to the extent that the UR remains unaffected (over a range of US intensities). Sensory deficits can be eliminated to the extent that CRs are disrupted with different CS modalities. Attentional deficits can be ruled out to the extent that CRs occur in the contralateral eye. Disruptions of timing can be assessed by varying the CS-US interval (Desmond & Moore, 1982).

Researchers do not agree on whether any brain region identified as crucial for the performance of robust CRs, such as the cerebellum, is actually involved in the learning process or, indeed, whether learning also depends on the integrity of other structures (Desmond & Moore, 1986). One possibility is that "learning the connection between a CS and the US" occurs within Purkinje cells (PCs) of cerebellar cortex, specifically at parallel fiber synapses. Contending views are that the critical connections are formed within the deep cerebellar nuclei or brain stem. It is well to get these issues in front of us before introducing implementation schemes. Thus armed, the reader can better judge their strengths and weaknesses. Some unresolved issues regarding the lesion data are reviewed here.

• Do the neural commands (motor programs) that result in a CR originate in the cerebellum? That is, can it be said that the cerebellum is a proximal *cause* of CR topography? We have recently performed a linear systems analysis of the relationship between the firing rate of single neurons in NIA and the position of the NM during conditioning trials (Berthier, Barto, & Moore, 1988). Regarding a causative link, we find that in some NIA cells the relationship between rate of firing and position of the NM can be modeled by a nonrecursive (causal) digital filter (Hamming, 1983). Using a related approach, we estimate the transfer function between neural activity and NM position, and from this it is possible to write a differential equation relating the two variables. Some NIA cells with CR-predictive firing yield equations for second-order systems. These equations relate a cell's firing rate to the acceleration, position, and velocity of NM movement. It is worth noting that NM movement is linearly related to eyeball retraction. In fact, the sweep of the NM over the eye is caused by retraction of the globe. The mechanics of eyeball retraction are those of a Voight element consisting of elastic and viscous components, a classic textbook example of a second-order linear system. Hence, it is perhaps not surprising that equations relating the firing of some NIA cells to the conditioned NMR are those of second-order linear systems.

• Do lesions of cerebellar cortex produce a complete and permanent loss of a previously acquired CR? This question is important in itself, but especially so because of the well-known theories of Marr (1969) and Albus (1971), who independently suggested that motor learning, including classical conditioning, involves cerebellar cortex in a fundamental way. Yeo and his colleagues were the first to show that HVI lesions disrupt CRs, typically by either eliminating them altogether or greatly reducing their amplitude (Yeo, Hardiman, & Glickstein, 1984). These studies have been extended to show that lesions of HVI must be complete in order to eliminate small amplitude responses that would normally be counted as CRs and to prevent recovery after extended post-operative training (Yeo & Hardiman, 1988). Lavond, Steinmetz, Yokaitis, and Thompson (1987) report small amplitude CRs and recovery following extended training in cases where HVI removed appears to have been complete, and so the issue remains in question.

• Do lesions of the inferior olivary nuclei cause a progressive decline of the CR resembling extinction as some investigators claim (McCormick, Steinmetz, & Thompson, 1985), or is it the case that such lesions merely disrupt normal cerebellar functioning and in this way bring about an immediate detrimental effect on performance (Yeo, Hardiman, & Glickstein, 1986)? This issue is as yet unresolved because experiments have yielded conflicting results. If lesions of the inferior olivary nuclei, particularly DAO, do result in experimental extinction, then this would support models built on the idea that climbing fibers from DAO to cerebellum carry the reinforcement signal from the US to sites of learning.

• Does stimulation of the DAO provide a reinforcing signal for learning in cerebellar cortex (Steinmetz et al., 1989)? Or is any learning in the cerebellum confined to NIA, which receives climbing fiber collaterals from DAO? Related to this question is the possibility that stimulation of DAO does not reinforce learning in the cerebellum at all. It may merely stimulate brain stem elements of the reflex pathway underlying the UR where learning might occur (Bloedel, 1987; Yeo, 1989).

There are other issues to be resolved among workers in the extended laboratory of the conditioned NMR. I have mentioned only those that bear on evaluating the implementation schemes suggested for each of the models outlined in subsequent sections. Each model has its own implementation scheme. They both assume that learning occurs through Hebbian mechanisms that involve synaptic modification through convergence of CS and US information onto single neurons (Byrne, 1987). They both assume that these synaptic modifications occur in cerebellar cortex (HVI) and that CRs are initiated by the action of PCs on NIA cells to which they project. The implementation for the SBD model assumes that learning occurs in cerebellar cortex, but one synapse before the PC or output stage (Moore & Blazis, 1989a, 1989b, 1989c). The other model, designated VET, is more complex and assumes that learning occurs within the brain stem as well as in cerebellar cortex. The cortical component assumes that learning occurs through modification and parallel fiber/PC synapses. In addition, it assumes that learning also occurs at the synapses of parallel fibers and Golgi cells (Moore, Desmond, & Berthier, 1989).

THE SBD MODEL

The SBD model has been described in detail elsewhere (e.g., Moore et al., 1986; Moore & Blazis, 1989a, 1989b, 1989c), and so a brief summary will suffice. The model is based on a single neuron-like processing unit that receives input from many potential CSs and a US. The processing unit adjusts the weights (synaptic efficacies) of the CS input so that future output of the unit matches its current output.

Equation 1 specifies that the output of this system, denoted $s(t)$, equals the weighted sum of its input from potential CSs and the US. The variable $s(t)$ is a linear function of its weighted input only within an allowed range imposed by the fact that the NM can only move so far (about 10 mm) as the eyeball retracts. A CS's contribution to the output of the element is computed as the product of the current strength of its representation, denoted $x(t)$ in the model, and a corresponding "synaptic" weight denoted $V(t)$. Formally, the output of the system at time t, denoted $s(t)$, equals the weighted sum of input from all CSs, where $x_i(t)$ refers to the magnitude of CS_i, $i = 1, \ldots, n$, at time t:

$$s(t) = \sum_{i=1}^{n} V_i(t)\, x_i(t) + \lambda'(t). \tag{1}$$

$\lambda'(t)$ is the US's contribution to $s(t)$.

In Equation 1, $x_i(t)$ represents the activation level of the ith member of a set of potential CSs at discrete times t after onset (t represents successive time steps of 10-msec duration). The following specifications for x_i were dictated by two contraints: (a) generation of realistic response topography for a forward-delay paradigm with a favorable CS-US interval, and (b) generation of realistic interstimulus interval (ISI) functions. The optimal CS-US interval for NMR conditioning is generally taken to be 250 msec (Gormezano, 1972). When the CS_i begins, $x_i = 0.0$. It remains at 0.0 until $t = 7$, i.e., 70 msec after CS_i onset. At this point, x_i begins to increase in an S-shaped fashion. It levels off to a maximum value of 1.0 by $t = 30$ (300 mec after CS_i onset) and remains at this value until CS_i offset, at which time x_i begins to fall exponentially back to 0.0. Thus, according to Equation 1 the output of the model, $s(t)$, to CS_i conforms to the temporal map or *template* provided by x_i. As the number of training trials increases, the variable $V_i(t)$ increases, and the CR becomes increasingly robust. This process is reversed over a series of extinction trials.

Learning in the SBD model follows a modified Hebbian rule that states that changes of the synaptic weight of CS_i, ΔV_i, are proportional to the *difference* between the current output, $s(t)$, and the trace of preceding outputs, $\bar{s}(t)$. At time t, ΔV_i is computed as follows:

$$\Delta V_i(t) = c[s(t) - \bar{s}(t)] \, \bar{x}_i(t), \qquad (2)$$

where c is a learning rate parameter, $0 < c \leq 1$.

The factor $\bar{x}_i(t)$ in Equation 2 specifies the degree to which the "synaptic junction" corresponding to CS_i is *eligible* for modification (Sutton & Barto, 1981). $\bar{x}_i$ is driven by the variable $x(t)_i$: After CS_i onset, it increases with the x_i, but with a lag of 30 msec. It remains at full strength as long as the CS_i is on and begins to decay to a baseline value of zero 30 msec after Cs_i offset. The rate of this decay is inversely related to CS_i duration whenever CS_i exceeds 250 msec.

Equation 2 does not contain an explicit term for the reinforcing action of the US. The US is important only insofar as it affects the term $s(t) - \bar{s}(t)$. The interaction of the two time-dependent variables associated with CS_i, $x_i(t)$ and $\bar{x}_i$, together with $s(t) - \bar{s}(t)$, govern the rate of learning and shape of ISI functions. Trial-wise learning curves reflect accumulated net changes in V_i occurring *within* each trial. Such changes occur before and after the occurrence of the US. For example, given the 10 msec time step used in our simulations (e.g., Moore et al., 1986), a trial with an ISI of 350 msec might involve over 400 computations of ΔV_i.

The term $\bar{s}(t)$ in Equation 2 can be thought of as a short-term decaying trace of the system's output from previous time steps. Alternatively, it can be interpreted as a prediction or expectation of output based on previous output. It is computed as follows:

$$\bar{s}(t + 1) = \beta \bar{s}(t) + (1 - \beta)s(t), \qquad (3)$$

where $0 \leq \beta < 1$.

Simulation studies indicate that, with a 10 msec time step, optimal performance of the model requires that β be on the order of 0.6–0.7 (Blazis & Moore, 1987). With β in this range, Equation 3 approximates the inhibitory action of cerebellar Golgi cells on information flow through the granular layer of cerebellar cortex (Eccles, Sasaki, & Strata, 1967). This coincidence was exploited in implementing the model (Moore & Blazis, 1989a, 1989b, 1989c).

The foregoing assumptions enable the SBD model to simultaneously generate response topographies and ISI functions for both trace and forward-delay conditioning paradigms (Moore et al., 1986). In addition, it retains the original Sutton-Barto model's ability to describe multiple-CS phenomena such as blocking, higher-order conditioning, and conditioned inhibition (Barto & Sutton, 1982; Sutton & Barto, 1981). In agreement with experimental literature (Miller & Spear, 1985), the model does not predict extinction of conditioned inhibition.

Implementing the SBD model

Moore and Blazis's (1989a, 1989b) implementation of the SBD model is summarized in this section. We required that the implementation scheme meet the following criteria: (a) It had to involve the cerebellum and be consistent with its anatomy and physiology. (b) It had to account for the lesion data, namely the fact that lesions of cerebellar cortex (HVI) or associated brain stem circuits virtually eliminate the CR. (c) It had the propose neuronal loci where the learning rule (Equation 2) might be implemented. This meant proposing sites where CS information, the variable $\bar{x}$, converges with the reinforcement signal, $s - \bar{s}$. (d) It had to propose schemes for computing $\bar{s}$ via Equation 3 and $s - \bar{s}$. The last item held the key. Once overcome, the rest of the implementation fell into place.

Turning now to the details of the implementation, the output variable s, which expresses the form of the CR and is used in the learning rule, is generated by the action HVI PCs on neurons in NIA to which they project. Evidence supporting this construction was reviewed earlier. Next, s is transmitted with high fidelity through each of the synaptic links leading to AAN motoneurons and generation of the peripherially observed CR. One synapse before this stage, within SpoV, an efference copy of s peels off and ascends back to HVI via mossy fibers. This efference copy of s consists of two streams. One stream passes through the granular layer without modulation by Golgi cells. This stream gives rise to parallel fibers (axons of granule cells) that carry s information to other circuit components, including Golgi cells that modulate the other s stream. This modulation computes $\bar{s}$ (see Moore & Blazis, 1989a, 1989b, 1989c) and gives rise to parallel fibers carrying $\bar{s}$ information.

The existence of separate parallel fibers that carry s and $\bar{s}$ information, respectively, allow for computation of the reinforcement factor in Equation 2, $s - \bar{s}$, by Golgi cells. These Golgi cells are distinct from those that compute $\bar{s}$. The computation of $s - \bar{s}$ occurs as follows: Golgi cells receive two simultaneous inputs from parallel fibers: an excitatory (depolarizing) input from the $\bar{s}$-carrying parallel fibers mentioned above and an inhibitory (hyperpolarizing) input from axon collaterals of PCs activated by s-carrying parallel fibers mentioned above. Notice that the algebraic sum of these inputs is $\bar{s} - s$, but because Golgi cells are inhibitory neurons, the effect of their axonal output on granule cells is proportional to $s - \bar{s}$, as required by the model.

In the implementation scheme, this output is directed to granule cells, different from those mentioned previously, which are activated by CSs via mossy fibers. CS information has been preprocessed such that the input to these granule cells from mossy fibers is proportional to the variable x.[3] The eligibility factor used in the learning rule, $\bar{x}$, presumably resides within x-activated granule cells (Sutton & Barto, 1981). Thus, these granule cells receive convergent input from mossy fibers carrying x-information and Golgi cell input carrying $s - \bar{s}$-information. This convergence allows for implementation of the learning rule for updating V by modified Hebbian mechanisms assumed by the model. The output of these granule cells is proportional to Vx, which in the one-CS case is equal to s.

Parallel fibers arising from these granule cells convey Vx to basket cells and PCs. The basket cells, which are inhibitory neurons, send axons to PCs on an adjacent, previously unmentioned group of parallel fibers. As Vx increases after CS onset, the accompanying increase in activation of basket cells causes the firing of PCs on the adjacent group of parallel fibers to decreases below their baserate of firing. This decrease causes disinhibition of neurons in NIA, to which they project. This, in turn, causes NIA cells to increase their firing rate and send an excitatory pulse of activation proportional to s through the efferent pathway for expression of the CR described previously. The penultimate link in this sequence are neurons in SpoV that return the current value of s to HVI for the next computation of ΔV.

Implications of the SBD Model

The complexity of the implementation scheme stems from considerations of anatomy and physiology. These considerations are elaborated elsewhere (Blazis & Moore, 1987; Moore & Blazis, 1989a, 1989b, 1989c). The most compelling evidence for the scheme comes from Berthier and Moore's (1986) study of how PCs in HVI respond during NMR conditioning.

Berthier and Moore (1986) recorded from single PCs in HVI during the asymptotic stages of two tone differential conditioning. PCs with CR-related firing patterns could be classified as either increasing or decreasing their baserate of firing whenever a CR occurred. Three cells increased firing for every one that decreased firing. Firing patterns of PCs could also be classified as either preceding or occurring simultaneously with CRs. All of the PCs that decreased their firing

---

[3] For discussion of this point, see Moore and Blazis (1989c).

did so before the CR occurred, as would be necessary if their activity were responsible for initiating CRs. PCs that increased their firing before the occurrence of CRs are implied by parallel fibers carrying Vx information. PCs that increase their firing simultaneously with CRs are implied by the two streams of s-carrying efference from SpoV used in the computation of $\tilde{s} - s$. It remains to be determined whether we can objectively discriminate increases in PC firing that mirrors s from that mirroring $\tilde{s}$. This caveat aside, the frequency distribution of firing patterns observed by Berthier and Moore (1986) are accounted for by the implementation scheme.

Behavioral Predictions

The SBD model predicts that the rate of CR acquisition in a trace conditioning paradigm (but not a forward-delay paradigm) is an increasing function CS duration, provided the interval between CS offset and US onset (trace interval) is sufficiently long, e.g., 300 milliseconds or more. The prediction follows directly from the model's assumption that the eligibility factor in the learning rule, $\tilde{x}$, decays at a rate that is inversely related to the duration of the CS. The prediction is counterintuitive because it states that acquisition can be faster with a longer-than-optimal ISI than one nearer to optimal. Preliminary studies with an acoustic CS (Blazis & Moore, 1989) have born out this prediction, but only when the CS is sufficiently intense (e.g., 80 dB). The prediction is not supported with a CS on the order of 60 dB. In this case, acquisition rate appears to be dominated by ISI instead of CS duration, as in forward-delay conditioning.

I mention this particular experiment in order to point out that the SBD model does not yet provide a complete account of NMR conditioning. In this case, the model is incomplete because it says nothing about how CS intensity affects the variables x or $\tilde{x}$. Part of the motivation for the experiment discussed earlier was to obtain information on how to incorporate these effects into the model. The SBD model also fails to take account of processes that might occur within intertrial intervals, for example, the "consolidation" effects from studies showing that rate of conditioning is a direct function of intertrial interval (Moore & Gormezano, 1977).

Physiological Predictions

The most important prediction from the implementation scheme is that learning occurs within the granular layer of cerebellar cortex. Testing this prediction will require experiments on cerebellar slices using designs similar to those employed by Coulter and Disterhoft to investigate long-term effects of NMR conditioning on hippocampal pyramidal cells (e.g., Coulter et al., 1989). It also remains to be proven that SpoV cells that fire in relation to CRs actually send *collateral* efference copy mirroring CR waveform to HVI. All we can say with confidence at this point is that cells in SpoV exist that show CR-related firing of the kind needed to provide HVI with a current copy of the variable s (Ricciardi, Richards, & Moore, 1989). We also know that some SpoV cells send mossy fibers to HVI. We do not know that the two categories actually overlap.

The implementation scheme does not assign a role in learning to climbing fiber inputs from DAO. Additional studies are needed to determine whether this is justified. One reason for discounting the possible contribution of climbings fibers is that their low firing rate makes them poor candidates for providing efference copy about s for implementation of the learning rule. In addition, very few PCs observed by Berthier and Moore (1986) responded with a complex spike, indicative of climbing fiber input, when the US occurred. Nevertheless, the possibility that climbing fibers are important for learning cannot be ruled out without further experimental work (Ito, 1989; Moore & Berthier, 1987). In fact, Moore et al.'s (1989) implementation of the VET model, reviewed in the next section, assumes that learning in the cerebellum is reinforced by climbing fiber input elicited by the US.

THE VET MODEL

The SBD model is able to simulate response topography because it assumes that every potential CS provides the system with a fixed pattern of activation that serves as a template for the CR. We have speculated about how templates might be formed and the possible contributions of other brain regions, especially the hippocampus, to this process (Blazis & Moore, 1987). The SBD model basically sidesteps this issue. In addition, it is incapable of adaptively changing CR topography so as to simulate a number of paradigms. The SBD model cannot yield appropriate CR waveforms for the following cases:

• Trace conditioning. CR waveforms should peak just before the occurrence of the US, as in forward-delay paradigms, but the dynamics of the variable x do not permit this to happen. Instead, CR waveforms begin to fall toward baseline when CS offset occurs.

• Forward-delay conditioning with long CS-US intervals. The SBD model cannot simulate inhibition of delay. Since CR-waveform mirrors the template provided by the variable x, its latency and form are not influenced by the CS-US interval.

• Multiple CS-US intervals. Training with multiple CS-US intervals yields complex CR-waveforms. For example, a study by Millenson, Kehoe, and Gormezano (1977) showed that training with randomly mixed trials having CS-US intervals of 200 and 700 milliseconds gives rise to CRs with two peaks, each centered at a time of US onset.

The VET model overcomes these deficiencies. It is able to simulate appropriate CR waveforms for these cases because of features absent in the SBD model. These features include the following:

- CSs are provided with a temporal dimension through tapped delay lines that encode, not only the source of the stimulus, but also the time since the stimulus began. Another set of tapped delay lines encodes the time since the stimulus ceased. Hence, the model has timed-tagged input elements for both stimulus onset and offset (Equation 4).

- There are two neuron-like processing units that receive convergent input from CSs and the US. One unit (designated V) is the output device (Equation 5). It has modifiable synaptic weights that are changed according to an LMS rule resembling the Rescorla-Wagner model (Equation 6). The main difference from the Rescorla-Wagner model is that weight changes depend on local eligibility factors (Equation 7), a global ISI parameter (Equation 8), and on an additional reinforcement signal reflecting the expected time of occurrence of the US (Equation 9).

- This additional reinforcement signal is computed by the other processing unit, designated the E unit, which learns when the US occurs. Like the V unit, the E unit receives convergent input from CSs and the US, and it has modifiable synaptic weights that are changed according to a simple linear difference equation (Equation 11) that includes local eligibility factors (Equation 10) and the global ISI parameter defined in Equation 8.

A more formal treatment follows:

Architectural assumptions for the network have been described in detail elsewhere (Desmond & Moore, 1988; Moore et al., 1989). Basically, the onset and offset of each CS begins activation of separate tapped delay lines. The elements in the delay line are referred to as x_{ijk} elements because each element can be referenced by (a) its CS (i), (b) whether it is activated by the onset ($j = 1$) or offset ($j = 0$) of the CS, and (c) its number within the delay line (k). For example, element x_{208} belongs to the offset delay line for CS2, and is the eighth element activated in the delay line. The output of an x_{ijk} element (which is either 1 or 0) at time t is designated $x_{ijk}(t)$.

The x_{ijk} tapped delay line elements are activated sequentially, with a new element recruited every time step (10 ms). When activated, an x_{ijk} element changes value from 0 to 1, and remains at 1 for 10 time steps. Thus, for each trial beginning at time $t = 1$:

$$x_{ijk}(t) = \begin{cases} 1 & \text{if } \tau_{ij} + k - 1 \leq t < \tau_{ij} + k + 9; \\ 0 & \text{otherwise} \end{cases} \qquad (4)$$

where τ_{ij} is the onset time ($j = 1$) or offset time ($j = 0$) of CS i ($x_{ijk} = 0$ is CS i is not presented).

In addition to the x_{ijk} elements, the network has two higher order processors designated the V unit and E unit. Each x_{ijk} element gives off two taps, one of which projects to the V unit and the other to the E unit. The latter connections are modifiable, and the weights for these connections are referred to as V_{ijk} and E_{ijk}. All other connections are non-modifiable; these include: US connections with the V and E units, an E-unit projection to the V unit, and connections between adjacent x_{ijk} elements in the delay line.

The output of the network, $s(t)$, is derived from the US input and from the weighted sum of the V unit inputs, and is defined as:

$$s(t) = \sum_i \sum_j \sum_k V_{ijk}(t)x_{ijk}(t) + L(t) \tag{5}$$

where $s(t)$ is confined to the closed unit interval.

Changes in the V_{ijk} weights are given by the following expression:

$$\Delta V_{ijk}(t) = c\{L(t) - \hat{s}(t)\}h_{ijk}(t)\bar{x}_{ij}(t)r(t). \tag{6}$$

where:

c is a rate parameter, $0 < c \leq 1$.

$L(t)$ is the reinforcement, $0 \leq L(t) \leq \lambda$, where λ is analogous to the strength of the reinforcement during conditioning, $0 < \lambda \leq 1$. $L(t) = 0$ until the US occurs, at which time $L(t) = \lambda$.

$\hat{s}(t) = \sum\sum\sum V_{ijk}(t)x_{ijk}(t)$, and is confined to the closed unit interval.

$h_{ijk}(t)$ constitutes an eligibility trace for each V_{ijk} synapse, $0 \leq h_{ijk}(t) \leq 1$. This term has maximum value at the onset time of the element and decays geometrically. It is computed as follows:

$$h_{ijk}(t) = \begin{cases} 1.0 & \text{if } t = \tau_{ij} + k - 1; \\ (0.8)h_{ijk}(t - 1) & \text{if } t > \tau_{ij} + k - 1; \\ 0.0 & \text{otherwise,} \end{cases} \tag{7}$$

$\bar{x}_{ij}(t)$ is an *overall CS eligibility* for onset and offset processes, $0 \leq \bar{x}_{ij}(t) \leq 1$. This function governs the rate of conditioning that occurs at a given interstimulus interval. $\bar{x}_{ij}(t)$ is globally available to all V_{ijk} and E_{ijk} synapses. The equations described below approximate the empirically observed inverted-U-shaped function found in rabbit nictitating membrane conditioning:

$$\bar{x}_{ij}(t) = \begin{cases} (.05)(t - \tau_{ij}) - 0.25 & \text{if } \tau_{ij} + 6 < t < \tau_{ij} + 25; \\ (-1/475)(t - \tau_{ij}) + (500/475) & \text{if } \tau_{ij} + 25 \leq t < \tau_{ij} + 500; \\ 0.0 & \text{otherwise.} \end{cases} \tag{8}$$

$r(t)$ is the output of the E unit, and represents the temporal expectation of reinforcement, $0 \leq r(t) \leq \lambda$. It is defined as follows:

$$r(t) = \max\{E_{ijk}(t)\Delta x_{ijk}(t) \mid i = 1, \ldots, n; j = 0, 1; k = 1, \ldots, N\} \tag{9}$$

where:

$$\Delta x_{ijk}(t) = \begin{cases} 1 & \text{if } x_{ijk}(t) - x_{ijk}(t-1) = 1; \\ 0 & \text{otherwise,} \end{cases} \tag{10}$$

and E_{ijk} are the connection weights of the input elements onto the E unit. Changes in these weights are given by:

$$\Delta E_{ijk}(t) = c[L(t) - r(t)]\Delta x_{ijk}(t)\tilde{x}_{ij}(t) \tag{11}$$

Implementing the VET Model

The implementation criteria for the VET model were the same as for the SBD model. The two schemes share many common features, e.g., they both assume that CRs are generated by the disinhibiting action of HVI PCs on NIA neurons, and Golgi cells play a crucial role in both. The scheme assumes that the time-tagged input elements associated with a CS ascend to cerebellar cortex via mossy fibers. They synapse within the granule layer, and with no modulation are assigned to a corresponding set of parallel fibers. The parallel fibers that carry CS information synapse on both PCs and Golgi cells. Notice that the implementation assumes that the tapped delay line architecture exists outside the cerebellum, possibly within the brain stem reticular formation (Scheibel & Scheibel, 1967). The anatomical justification for this assumption is discussed elsewhere (Desmond & Moore, 1988; Moore et al., 1989).

These synapses are modifiable. Synapses on PCs are sites where Equation 6 is implemented, and hence these cells are the V units. Synapses on Golgi cells are sites where Equation 11 is implemented, and hence these cells are the E units. Both sets of modifiable synapses are changed when the US occurs to the extent that they are eligible. The US triggers a climbing fiber volley that causes these synapses to undergo synaptic depression via mechanisms of long term dression (LTD) identified by Ito (1989) and others (e.g., Crepel & Krupa, 1988). With learning, CS input causes V and E units to decrease their baserate of firing. For V units, this decrease initates a CR. For E units, this decrease disinhibits certain granule cells. These granule cells receive input from brain stem neurons, possibly in SpoV, which also undergo learning due to the convergence of CS and US information. Although no learning rule is specified, the SBD (or SB) model suffices provided the variable x in Equation 1 increases rapidly to a plateau at CS onset and decays slowly at CS offset.

With learning, the output (rate of firing) of these brain stem units consists of a short latency plateau that extends well beyond the offset of the CS. This output provides a uniform and long lasting stream of excitatory input to granule cells. The E units, however, gate this excitation and prevent it from exciting parallel fibers that synapse on the V units. As learning builds up in the E units, this gate is opened, but only during times near the expected occurrence of the US. Thus, E units provide the second reinforcing event necessary for modification of CS-input synapses on the V units so that they express appropriately timed CR waveforms.

Implications of the VET Model

The VET model simulates virtually all of the behavioral phenomena encompassed with the SBD model, but unlike that model, it has the added feature of predicting realistic S-shaped acquisition curves. Unlike the SBD model, however, it cannot simulate second-order conditioning. As in the Rescorla-Wagner model, connection weights are strengthened only in the presence of the US. Desmond and Moore (1988) describe several novel predictions of the VET model. One prediction is that lengthening the duration of a CS after trace conditioning should result in double peaked CR waveforms, reflecting contributions of both onset and offset tapped delay elements. The weights of offset elements are normally masked by those of the onset elements, but lengthening the CS exposes these offset elements, and CRs with two peaks emerge.

The physiological implications of the implementation scheme are many. The most interesting implication is the possibility that cerebellar Golgi cells express LTD at parallel fiber synapses in a manner analogous to that of PCs. There appears to be no evidence on this point in the literature. Another implication of the scheme is that climbing fibers from DAO do, in fact, reinforce learning. In order for this possibility to be taken seriously, it would be necessary to record from PCs (and Golgi cells) in HVI during the initial stages of CR acquisition. This was not done in the Berthier and Moore (1986) study, nor in any other study, for methodological reasons: It is virtually impossible to record from single neurons in an awake animal for the hundreds of trials normally required to obtain robust CRs. Nevertheless, there are ways around this problem we are pursuing in our laboratory. Finally, the scheme implies the existence of neurons in the brain stem that project to HVI and fire in accordance with the scheme's requirements. We have seen a number of candidate neurons in recordings from cells in NIA, RN, SpoV. Cells in RN and SpoV project to HVI, and it is possible, though not proven in rabbit, that NIA cells send mossy fibers to HVI as axon collaterals.

CONCLUDING REMARKS

The two models and implementation schemes are quite different, but not mutually exclusive. The VET model implies a neural network architecture well suited for forming appropriate CR waveforms for a wide range of circumstances. The output of this network need not be regarded as input to motoneurons. Instead, it could be regarded as a template used by another learning system responsible for such things as generating CRs and implementing second-order conditioning. Such a system might resemble the SBD model or some other supervised learning (error correction) algorithm.

Such a hybrid model would not encompass phenomena that both models fail to address: intertrial interval effects, CS intensity, stimulus generalization and discrimination. These phenomena require a richer representation of CS input to learning networks than have been considered by either model to date. Desmond (1988) has developed one approach that can potentially address these topics. It represents CSs as planar arrays of elements through which activation spreads and decays in an orderly, yet stochastic manner. The planar array approach follows directly from the foundations provided by the SBD and VET models, and we might anticipate as abundant a harvest of interesting implications as have sprung from its forerunners.

ACKNOWLEDGMENT

Preparation of this chapter and the author's research program were supported by grants from the Air Force Office of Scientific Research and the National Science Foundation.

REFERENCES

Albus, J.S. (1971). A theory of cerebellar function. *Mathematical Bioscience, 10*, 25–61.

Barto, A.G., & Sutton, R.S. (1982). Simulation of anticipatory responses in classical conditioning by a neuron-like adaptive element. *Behavioural Brain Research, 4*, 221–235.

Berthier, N.E., Barto, A.G., & Moore, J.W. (1988). Linear systems analysis of cerebellar deep nuclei cells during performance of the classically conditioned eyeblink. *Society for Neuroscience Abstracts, 14*, 1239.

Berthier, N.E., & Moore, J.W. (1986). Cerebellar Purkinje cell activity related to the classically conditioned nictitating membrane response. *Experimental Brain Research, 63*, 341–350.

Blazis, D.E.J., & Moore, J.W. (1987). *Simulation of a classically conditioned response: components of the input trace and a cerebellar neural network implementation of the Sutton-Barto-Desmond model.* Computer and Information Science technical report 87–74, University of Massachusetts, Amherst, MA.

Blazis, D.E.J., & Moore, J.W. (1989). Conditioned stimulus duration: Behavioral assessment of a prediction of the Sutton-Barto-Desmond model. *Society for Neuroscience Abstracts, 15*, 506.

Bloedel, J.R. (1987). Technical comment. *Science, 238*, 1728–1729.

Byrne, J.H. (1987). Cellular analysis of associative learning. *Physiological Reviews, 67*, 329–439.

Churchland, P.S. (1986). *Neurophilosophy: Toward a unified science of the mind-brain.* Cambridge, MA: MIT Press.

Coulter, D.A., Lo Turco, J.J., Kubota, M., Disterhoft, J.F., Moore, J.W., & Alkon, D.L. (1989). Classical conditioning reduces amplitude and duration of calcium dependent afterhyperpolarization in rabbit hippocampal pyramidal cells. *Journal of Neurophysiology, 61*, 971–981.

Crepel, F., & Krupa, M. (1988). Activation of protein kinase C induces a long-term depression of glutamate sensitivity of cerebellar Purkinje cells. An in vitro study. *Brain Research, 458,* 397–401.

Desmond, J.E. (1988). *Temporally adaptive conditioned responses: Representation of the stimulus trace in neural-network models.* Computer and Information Science technical report 88–80, University of Massachusetts, Amherst, MA.

Desmond, J.E., & Moore, J.W. (1982). A brain stem region essential for the classically conditioned but not unconditioned nictitating membrane response. *Physiology & Behavior, 28,* 1029–1033.

Desmond, J.E., & Moore, J.W. (1986). Dorsolateral pontine tegmentum and the classically conditioned nictitating membrane response: analysis of CR-related single-unit activity. *Experimental Brain Research, 65,* 59–74.

Desmond, J.E., & Moore, J.W. (1988). Adaptive timing in neural networks: The conditioned response. *Biological Cybernetics, 58,* 405–415.

Eccles, J.C., Sasaki, K., & Strata, P. (1967). A comparison of the inhibitory action of Golgi and basket cells. *Experimental Brain Research,* 81–94.

Gabriel, M. (1988). An extended laboratory for behavioral neuroscience: A review of *Classical conditioning* (third edition). *Psychobiology, 16,* 79–81.

Gormezano, I. (1972). Investigations of defense and reward conditioning in the rabbit. In A.H. Black & W.F. Prokasy (Eds.), *Classical conditioning II: Current research and theory* (pp. 151–181). New York: Appleton-Century-Crofts.

Gormezano, I., Prokasy, W.F., & Thompson, R.F. (Eds.) (1987). *Classical conditioning* (third edition). Hillsdale, NJ: Lawrence Erlbaum Associates.

Hamming, R.W. (1983). *Digital filters.* New York: Prentice-Hall.

Ito, M. (1989). Long-term depression. *Annual Review of Neuroscience, 12,* 85–102.

Lavond, D.G., Steinmetz, J.E., Yokaitis, M.H., & Thompson, R.F. (1987). Reacquisition of classical conditioning after removal of cerebellar cortex. *Experimental Brain Research, 67,* 569–593.

Marr, D. (1969). A theory of cerebellar cortex. *Journal of Physiology, 202,* 437–470.

McCormick, D.A., Steinmetz, J.E., & Thompson, R.F. (1985). Lesions of the inferior olivary complex cause extinction of the classically conditioned eyeblink response. *Brain Research, 359,* 120–130.

Millenson, J.R., Kehoe, E.J., & Gormezano, I. (1977). Classical conditioning of the rabbit's nictitating membrane response under fixed and mixed CS-US intervals. *Learning and Motivation, 8,* 351–366.

Miller, R.R., & Spear, N.E. (Eds.) (1985). *Information processing in animals: Conditioned inhibition.* Hillsdale, NJ: Lawrence Erlbaum Associates.

Moore, J.W. (1979). Brain processes and conditioning. In A. Dickinson & R.A. Boakes (Eds.), *Mechanisms of learning and motivation: A memorial volume to Jerzey Konorski* (pp. 111–142). Hillsdale, NJ: Lawrence Erlbaum Associates.

Moore, J.W., & Berthier, N.E. (1987). Purkinje cell activity and the conditioned nictitating membrane response. In M. Glickstein, C. Yeo, & J. Stein (Eds.), *Cerebellum and neuronal plasticity* (pp. 339–352). New York: Plenum.

Moore, J.W., & Blazis, D.E.J. (1989a). Cerebellar implementation of a computational model of classical conditioning. In P. Strata (Ed.), *The olivocerebellar system in motor control* (pp. 387–399). Berlin: Springer-Verlag.

Moore, J.W., & Blazis, D.E.J. (1989b). Simulation of a classically conditioned response: A cerebellar neural network implementation of the Sutton-Barto-Desmond model. In J.H. Byrne & W.O. Berry (Eds.), *Neural models of plasticity: Experimental and theoretical approaches* (pp. 187–207). San Diego: Academic Press.

Moore, J.W., & Blazis, D.E.J. (1989c). Conditioning and cerebellum. In M.A. Arbib & S. Amari (Eds.), *Dynamic interactions in neural networks: Models and data* (pp. 261–277). New York: Springer-Verlag.

Moore, J.W., Desmond, J.E., & Berthier, N.E. (1989). Adaptively timed conditioned responses and the cerebellum: A neural network approach. *Biological Cybernetics, 62,* 17–28.

Moore, J.W., Desmond, J.E., Berthier, N.E., Blazis, D.E.J., Sutton, R.S., & Barto, A.G. (1986). Simulation of the classically conditioned nictitating membrane response by a neuron-like adaptive element: Response topography, neuronal firing, and interstimulus intervals. *Behavioural Brain Research, 21,* 143–154.

Moore, J.W., & Gormezano, I. (1977). Classical conditioning. In M.H. Marx & M.E. Bunch (Eds.), *Fundamentals and applications of learning* (pp. 87–120). New York: Macmillan.

Ricciardi, T.N., Richards, W.G., & Moore, J.W. (1989). Single unit activity in spinal trigeminal oralis and adjacent treticular formation during classical conditioning of the rabbit nictitating membrane response. *Society for Neuroscience Abstracts, 15,* 507.

Robinson, D.A. (1981). The use of control systems analysis in the neurophysiology of movement. *Annual Review of Neuroscience, 4,* 463–503.

Robinson, D.A. (1989). Integrating with neurons. *Annual Review of Neuroscience, 12,* 33–45.

Scheibel, M.E., & Scheibel, A.B. (1967). Anatomical basis of attention mechanisms in vertebrate brains. In G.C. Quarton, T. Melnechuk, & F.O. Schmitt (Eds.), *The neurosciences. A study program* (pp. 577–602). New York: Rockefeller University Press.

Steinmetz, J.E., Lavond, D.G., & Thompson, R.F. (1989). Classical conditioning in rabbits using pontine nucleus stimulation as a conditioned stimulus and inferior olive stimulation as an unconditioned stimulus. *Synapse, 3,* 225–233.

Sutton, R.S., & Barto, A.G. (1981). Toward a modern theory of adaptive networks: Expectation and prediction. *Psychological Review, 88,* 135–170.

Yeo, C.H. (1989). The inferior olive and classical conditioning. In P. Strata (Ed.), *The olivocerebellar system in motor control* (pp. 363–373). Berlin: Springer-Verlag.

Yeo, C.H., & Hardiman, M.J. (1988). Loss of conditioned responses following cerebellar cortical lesions is not a performance deficit. *Society for Neuroscience Abstracts, 14,* 3.

Yeo, C.H., Hardiman, M.J., & Glickstein, M. (1984). Discrete lesions of the cerebellar cortex abolish the classically conditioned nictitating membrane response of the rabbit. *Behavioral Brain Research, 13,* 261–266.

Yeo, C. H., Hardiman, M. J., & Glickstein, M. (1986). Classical conditioning of the nictitating membrane response of the rabbit. IV. Lesions of the inferior olive. *Experimental Brain Research, 63,* 81–92.

MODELS OF
INSTRUMENTAL
CONDITIONING

8 Models of Acquisition and Preference

Michael L. Commons,
Eric W. Bing,
Charla C. Griffy,
and Edward J. Trudeau
Harvard University

ABSTRACT

Previously, preference and acquisition studies have only been concerned with developing acquisition models based on observed behavior for known reinforcement parameters. Schedules of reinforcement were taken into account when models were constructed, and learning curves were developed as if the subject knew the reinforcement parameters. The learning parameters were based upon a response strengthening model derived from Thorndike's Law of Effect (Thorndike, 1898, 1911).

This study attempts to predict acquisition based on probabilistic values of reinforcers from the subject's limited perspective. We assert that underlying instrumental conditioning, two respondent conditioning steps take place. We also assert the effect of having a response predict the occurrence of a following reinforcer decreases in a hyperbolic fashion as the time between them increases. Thus, a model is constructed in which the learning curve changes over time as the subject's behavior progresses. A final predicted learning curve becomes apparent data point by data point, as each is processed by this model. The resulting predicted behavior for one such model is compared to the observed behavior, and an analysis is then carried out for accuracy of fit.

MODELS OF ACQUISITION AND PREFERENCE

Quantitative acquisition of preference studies with nonhumans essentially began in the early 1980's (e.g., Commons, Woodford, Boitano, Ducheny, & Peck, 1982; Herrnstein, 1982; Herrnstein & Vaughan, 1980; Myerson & Mizzen, 1980). In preference studies, organisms select among schedules of reinforcement by either responding on one operandum (key) more often than on others or by responding on an operandum that selects one of the schedules. The reinforcement schedules deliver reinforcers (S^{R+}) to the organisms based upon some rule, for example, following the first response (R) after one minute has elapsed since the last reinforcer had been delivered (Fixed Interval, FI). In the late 1980s there have been a few such studies (Myerson & Hale, 1988; Vaughan & Herrnstein, 1987). Preference experiments, in which organisms make real choices and experience real outcomes form an even smaller subclass of such studies (e.g., Bailey, 1988; Bailey & Mazur, submitted; Commons, Woodford, Boitano, Ducheny, & Peck, 1982; Myerson & Mizzen, 1980). Corresponding theories and experimental results are so few that a fairly detailed history of them is possible.

Preference situations can be characterized as discrimination situations, in which the stimuli associated with the responses are quite easily distinguishable. For example, pecks on the red-left key (R_L) are reinforced on one schedule. Pecks on the green-right key (R_R) are reinforced on another schedule. An acquisition experiment studies how an organism changes its behavior in response to a changing set of stimuli. At the beginning of acquisition, one set of schedules has been in effect for a long time. Organisms show stable preferences reflected by relatively constant choice probabilities or rates. The schedules are then changed and the resulting changes in response frequency or rate as performance restabilizes constitute the acquisition data.

In the short history of data based quantitative preference studies, simple concurrent, schedules have been examined by Myerson (Myerson & Hale, 1988; Myerson & Mizzen, 1980). He placed pigeons on two concurrent schedules, whose values (programmed reinforcement frequency) were periodically switched. In such a concurrent schedule, each response (R) that met the reinforcement contingency was reinforced (SR+). The clock for one schedule continued to run even though responding on the other schedule was taking place. Therefore, for time-based reinforcement schedules, responding on the left increased the likelihood that responding on the right would be reinforced.

Myerson and his associates proposed the Kinetic Model to explain their acquisition data. For example each of the schedules could be random ratio schedules:

$$R_L \rightarrow S^{R+} \text{ with p} = .1$$
$$R_R \rightarrow S^{R+} \text{ with p} = .2 \tag{1}$$

The word ratio reflects that fact that there is a ratio of the number of responses, n, to the 1 response that is reinforced (Reynolds, 1968). With a random ratio schedule, a response is reinforced with probability p, a Poisson process (Schoenfeld & Cole, 1972).

At the same time, two other groups of researchers were examining acquisition of preference. First, (Herrnstein, 1982; Herrnstein & Vaughan, 1980; Vaughan & Herrnstein, 1987) set forth their notion of melioration. Herrnstein had previously established his matching law as a model describing the relationship between responding and reinforcement after responding has stabilized. Melioration is a model describing what an organism will do if reinforcement conditions are changed. Vaughan (1981, 1985) made clear that the "value" of the reinforcement schedules being preferred was carried by the stimuli that preceded the responses. Otherwise, response rate would not have momentarily dropped when key color was changed but programmed reinforcement rate stayed constant. Another aspect of their theory is that the mechanism of change depended on the obtained rate of responding and obtained rate of reinforcement rather than the programmed rate of reinforcement. No specific experiments were analyzed at the time, and few details in their model were specified.

Second, Commons, Woodford, Boitano, Ducheny, and Peck (1982) examined the acquisition of preference by using a concurrent chain procedure as shown next:

Pattern of reinforced pecks across cycles, C_1

	C_1	C_2	C_3	C_4
Completing requirement 1, VI-12 seconds for R_L, leads to:	R-0	R-S^{R+}	R-0	R-0
Completing requirement 2, VI-12 seconds for R_R, leads to:	R-0	R-S^{R+}	R-S^{R+}	R-0

$$(2)$$

In the concurrent-chain schedule they used, completing requirement 1 (a VI-12 seconds component) led to four trials (C_1 through C_4) described by the pattern of reinforced pecks. If a response was reinforced, it was delivered at the end of the cycle, C_i; no reinforcement is indicated by 0. The length of C_i was 3 seconds. In a variable interval schedule (VI), after a variable length with a mean of t seconds of time after the last reinforcer, the first response is reinforced. Commons, Woodford, et al. were interested in relating the value of reinforcement obtained in a preference situation to the value obtained in the situation where reinforcement schedules were discriminated. In a discrimination situation, some form of responding indicates which schedule has been in effect previously. Correct indications are reinforced. Concurrent chains, in which completing one schedule leads to another, were adopted at the suggestion of Nevin (1978, personal communication). Nevin suggested that concurrent schedules were better understood than simple choice procedures for assessing preference. The Commons, Woodford, et al's model was midway between others in that the brief obtained rates of responding were used, but steady-state values of obtained reinforcement were also used. In any case, a titration procedure was used, in which reinforcer values were switched for choice outcome, and shifts toward stability were recorded.

We know of only three other data based studies. Dreyfus (1985, March; 1985, April) presented some data on what happens with concurrent schedules when values are switched. A final report of that data is forthcoming. Myerson and Hale (1988) also studied acquisition of preference using concurrent schedules. They suggest that their data is inconsistent with a melioration model and consistent with their kinetic model.

Most recently Bailey and Mazur (Bailey, 1988; Bailey & Mazur, submitted) ran the simplest experiment. They first stabilized pigeons pecking on a simple two choice situation. They then examined a number of theories including melioration (Vaughan, 1981), probability learning (Estes, 1959) as well as the kinetic model.

Relating Instrumental Preference to Classical Conditioning Neural Networks

This paper presents a conceptual reduction model of how preference can be conceived of as two steps of respondent (classical) conditioning. It then presents pilot models integrating the reduction and the Linear Noise Model (Commons, Woodford, & Ducheny, 1982), a model on effect of the delay of reinforcement. These models are of a behavioral nature and do not compete with neural network models of the same processes (Grossberg, 1987). They may be useful, however, to network modeling of the acquisition of preference. One of the most important aspects of neural networks is that they are formed by not only modeling behavioral data but also modeling neural network processes. As most of the papers appearing

Flowchart of Experiment

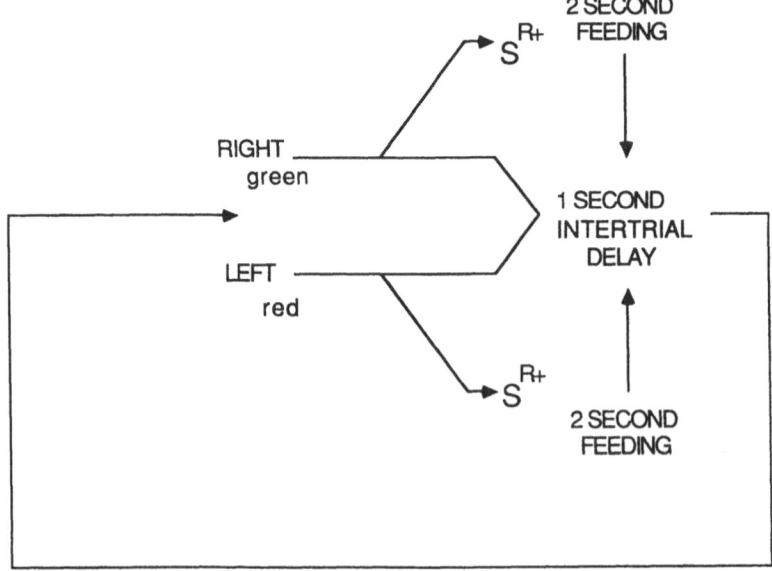

FIG. 8.1. Flowchart of experiment.

in this volume attest, models of conditioning generally refer to classical (respondent) conditioning rather than instrumental (operant) conditioning. Grossberg (1971a, 1971b, 1974, 1982a, 1982b, 1987) is the most prominent exception because he integrates operant and respondent conditioning and develops neural networks of them. There is a long history of relating these two forms of conditioning. Two factor theories suggest that operant and respondent condition are separate processes. Single factor theories suggest that although there are surface differences in the procedures, the underlying processes are the same. Our strategy is to show how operant conditioning is related to respondent conditioning and how they differ, so that the methods used for respondent conditioning may be applied to operant conditioning.

The Difference Between Operant and Respondent Conditioning

In order to model the acquisition of operant preference, one must draw upon the neural network studies of respondent (classical) condition. We identify and examine differences between the two to show why our reduction of operant to respondent conditioning is necessary in order to use respondent conditioning neural network results.

Operant conditioning and respondent conditioning cannot be immediately reconciled. First, in operant conditioning, an environmental stimulus (S) followed by an operant reinforcer never produces a response, whereas in respondent conditioning it does. Second, in classical conditioning there is no necessity to follow the conditioned response (CR) with an unconditioned stimulus US or operant reinforcer (S^{R+}). Third, simple classical conditioning between the environmental stimulus and the stimulus that elicits the operant response fails. The unobserved but inferred stimulus that elicits the operant response is an unconditioned stimulus (us). In classical conditioning, presentation of the environmental stimulus followed by an unconditioned stimulus leads to conditioning. In the operant case, presenting the environmental stimulus followed by the (us) that elicits the operant response leads to extinction of the operant response. Although the differences are not limited to these three, these distinctions are specifically addressed by reduction.

Although operant conditioning and respondent conditioning cannot be immediately reconciled, they can be united by reducing operant conditioning to respondent conditioning. For background on how we conceive of the contingencies in operant conditioning, the following reduction of operant to respondent conditioning is presented. Pilot behavioral models of operant preference acquisition are also introduced that update and combine features of the Commons and Woodford (Commons, Woodford, & Ducheny, 1982) model with features of the Herrnstein and Vaughan (1980) model. Together, the reduction and behavioral models provide a conceptual springboard, from which a neural network of preference acquisition might be built.

The Reduction of Operant Conditioning
to Respondent Conditioning

It is argued here that the functions of operant reinforcement can be derived from the functions and properties of respondent pairing operations. Two important functions of operant reinforcement are the strengthening of responding and the establishment of discriminative control by events. We propose that the operant response is not spontaneously emitted behavior, but rather is an acquired response to certain appropriate stimuli.

The two extra requirements that preserve the explanatory power of two factor theories are the two respondent pairing steps in the present reduction: (a) the salience or "what to do" pairing step, in which the internal causal events (US) that precede the operant response (R/UR_1) become salient to the subject, who thereby learns "what to do" to receive reinforcement; and (b) the environmental control or "when to do" pairing step, in which the now salient internal stimulus/conditioned response complex $(US\text{-}CR_2)$ is paired with the neutral environmental stimulus (S_1/NS). The subject thereby learns "when to do" the operant behavior and under what circumstances.

In respondent conditioning the first step is unnecessary, because the unconditioned stimulus (US) is already salient to the subject. Because of requirement (b), the organism cannot learn to make the operant response under different circumstances unless it has been exposed to various situations. More reinforced responding within one situation will not lead to generalization of learned behavior in other situations.

We posit the existence of an event of unknown origin within the brain called the internal unconditioned stimulus (us). This us is a behavioral-related, analytic way of writing the stimulus properties of the brain event that precedes the operant response (R). It is the inferred cause of the operant response. After this us is paired with a reinforcer (S^{R+}/US), the response (cr) of excitement anticipates the excitement elicited by the (S^{R+}/US). It is a simple classical conditioning step that makes the us salient. The new compound is written $us\text{-}cr$, but in the brain, we do not differentiate the stimuli and responses. A stimulus for one layer is a response in another layer. The salience of the us is maintained by its continued pairing with the final reinforcer (S^{R+}/US). Thus, there is no extinction for responses to the us while these pairings are occurring. Over long periods of time a constant reinforcement rate loses its salience and responding becomes automatic.

The value associated with an environmental stimulus outside (a red or a green light or a key) and the cr that represents the value associated with the density of past reinforcement are compared to the density of presently obtained reinforcement. If the values are discrepant, this elicits an emotional response that again is associated with each stimulus, the us and the S_1. This raises the salience of the us leading to changes in the vigour of one of the responses.

Partly from these observations, we make the following claim. Provided the cr

is salient, the animals are "aware" of what they are going to do before they do it. We speculate, and there is good evidence, that as one goes up phylogenetically, the planning step, or internalization of this unconditioned response, is what lengthens and becomes more regularized.

Once this pairing step occurs regularly, the environmental aspect elicits the response directly. This is why the response occurs more and becomes habituated. If there is no change in the value of the reinforcer, the subject no longer detects it unless there is a mismatch. For example, if you drive your car a particular route every day, you will not recall many of the details of the surroundings through which you drive. The drive is chunked into large segments, and the details control automatic driving but do not require reflective attention. If, however, you become lost, you suddenly become much more aware of your surroundings. This does not differ from having higher order verbal terms, such as long words, in which one does not necessarily plan or reflect upon each syllable. If the spelling is to be checked, the syllables are then examined.

An example will serve as an anchor for the theory to follow. A pigeon is presented with two keys in a Skinner Box for the first time. The pigeon sees grain in the lit hopper, and hears the click associated with its activation (an external stimulus, S_1). The sound and sight of the lit hopper (NS/S_1) are *paired* with the grain's consumption (US_2). The key peck is the response, and each key color becomes associated with one rate of reinforcement. When a change in reinforcement density takes place, the value of the key color changes. The conditioning of the sight of the grain (CS) comes to produce the conditioned response, that is, excitement (CR_2).

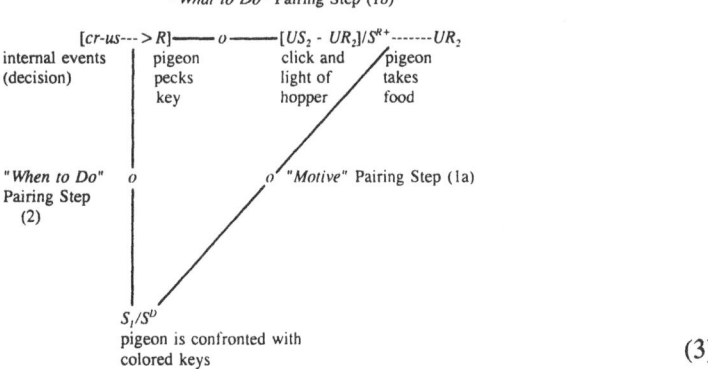

$$\tag{3}$$

The "What to Do" Pairing Step (1b): Salience

Return to the example cited earlier just before the *US* becomes salient as diagrammed in expression three. The salience of the decision to peck the key (*US*) is established by pairing it with the arrival of the hopper and ingesting the grain (US_2/S^{R+}). The result of this pairing is that the decision to peck the key

(US) is now salient, because the pairing elicits the conditioned response and the stimulus it produces, (CR_2/CS_2) as diagrammed in expression four.

$cr-(US-CR_2/CS_2)-R$

              excitement,   (pecks the colored key,
                                     possibly more frequently)
              image of the reinforcer   (4)

Before conditioning, these responses (CR_2/CS_2) were elicited only by the arrival of the hopper and the ingestion of food (US_2).

This proposed pairing step 1b has been anticipated by others. In Guthrie's (1952) theory, one role of the final reinforcing US_2 is to increase salience. In Tolman's (1932) theory, the US comes to elicit a portion of the responses to the reinforcing US_2. That is, the activity of preparing to act (US) elicits a set of memories (CR_2/CS_2) of past reinforcements and of the relationship (a relational event) between that previous activity and the reinforcements. For Tolman, this set of recalled events leads to the action being experienced as *purposeful* action. Our interpretation is slightly different. As subjects prepare to act (decide to peck the colored key), they recall the outcome of similar actions in the past (possibly get excited, salivate, and have an image of the food and its taste). These recalled events and the sense of the relationship between the previous similar actions and previous outcomes is experienced as the reason to act. Thus, in Tolman's terms, subjects have the sense that their acts are purposeful and voluntary, although this is actually an illusion, because memory is elicited, but is not in fact the cause of the act.

The "When to Do" Pairing Step: Environmental Control

In order for respondent conditioning to occur (i.e., so that pecking a colored key, R, follows the environmental stimulus of introducing the grain, (S_1), both the internal stimulus that elicits the reaching for the grain, (US), and the environmental stimulus, (S_1), must be salient. After a sufficient number of pairings during the "What to Do" step, the internal unconditioned stimulus becomes the salient complex, $(US-CR_2/CS_2)$, as modern theories of classical conditioning (Mackintosh, 1974; Rescorla & Wagner, 1972) would suggest. The environmental stimulus, (S_1), will also be salient because it has previously been paired (pairing 1a) with (US_2/S^{R+}). Thus, the internal event, $us\text{-}CR_2/CS_2$, can be effectively paired with the neutral environmental stimulus, S_1/NS, changing that environmental stimulus into a conditioned stimulus, CS_1.

CS/S^D -------------CR_2 (R) -------------US_2/S^{R+}
perception of   arrival of   frequent key pecks
colored keys   hopper, pigeon gets food   (5)

The "When to Do" pairing results in a response following the environmental stimulus, S_1. Thus, the probability of pecking a given key is increased because the complex that elicits the response is now elicited by the environmental stimulus. Changes in preference when reinforcement schedules are altered should depend on this step.

Results of the Three Pairings

The three pairings result in the following chain of events:

$$CS_1/S^D - (us - CR_2/CS_2) - R. \ldots \ldots US_2/S^{R+} \qquad (6)$$

After conditioning, the next time that the pigeon is confronted with the lit keys, CS_1, it will elicit the complex us-CR_2/CS_2, of repeated pecks to a given key. The model allows for only strengthening of responding. For example, this model predicts that punishment strengthens responding. The responses, however, are those that compete with the behavior that is being punished (Alkon, personal communication, January 23, 1990; Hull, 1952). Increasing the rate of making the competing response, increases the value of the outcome by decreasing the amount of punishment.

Extension to Two-key Preference

The reduction of the acquisition of a single operant response to respondent condition can be extended to the acquisition of preference. In the more complex two key situation, the role of value of the final reinforcer is emphasized rather than just its existence.

With this view of conditioning in mind, the Bailey and Mazur data were examined with the hope of developing a pilot model that could be simulated by a neural network. These pilot models are our first attempt to model preference acquisition and do not test the notions that they are based on. Commons and Hallinan (1990) use a very general notion of such neural nets. The interest here is to present a set of models that are restricted to the "pigeon view of preference" removing the experimenter perspective. They postulated that any frequently repeated response to a stimulus will lead to the slow development of a "cell assembly" within parts of the brain. Such cell assemblies are capable of acting briefly as a closed system and interacting with other such systems. A series of such events constitutes a "phase sequence." The phase sequence constitutes the melioration (Herrnstein, 1982; Herrnstein & Vaughan, 1980) or learning process.

In the two choice preference situation here, learning is postulated to occur when the amount of reinforcement delivered for responding on each key is altered. As with Grossberg's Adaptive Resonance Theory (ART) model (1980; chap 4 in this volume), the change in local reinforcement density creates a discrepancy between the value of reinforcement obtained historically over the short range and that obtained momentarily. When the reinforcement density changes, there is a discrepancy between the CR, the reinforcement predicted by the us, and the obtained density. The us elicits some motivational CR, which some consider an expectation of a given reinforcement density. Tendency to respond to the little (*us*) also reflects an expectation of what the payoff for the response should be. The discrimination of this discrepancy makes the *us* salient again. This *us* is then paired again with the environmental S_1 that is more potent because it has been paired with the (US_2/S^{R+}). Together, the more-often-reinforced response is more vigorously elicited (Estes, 1969; Grossberg, 1982a, 1982b).

There is substantial evidence for the relativization of responding—that is the ratio of response allocation to one key to the response allocation of the sum of the two keys (e.g. Commons, Herrnstein, & Rachlin, 1982) as well as for individual strengthening of each response in the two key situation (Myerson & Mizzen, 1980). The sum of the rates increases with an increase in total reinforcement. The relative rates stay constant if the relative rate of reinforcement stays constant.

The Effect of the Time Between Reinforcement and a Response

Three time-related variables (Commons, Woodford & Trudeau, in press) affect acquisition. Each is identified with the effect that reinforcers have on a choice of behavior over time. The first is time scheduled associativity (Commons, Woodford, Boitano, Ducheny, & Peck, 1982), or the time lapse from each reinforcer to the choice that it affects. Another is the relative time between reinforcers, or the change in time lapse between reinforcers over several consecutive reinforcer/choice cycles. The assumption here is that the subject would choose more often the key with the shorter time delay between reinforcement. The last is based on the number of intervening events between reinforcers and choice. Time in this case is measured in terms of events passed rather than seconds.

Time allocation for responding to changes in scheduled reinforcement is a mechanism for characterizing melioration or learning. The value of events to a subject is reflected in how it spends its time (Commons, Woodford, & Ducheny, 1982). This model hopes to demonstrate that events in time and their value interact to determine allocation according to the matching law. The level of this allocation can be determined by response to scheduled reinforcement.

Our pilot models for acquisition are also based upon the General Additive Noise Model developed by Woodford (Commons, Woodford, & Ducheny, 1982). They describe the memory of an event, such as reinforcement, as some memory value plus some random noise term. For each subsequent event, an additional noise term is added to previous memories, so that over time, memories become more indistinct. For a four cycle experiment, for example, the model might predict the following memory values when the time came to make a choice. Here, M_i, is the memory of whether or not there was a reinforcer on cycle i, and n_i is the noise term associated with that cycle.

Commons, Woodford, and Trudeau (in press) have shown that this linear noise model predicts the hyperbolic decrease of the effect of reinforcers over time.

Cycle: C_1 C_2 C_3 $C_4 \rightarrow$ choice
Reinforcer: 1 (M_1) 1 (M_2) 1 (M_3) 0 (M_4)
Memory 1: M_1+n_1 $M_1+n_1+n_2$ $M_1+n_1+n_2+n_3$ $M_1+n_1+n_2+n_3+n_4$
Memory 2: M_2+n_2 $M_2+n_2+n_3$ $M_2+n_2+n_3+n_4$
Memory 3: M_3+n_3 $M_3+n_3+n_4$
Memory 4: M_4+n_4 (7)

Further analysis has been carried out, examining the effect of the length of a cycle, measured in seconds. With that variable, the rate at which the remembered value of a reinforcement decreases, can be adjusted according to the average rate that events occur in a particular experimental situation. Hence, it is not how much time has passed that causes the decrease in value, but simply length of time relative to the average time between events in the particular environment. According to this hypothesis, a change in the length of the average reinforcement delay would have no effect, but a particularly long delay after the subject has already calibrated its rate of memory decay to a different rate of occurrence of events will affect the size of the reinforcer. In other words, the subject will give more weight to a reinforcer given after a relatively large time delay between reinforcers, remembering the most recent reinforcer for a longer period of time. In the interests of minimizing delay between reinforcement, the subject would develop a preference of the schedule that delivered reinforcement with the least amount of intervening relative time.

Using this model for decrementation of reinforcers over time, this simulation attempts to develop a learning algorithm that captures the time allocation predicted by melioration (Herrnstein, 1982; Herrnstein & Vaughan, 1980). Melioration predicts that the subject will modify its behavior so that the relative allocation of behavior matches the relative obtained reinforcement. Here, with two ratio-like schedules, subjects' behavior will stabilize on the richest reinforcement schedule over time.

Our study focuses on determining what happens when the density of reinforcement is changed and how the time allocation of choices that result can be predicted by the acquisition and delay functions (Commons, Mazur, Nevin, & Rachlin, 1987). The analysis of Bailey and Mazur's data presented here deviates from all past models including their own. Our pilot models examine reinforcer value over time instead of developing that fit based on the entire data set. This general model is used to calculate the total value of each reinforcer, each decremented for time, that has been delivered up to a given point in the experiment. The sum of the decremented values of all previous reinforcers yields the total reinforcement at any given point in time. This sum of all previous reinforcer values is then used to predict the behavior of the subject on the succeeding choice period. This process is repeated for each choice period, reassessing the value of reinforcers at each succeeding point in the experiment. We compared these predictions to data.

Method

Four pigeons were run on several 800 trial sub-experiments in a standard Skinner box. In each experiment, two keys were transilluminated, one with a red light and one with a green light. A peck on either key was reinforced with a single pellet of food, or not reinforced at all, with a probability based on which key was selected.

The probability of reinforcement on either key varied from trial to trial. Five different probability pairs were each run twice, with the rich or higher probability on one key first and then on the other. Probabilities were selected to test the effects of greater or lesser difference in probability, ratio between probabilities, and discrimination between values of probabilities. Before each trial, each bird was run on a special series of trials designed to bring the probability of pecking a given key as close to 0.5 as possible, and then a change in reinforcement density was slowly introduced. The data we are using are the 800 trial experimental sessions that were run after the transition sessions.

The Pilot Models

In the Linear Noise Model, the effect of any given reinforcer decreases over time as a decrementing hyperbolic function, as shown in figure 8.2. The curves represent the decrease of the two reinforcer values of R_1 and R_2, delivered at time t_i and t_{i+1}, respectively. The total value of both reinforcers at time T is the sum of the two decrementation functions at time T. The difference in time between time t_i and t_{i+1} is the relative time between reinforcers, denoted $delta t_i$. Then the total value at time t_{i+1} of R1 can be expressed:

$$R_1(t_{i+1}) = \frac{R_1(t_i)}{(t_{i+1} - t_i)};$$ (8)

where $R_1(t_i)$ is the weight assigned to the reinforcer at the time it is obtained, and $(t_{i+1} - t_i)$ is the time elapsed at t_{i+1} since the reinforcer R1 was delivered. One possible value for $R_1(t_i)$ is the inverse of the time delay between R_1 and the previous reinforcer, or $delta t_{i-1}$. This inverse would place a greater weight on reinforcers that were delivered with little intervening time. In this manner, reinforcers preceded by large time delays contribute less to memory, as one model for decrementation (Commons, Woodford, & Trudeau, in press) suggests.

Similarly, the total value of all reinforcers at time T can be expressed as the hyperbolically decremented sum of the values of all reinforcers delivered before T. That is:

$$\text{Total reinforcement at time T} = \sum_{j=1}^{n} \frac{R_j(t_j)}{(T - t_j)};$$ (9)

LINEAR NOISE MODEL

The Effects of Any Given Reinforcer Decreased Over Time Hyperbolically

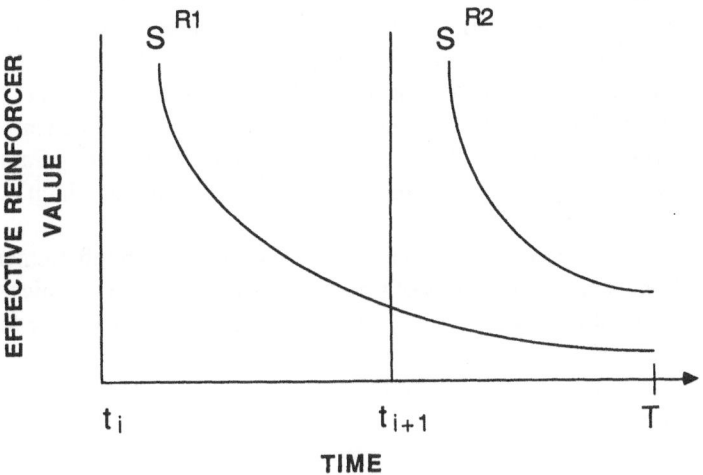

FIG. 8.2. Linear noise model. The effects of any given reinforcer decreased over time.

where n is the number of reinforcers that have been delivered up to time T, each $R_j(t_j)$ is the value of the jth reinforcer when it was delivered at time t_j, and the inverse of $(T - t_j)$ is the hyperbolic decrementation of the jth reinforcer since it was delivered at time t_j.

Because the value of each reinforcer decrements hyperbolically, the contribution of any reinforcer at the moment it is delivered is infinite. This reflects the certainty of how the subject will choose at the moment the choice is made. Immediately thereafter, the contribution begins to drop off. Because the contribution is infinite at the moment the reinforcer is delivered, its value cannot be calculated. Thus, the difference of the values of the reinforcer at the moment of delivery to its contribution to behavior one time unit later, is similar to Mazur's (1987) adding a constant.

The simplest model assumes that the value of any reinforcer when it is given $(R_1(t_i))$ is 1. Then 1 time unit after the reinforcer is given the value of the reinforcer can be written:

$$R_1(t_{i+1}) = \frac{R_1(t_i)}{(t_{i+1} - t_i)} = \frac{1}{(x + 1 - x)} = 1; \qquad (10)$$

where x represents the time that the reinforcer was delivered. Two time units away, this value would be 1/2, three time units away 1/3, and so on. The total value of several such reinforcers would be the sum of the values of each of these reinforcers at some point in time.

This method of evaluating reinforcers is the unweighted sum of decrementation, so called because the value is unweighted, or simply 1. Graphically, a possible reinforcement schedule might produce the sums shown in Figure 8.3. Here, units of time are shown along the x axis, whereas values of reinforcers are shown along the y axis. The total value of all previous reinforcement at a given point is indicated by the dotted line, which, until t_{i+1}, is coincident with the decrementation of reinforcer R_i (since that is the only reinforcer operating). The total value of reinforcement 10 time units along is the point at which the dotted line crosses the $x = 10$ line.

As can be seen by the graph in Figure 8.3, the sum of all reinforcement yields a noncontinuous curve over units of time that can be described graphically as the sum of each reinforcer graph, but cannot be accurately evaluated for any given instant of reinforcement. For this reason, the total value of reinforcement is calculated in discrete time intervals, and the value of reinforcement at the time period, for which the value is being calculated never includes the reinforcer (if any) that will be delivered on that instant, since the subject has no way of knowing whether reinforcement will occur until immediately after

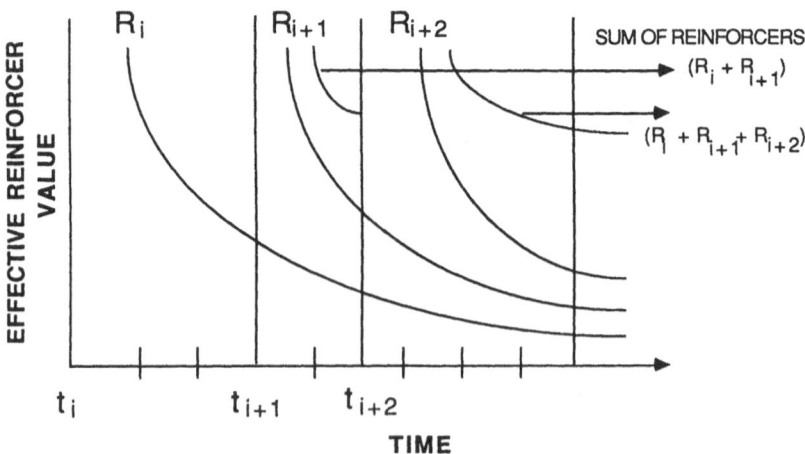

FIG. 8.3. Sample schedule for unweighted sum of decrementation.

the choice is made. This yields a smoother, discrete step function with an overall tendency of responding to increase as long as reinforcers are delivered with some regularity, but a local tendency to decrease as reinforcers decrement over discrete time intervals.

Graphically, a possible total reinforcement function is shown later for unweighted reinforcers. The bumps indicate that a reinforcer was delivered just prior to it, and give an overall increase to the function.

Note that if reinforcement were discontinued, the function would drop off, approaching but never reaching 0. Imagine a situation where the subject ceased to receive reinforcement for some type of behavior. Eventually, the subject would stop that type of behavior in favor of some type that did deliver reinforcement with some regularity, if possible. The subject would, however, always remember that the first type of behavior did at one point elicit reinforcement, and thus the memory of reinforcement would never theoretically reach 0.

The discrete time interval that was chosen for the task of calculating these curves in the pilot models we evaluated was the number of intervening events. For this experiment, this is equivalent to the number of key pecks between reinforcers. The use of this time scale solves two problems. First is the question of which time scale most accurately represents decrementation, and second is the role played by the number of intervening events from the original decrementation model. Both of these questions will be more fully addressed later.

HOW THE OCCURENCE OF REINFORCERS AFFECT RESPONSE STRENGTH

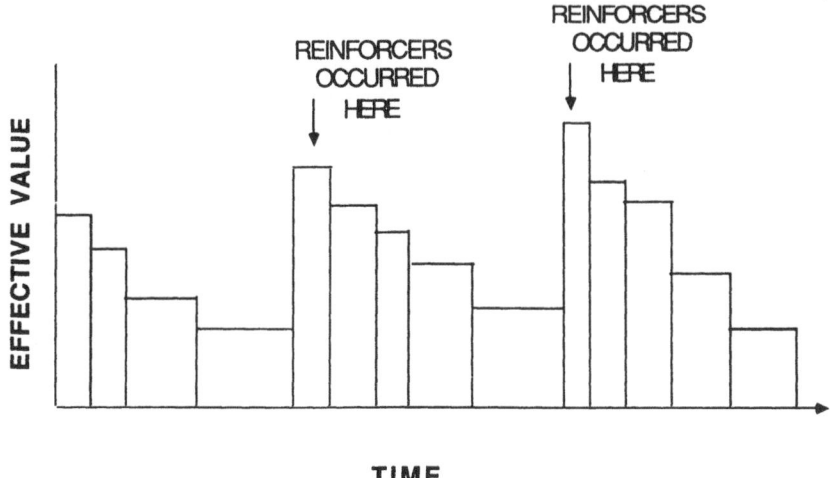

FIG. 8.4. How the occurrence of reinforcers affect response strength.

Calculating the Probability to Peck Right

The calculation of the probability to peck right does not include any term for the reinforcement schedules, their differences, ratios, or which side is rich or lean (that is, which key has the higher reinforcement schedule). All probabilities are calculated solely from the history of reinforcement that the subject has received, and expressed as· a probability that at any discrete point in time, the subject will choose the right hand key. If the right hand key carries the rich schedule, we assume by the matching law that the probability of selecting right should increase. Conversely, if the left key carries the rich schedule, we expect that the probability of selecting right should decrease.

The primary assumption behind our calculation of the probability of choosing right is that the sum of all reinforcers can be expressed as the sum of right hand reinforcers plus the sum of left hand reinforcers. This allows us to express the value of the sum of reinforcers to a key as a probability of choosing that key. If ΣR and ΣL represent the sum of reinforcers on the right and left key, respectively, the $P(R)_i$, the probability of choosing the right hand key at time i is:

$$P(R)_i = \frac{\Sigma R}{\Sigma R + \Sigma L}; \tag{11}$$

And P(L)Fi, the probability of choosing the left hand key at time i, is one minus this value:

$$P(L)_i = \frac{\Sigma L}{\Sigma R + \Sigma L} = 1 - P(R)_i; \tag{12}$$

To estimate the accuracy of these predictions, a local probability of selecting the right or left key should be compared to the results of these equations. However, there is no way to obtain the actual local probability that the subject will choose a particular key. The actual behavior of the bird on that action, a binary value of right or left, best not be regressed with the probability we propose. We estimate this probability by a ratio of the actual number of pecks to the right or left key to total pecks over the last 50 trials. Because behavior changes over time, error is introduced into this estimate the farther back in time it is calculated. However, we assume the last 50 trials do not produce too great an error, since transition sessions run with 100 trials of correction procedure and 500 trials of experimentation procedure are considered the minimum necessary to insure a change of behavior. On the other hand, at least 50 trials are required to gain an accurate probability. Changes often only occur in the hundredths place or lower.

Results

For each reinforced action, the predicted probability that the choice would be a right key peck, as described earlier in equations 2, 4, and 5 was calculated. The program also produced a probability for each of these actions, which represented the proportion of right key pecks the subject had delivered through the last fifty trials. This proportion, as described earlier, is an approximation of the subject's probability of choosing the right key on that action. To test the accuracy of the predicted probability, these two lists were regressed against each other using least-squares linear regression. Recall our general equation:

$$
\begin{array}{c}
\text{Total} \\
\text{reinforcement} \\
\text{at time T}
\end{array}
= \sum_{j=1}^{n} \frac{R_j(t_j)}{(T - t_j)}.
\tag{9}
$$

This equation can, for individual models, be varied along two different parameters. The first parameter is the calculation of $R_j(t_j)$, the value or "weight" of the reinforcer at the time it is given. Some models varied this parameters, whereas others varied the time scale, along which the reinforcers were delivered.

The first model that was tested used a time scale measured in tenths of seconds, and a weight for the reinforcer, of:

$$
R_j(t_j) = \frac{1}{t_j - (t_{j-1})}.
\tag{13}
$$

This weighting placed greater influence on reinforcers that were closer together, although it was based on the assumption that the subject was attempting to obtain as many reinforcers as possible. The subject generally responded within 0.5 seconds, rarely exceeding 1 second. Often response was within 0.1 seconds, but time between reinforcers, being the sum of several of these 'latencies', varied between 0 and 15 seconds. When this value was placed into equation 2, the resulting equation had the form:

$$
\begin{array}{c}
\text{Total} \\
\text{reinforcement} \\
\text{at time T}
\end{array}
= \sum_{j=1}^{n} \frac{1}{(T - t_j)(t_j - t_{j-1})}.
\tag{14}
$$

The linear regressions obtained using this model indicated that our predictions accounted for 3 to 50 percent of the variance of the local probabilities of pecking right. F-statistics showed the regression to be significant at values ranging from .238 to .0001. The best fit was for the schedule of .15 left to .05 right. The worst fit was for the schedule of .05 right to .15 left. This seemed to indicate that either the model was random, or that there was some hidden bias that is unaccounted for.

That the model was completely random seems unlikely, given the high significance of the regression and its coefficients (for the best fit, .0001 significance level). A discussion of the bias for a given key, a problem that occurs in several models, will be given later.

The second pilot model examined the same type of weighting as described in equation (7) but measured the temporal variables in terms of events. Thus, 'latency' (the delay between lighting the right and left keys and the bird's choice of one) was ignored, and only the trial itself was counted as a time unit. In this manner, the weight of each reinforcer was given a larger overall value. This occurred because there were, as a rule, more tenths of a second between reinforcers than there were events, and the reciprocal of these values for tenths of a second was smaller:

Weighting term for model:

$$\overset{\#1}{\frac{1}{\text{tenths of a second}}} < \overset{\#2}{\frac{1}{\text{events}}} \tag{15}$$

These regressions on the whole varied little from those based on the first model because the weighting terms differed but not significantly.

More interesting are the pilot models run without the weighting terms. These simply count each reinforcer as a '1' one time unit after they are delivered, and sum the hyperbolic decrementation of the reinforcers as described earlier. The most significant of these used events as time units, as with the second weighted model, cited earlier. For nearly all schedules, this model yielded better fits and more significant regressions R^2 values for this model range between 8% and 52%, with an average R^2 of 36%. A comparison is shown between Bailey's fits for predicting response to the rich key over time and our fits for predicting probability to peck right over local probability for the same schedules.

Due to the rapid decrementation of each reinforcer in the hyperbolic model, there is a heavy bias towards the most recent reinforcers in the estimation of the probability to peck right. For example, based on this model, a large string of right reinforcers may be nullified by a single succeeding left reinforcer, especially if there is a significant delay between these reinforcers. Subjects have a much greater tendency to discount these sparse left reinforcers, even though the model suggests that each should have as much weight as a right reinforcer. For this reason, the pilot model tends to 'overreact' to the experimental conditions, with two noticeable affects; the model tends to fluctuate over a wider band than the subject actually does, and the probabilities tend towards the extremes. This can be seen in scatterplot 1 (see Figure 8.5), where the predicted values are along the x-axis and the local probabilities are along the y-axis. Note how the data ranges from 0 to .45 with fairly normal distribution, but predictions for this band have values appearing primarily between 0 and 0.15. In this way, a variation of probabilities on the part of the subject is overemphasized by the model because the probabilities all lie below 0.5. The model gives this marginal preference undue weight, and reduces the prediction still lower to hover around 0.1.

The counterpart of this overemphasis by the model can be seen in the same graph by examining the range of the values on both axes. Whereas values of the local probabilities range from 0 to 0.45, predicted values for these probabilities, though centering at 0.1, range from 0 to 0.8.

That the model seems to emphasize the most recent reinforcers may be a

COMPARISON OF BAILEY'S ANALYSIS
WITH THE PRESENT ANALYSIS

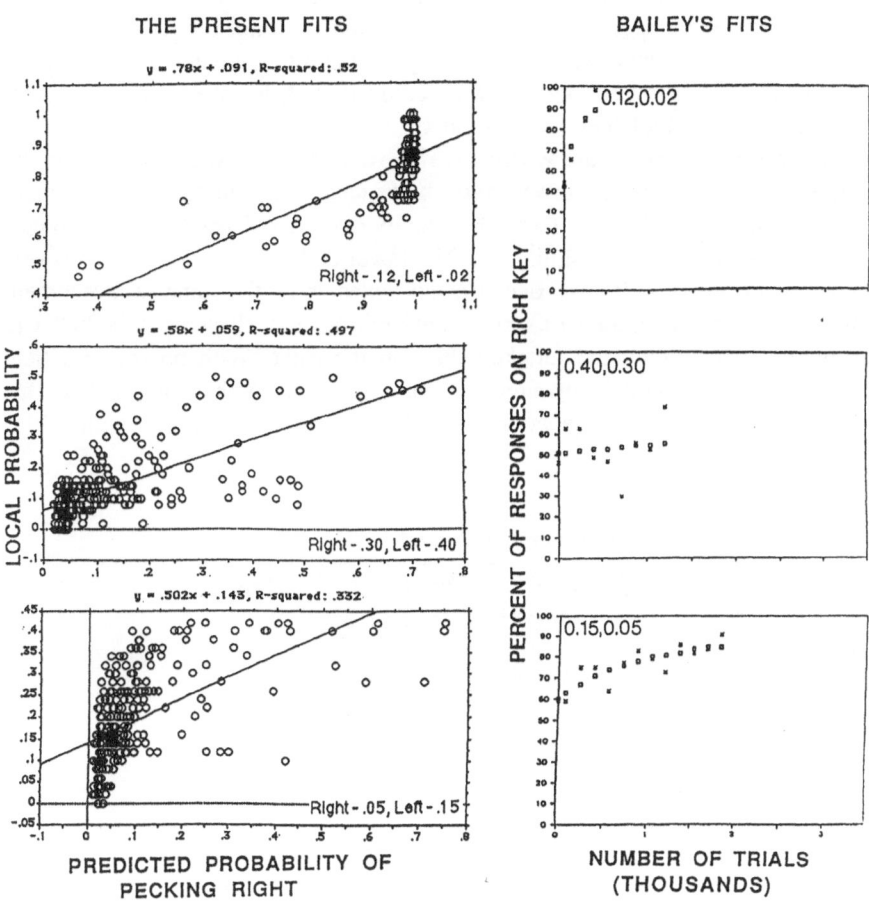

FIG. 8.5. Comparison of Bailey's analysis with present analysis.

function of the way the local probabilities are calculated. Because estimation of the local probability involves averaging over the behavior of the last fifty trials, local fluctuations in the subject's value assignments caused by recent reinforcement may not be given enough weight.

The fit was further complicated by two forms of bias—a global preference for a given key before reinforcement begins, and local patterns of behavior induced by the correction procedure. The first of these biases occurs when a tendency of the subject to prefer one key over the other is not totally eradicated by the correction procedure. The second occurs because the correction procedure normalizes behavior on a global, but not a local scale. Although the correction procedure effectively moves the subject back to evenly distributed response on both keys, the method of alternate reinforcements creates a local tendency for the pigeon to choose the key that was not reinforced in the last trial.

A χ^2 test for independence was performed on our pilot model's predictions and birds' actual performance. Right key probabilities greater than 0.5 were counted as 1. Values less than 0.5 were counted as 0. Right key response were counted as a 1 and left key response as 0.

The top table shows data where the rich reinforcement schedule was on the right key (N=269) and the lower table shows data where the rich reinforcement schedule was on the left key (N=309). No evidence for dependence was found, $\chi^2(1) = 0.84$ and was not significant. The lower table, was significant however, $\chi^2 (1) = 11.53$, p < 0.001. Because the tables show the same reinforcement probabilities, but with their rich keys reversed, our model seems to be better at predicting behavior on the left key than on the right. With no mathematical difference in our model between left and right, we might suppose that this anomaly is due to the subjects entering the experimental sessions with a bias

TABLE 8.1
Fit of the unweighted model to data

SUCCESS OF FIT OF THE UNWEIGHTED MODEL

Schedule with rich key on right
(right = .15, left = .05)

Actual Choice

Predicted Choice	1	0
≥.5	6	13
<.5	56	194

Schedule with rich key on left
(right = .05, left = .15)

Actual Choice

Predicted Choice	1	0
≥.5	249	37
<.5	14	9

toward the left key. In the upper table, where the rich schedule was on the right key, the majority of successful predictions were correct rejections. On the lower table, which displays predictions of behavior when the left key is rich, the majority of successful predictions were hits.

Conclusions

This investigation addressed questions on acquisition of preference. A proposed reduction of operant to respondent conditioning was extend to the two key preference situation. The Linear Noise Model was then combined with it to produce a pilot models explored here. The unique characteristics of these models included trial by trial analysis of the probability to peck a given key, analysis from the subject's view point and prediction of the future probability of pecking a given key.

ACKNOWLEDGMENTS

The data was collected by John Bailey for his undergraduate thesis at Harvard. He and James Mazur have submitted a report of the data to the Journal of the Experimental Analysis of Behavior. A condensed version of their method section is reproduced here with their permission. They also supplied the data analyzed here. We thank William Reynolds and Jared Jenisch for their editorial comments.

REFERENCES

Bailey, J. T. (1988). *Factors Affecting Pigeons' Development of Preference for the Better of Two Alternatives*. Harvard Honors Thesis.

Bailey, J. T., & Mazur, J. E. (submitted). Choice behavior in transition: Development of preference for higher probability of reinforcement.

Commons, M. L., & Hallinan, P. W. with Fong, W., & McCarthy, K. (in press). Intelligent pattern recognition: Hierarchical organization of concepts. In M. L. Commons, R. J. Herrnstein, S. M. Kosslyn, & D. B. Mumford (Eds.), *Models of behavior, 9, Computational and clinical approaches to pattern recognition and concept formation*. Hillsdale, NJ: Lawrence Erlbaum Associates.

Commons, M. L., Herrnstein, R. J., & Rachlin, H. (Eds.). (1982). *Quantitative analyses of behavior: Vol. 2, Matching and maximizing accounts*. Cambridge, MA: Ballinger.

Commons, M. L., Mazur, J. E., Nevin, J. A., & Rachlin, H. (1987). *Quantitative analyses of behavior: Vol. 5, Effect of delay and intervening events on value*. Hillsdale, NJ: Lawrence Erlbaum Associates.

Commons, M. L., Woodford, M., Boitano, G. A., Ducheny, J. R., & Peck, J. R. (1982). Acquisition of preference during shifts between terminal links in concurrent chain schedules. In M. L. Commons, R. J. Herrnstein, & A. R. Wagner (Eds.), *Quantitative analyses of behavior: Vol. 3, Acquisition* (pp. 391–426). Cambridge, MA: Ballinger.

Commons, M. L., Woodford, M., & Ducheny, J. R. (1982). How reinforcers are aggregated in reinforcement-density discrimination and preference experiments. In M. L. Commons, R. J.

Herrnstein, & H. Rachlin (Eds.), *Quantitative analyses of behavior: Vol. 2, Matching and maximizing accounts* (pp. 25–78). Cambridge, MA: Ballinger.

Commons, M. L., Woodford, M., & Trudeau, E. J. (in press). How each reinforcer contributes to value: "Noise" must reduce reinforcer value hyperbolically. In M. L. Commons, M. C. Davison, & J. A. Nevin (Eds.). *Models of behavior: Signal detection, 11*. Hillsdale, NJ: Lawrence Erlbaum Associates.

Dreyfus, L. R. (1985, March). *Choice in mixed concurrent schedules*. Paper presented at the annual meeting of the Eastern Psychological Association at Boston.

Dreyfus, L. R. (1985, April). *Reinforcement transitions and absolute reinforcement rate in concurrent schedules*. Paper presented at the annual meeting of the Southwest Psychological Association, Austin, TX.

Estes, W. K. (1959). The statistical approach to learning theory. In S. Koch (Ed.), *Psychology: A study of science* (Vol. 2.) New York: McGraw-Hill.

Estes, W. K. (1969). Outline of a theory of punishment. In B. A. Campbell & R. M. Church (Eds.), *Punishment and aversive behavior*. New York: Appleton-Century-Crofts.

Grossberg, S. (1971a). On the dynamics of operant conditioning. *Journal of Theoretical Biology, 33*, 225–255.

Grossberg, S. (1971b). Pavlovian pattern learning by nonlinear neural networks. *Proceedings of the National Academy of Science, 68*, 828–831.

Grossberg, S. (1974). Classical and instrument learning by neural networks. In R. Rosen & F. Snell (Eds.), *Progress in theoretical biology* (Vol. 3). New York: Academic Press.

Grossberg, S. (1980). How does the brain build a cognitive code? *Psychological Review, 87*, 1–51.

Grossberg, S. (1982a). A Psychophysiological theory of reinforcement, drive, motivation, and attention. *Journal of Theoretical Neurobiology, 1*, 286–369.

Grossberg, S. (1982b). *Studies of mind and brain: Neural principles of learning, perception, development, cognition, and motor control*. Boston: Reidel Press.

Grossberg, S. (Ed.). (1987). *The adaptive brain*, Volume 1. Amsterdam: Elsevier Science Publishers.

Guthrie, E. R. (1952). *The psychology of learning (rev. ed.)*. New York: Harper and Row.

Herrnstein, R. J. (1982). Melioration as behavioral dynamism. In M. L. Commons, R. J. Herrnstein, & H. Rachlin (Eds.), *Quantitative analyses of behavior: Vol. 2. Matching and maximizing accounts* (pp. 433–458). Cambridge, MA: Ballinger.

Herrnstein, R. J., & Vaughan, W., Jr. (1980). Melioration and behavioral allocation. In J. E. R. Staddon (Ed.), *Limits to action: The allocation of individual behavior* (pp 143–176). New York: Academic Press.

Hull, C. L. (1952). *A behavior system: An introduction to behavior theory concerning the individual organism*. New Haven, CT: Yale University Press.

Mackintosh, N. J. (1974). *The psychology of animal learning*. New York: Academic Press.

Mazur, J. E. (1987). An adjusting procedure for studying delayed reinforcement. In M. L. Commons, J. E. Mazur, J. A. Nevin, & H. Rachlin (Eds.), *Quantitative analyses of behavior: Vol. 5. The effect of delay and of intervening events on reinforcement value*. Hillsdale, NJ: Lawrence Erlbaum Associates.

Myerson, J., & Hale, S. (1988). Choice in Transition: A comparison of melioration and the kinetic model. *Journal of the Experimental Analysis of Behavior, 49*, 291–302.

Myerson, J., & Mizzen, F. M. (1980). The kinetics of choice: An operant systems analysis. *Psychological Review, 87*, 160–174.

Rescorla, R. A., & Wagner, A. R. (1972). A theory of Pavlovian conditioning: variations in the effectiveness of reinforcement and non-reinforcement. In A. H. Black & W. F. Prokasy (Eds.), *Classical conditioning* (Vol. 2). New York: Appleton-Century-Crofts.

Reynolds, G. S. (1968). *A primer of operant conditioning*. Glenview, IL: Scott, Foresman.

Schoenfeld, W. N., & Cole, B. K. (1972). *Stimulus schedules: The T-V systems*. New York: Harper and Row.

Thorndike, E. L. (1898). Animal intelligence: An experimental study of the associative process in animals. *Psychological Review Monographs, Supplement* 2(8), 28–31.

Thorndike, E. L. (1911), *Animal intelligence*. New York: Macmillan.

Tolman, E. C. (1932). *Purposive behavior in animals and man*. New York: Century (Appleton-Century-Crofts).

Vaughan, W., Jr. (1981). Melioration, matching, and maximization. *Journal of the Experimental Analysis of Behavior, 36*, 141–149.

Vaughan, W., Jr. (1985). A local analysis. *Journal of the Experimental Analysis of Behavior, 36*, 141–149 *43*, 383–405.

Vaughan, W., Jr., & Herrnstein, R. J. (1987). Stability, melioration, and natural selection. In L. Green & J. H. Kagel (Eds.), *Advances in behavioral economics* (Vol. 1, pp. 185–215). Norwood, NJ: Ablex.

9 A Connectionist Model of Timing

Russell M. Church
Hilary A. Broadbent
Brown University

ABSTRACT

An information-processing version of Scalar Timing Theory has been useful for explaining the ability of animals to estimate the duration of an interval. Unfortunately, this version contains some cognitive activities that are difficult to duplicate with known biological mechanisms. A new connectionist version of Scalar Timing Theory is now under development that requires only the standard assumptions about the operations of neural networks, and deals with many of the facts of duration discrimination. This connectionist version provides a good quantitative fit to data, whereas several modifications of it provide only a qualitative fit or fail to fit the data at all. Thus, quantitative fits to data are shown to constrain both the input representation and the architecture of a connectionist model of animal timing.

Two approaches to the study of animal timing have emerged quite independently of each other. They both use the metaphor of the internal clock; they both are concerned with biological and experiential determinants of timing; and they both involve mathematical models of the process. One of these approaches is primarily concerned with periodic timing and the other with interval timing. The properties of the internal clock that are assumed, the variables affecting timing, and the nature of the mathematical models of the process are quite different. Periodic timing has been investigated mostly by biologists whereas interval timing has been investigated mostly by experimental psychologists. Not many investigators of one type of timing are familiar with the detailed results of the other type of timing, and the amount of cross-referencing to the recent empirical literature is limited.

Periodic Timing

Periodic timing refers to the ability of animals to respond in a consistent way at particular times. In some cases, such periodic behavior persists in free-running conditions, that is, in the absence of any obvious external periodic cue. The most thoroughly studied of periodic behavior is the circadian rhythm, but there are many others both shorter and longer (Aschoff, 1981). For example, the fiddler crab *Uca minax* shows a circa-tidal rhythm in activity that persists in the laboratory in the absence of tidal cues and free-running digging activity of the ant lion, *Myrmeleon obscurus,* varies according to the lunar day (Neumann, 1981). Circannual rhythms that have been studied under free-running conditions include hibernation in ground squirrels, migratory restlessness of warblers, and reproductive behavior of domestic sheep (Gwinner, 1981). Periods shorter than circadian (ultradian) have been reported for core body temperature of human subjects in free-running conditions (Moore-Ede & Sulzman, 1981), and blood plasma cortisol levels are known to exhibit ultradian rhythms in a variety of mammals (Moore-Ede, Sulzman, & Fuller, 1982).

Consistent periodic behavior in the absence of periodic external stimuli has usually been attributed to endogenous periodic clocks (oscillators). Each of them may be characterized as having a regular oscillation of some relatively fixed period and the state of the clock can be described in terms of its phase (in degrees). Slower oscillating processes can be created by the beating of faster ones (Miall, 1989). Alternatively, consistent periodic behavior may be attributed to endogenous interval clocks. Each of these may be characterized by an accumulation process followed at some relatively fixed time by a rapid reset so that the state of the clock can be described by the amount of accumulation. In either case, stimuli external to the animal may serve to entrain the clock, and thus entrain the periodic behavior, but only within narrow entrainment limits.

If an animal could very accurately read the phase of any single periodic oscillator, such as the one associated with a circadian rhythm, it could learn to make a response at any particular phase. For example, a bee could learn to go to a source of nutrient at particular times of the day (Moore, Siegfried, Wilson, & Rankin, 1989). There are two ways an animal could achieve precision in a task of this sort. If a slow oscillator, such as the one associated with a circadian rhythm, is used, then there can be only a slight amount of variability in the phase and angular velocity of that oscillator. This situation is analogous to using a clock that has a single hand that sweeps once around the dial every 24 hours. To use such a clock to identify a particular time of day with reasonable precision, one must be able to read the position of the hand with great accuracy. Alternatively, an animal might have several oscillators available for timing, each with a different period. This situation is analogous to a more conventional clock that has several hands that sweep the dial at different rates (hours, minutes, seconds, etc.) Using a clock of this sort, one can glance at the positions of the various hands to get a reasonably precise reading of the time, without requiring so much accuracy in the position of the individual clock hands. Either mechanism is adequate for discrimination of a duration that begins at a fixed phase of an oscillation (synchronous timing), but not for the discrimination of a time that begins at some arbitrary phase of the oscillation (asynchronous timing).

Interval Timing

Interval timing refers to the ability of animals to deal with durations (Catania, 1970; Church, 1989), even when the interval begins at some arbitrary time. The interval may be the duration of a single stimulus, a response, or a reinforcement. For example, an animal can discriminate between stimuli of different lengths and it can produce responses of different durations. The interval may also be defined in terms of the time between two stimuli, two responses, or two reinforcements. Or it may be the time between a stimulus and a response, a stimulus and a reinforcement, or a response and a reinforcement. The principles of interval timing appear to be quite general and to apply to all of these cases.

The ability of animals to discriminate and produce relatively constant durations has usually been attributed to an endogenous internal clock. Such a clock may be characterized by an accumulation process (perhaps the integration of pulses from an oscillator) so that a duration is associated with a particular state of the clock, that is, the particular amount of accumulation. Such a clock, when embedded in an information processing description of a timing mechanism, accounts for many results of behavioral experiments (Church, 1984; Gibbon & Church, in press; Gibbon, Church, & Meck, 1984).

For a periodic clock to be useful for duration discrimination, either the beginning of the duration must reset the oscillators, or some differencing operation must occur between the state of the oscillators at the beginning of the duration and the state of the oscillators at end of the duration. In our simulations, we assumed that the relevant oscillators become entrained to the stimulus (i.e., that they are reset at the beginning of the duration). One purpose of this manuscript is to examine the usefulness of a periodic internal clock for duration discrimination.

The Possibility of a Connectionist Model of Timing

As described by Maki and Abunawass (chap. 10 in this volume) and Kehoe (1989), connectionist models are being used extensively in modelling associative processes and other cognitive processes in animals, and several connectionist models that can simulate some aspects of timing have been developed. These models are generally useful for periodic or interval timing, but not both, and are often restricted to particular time ranges. For example, Carpenter and Grossberg (1987) used mutually inhibitory oscillating units in a model of circadian rhythm. Although the same architecture could, of course, perform period timing in any other time range simply by changing the period of the oscillators, replication of hardware would seem to be necessary in order to provide flexibility of time range or to simulate multiple periodicities. Furthermore, the model is not intended to simulate interval timing. A model developed by Grossberg and Schmajuk (1989) to simulate interval timing in classical conditioning is similarly range-specific and cannot perform periodic timing. The temporal-difference model of Sutton and Barto (in press) captures some time-dependent aspects of classical conditioning. However, this model is specific to interval timing, and is limited to a time range determined by the rate of decay of a stimulus trace. Although all these models have value in their respective domains, a model that could perform both interval and periodic timing across a potentially unlimited range of times would be desirable.

One concern is that perhaps no connectionist model can provide an accurate fit to data. Many connectionist models have not been tested to determine whether or not they can do so. They often provide output that agrees qualitatively with established results, but a quantitative fit would provide additional constraints. The model should account for a high proportion of the variance in the data; the residuals should be unsystematic; the parameter values that reflect characteristics of the subjects should be approximately the same in different experiments; and the values of single parameters should change appropriately with variation in relevant intervening variables.

An opposite concern is that perhaps any connectionist model can provide an accurate fit to data. All such models contain computing elements with connections and modifiable connection weights, but the nature of the elements, the architecture of the system, and the rules of modification of the weights vary from application to application. A particular experimental result might be consistent with a large number of incompatible connectionist models. If many different connectionist models provide equally good fits of the data, then the behavioral data would not contribute to the selection of the appropriate connectionist model.

A connectionist model for interval timing is described in this chapter that provides a reasonable approximation to some facts of temporal generalization. Changes are then made in single features of the model, and each of them leads to failure to approximate the facts. Minor changes in a particular connectionist model can change a model that provided a reasonable approximation to the facts into one that is clearly in error. The major purpose of this manuscript is to examine the usefulness of a connectionist model for timing.

A DESCRIPTION OF A CONNECTIONIST MODEL
OF TIMING

The connectionist model shown in Figure 9.1 contains several separable modules, each of which is discussed in detail later. The oscillators provide the representation of time and the storage status indicators record the state of each oscillator. Working memory provides a temporary store of the information in the storage

A CONNECTIONIST MODEL

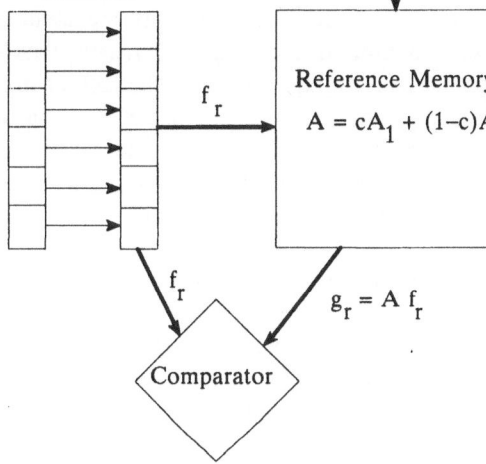

FIG. 9.1. A connectionist model for duration discrimination. From Church & Broadbent, in press.

status indicators. Information from working memory can be transferred to reference memory, a more permanent store of this information. The information in reference memory can be retrieved. For retrieval, the oscillators provide the representation of time and the retrieval status indicators record the state of each oscillator. A comparison is made between a retrieval status indicator and the output of memory, and a response occurs if the current time and the output of reference memory are sufficiently close.

The Oscillators

There are many oscillatory processes in animals, and they occur at various time periods (Gallistel, 1980). In the simulations, eleven oscillators were used with periods of .2, .4, .8, 1.6, 3.2, 6.4, 12.8, 25.6, 51.2, 102.4, and 204.8 s. The spacing of the oscillators at ratios of 2:1 probably is not an essential assumption, but it captures the idea of greater resolution at shorter periods. Greater temporal resolution would be possible with more closely spaced oscillators. With the present spacing of oscillators, 30 of them would be sufficient for the lifetime of a rat: The slowest oscillator would not complete one period for more than 3 1/2 years.

In the simulations, we assume that the oscillators are not completely accurate, but that they contain some random source of variability. The distribution of periods is normal, and the coefficient of variation of the period (the ratio of the standard deviation to the mean) is set to 0.3. On each trial, a random value from this distribution is taken to determine the periods. Thus, a particular physical time does not always produce the same oscillator values. As will be shown later, the assumption that the oscillators are not completely accurate is essential to the operation of the system. Sources of variability are assumed to affect all the oscillators in the same manner (increasing or decreasing their periods by the same percentage) so that they are tightly coupled. The lack of independence may not be an essential assumption, but it captures the idea that entrainment of one oscillator leads to entrainment of harmonics and that some factors affect the overall speed of all oscillators. Greater resolution might result if the variability of the oscillators were independent so that they would contain nonredundant information.

C. R. Gallistel (1990) has proposed that duration discrimination can be performed with oscillatory processes of this sort, and the present proposal has been influenced by his proposals for integrating periodic and interval timing. His specific proposal involves calculation of phase and phase difference. In the present version of the model, there is no assumption that the phase of an oscillator can be read with any precision. The record is simply a $+1$ or a -1 for the half-phase of an oscillator.

The Status Indicators

In the simulations, we attempt to calculate usable times without detailed phase information: only information about the half-phase is used. Each oscillator is connected to an element in a status indicator. Each element registers either a $+1$ or a -1, depending upon the half-phase of the oscillator. Thus, they form a binary counter. It should be noted, however, that the system deals with the similarities between binary representations in terms of the similarities of the bit configurations, not in terms of the magnitudes of the binary representations. In the circadian range, this would mean that this oscillator would indicate whether it was day or night, but would not represent the time of day. Some more rapid oscillator must be used to determine that food is available at various times within a day. As shown later, the representation of the oscillator half-phase as $+1$ or -1 is much more successful than a representation of $+1$ or 0. In the simulations, the status indicator used for storage, f_s, is the vector of 11 elements, plus one element that records whether or not reinforcement is delivered.

Working Memory

Working memory, A_1, is a matrix represented by the outer product of the storage vector, f_s, with itself. The auto-association matrix for the storage of information has been used extensively by Anderson and his colleagues, and the present proposal has been influenced by these applications (Anderson, Silverstein, Ritz, & Jones, 1977). We show later that with the present representation of time, the use of an auto-association representation of working memory, rather than a vector representation, is essential.

Reference Memory

Working memory, A_1, is represented as a matrix containing the most recent information; reference memory (A) is represented as a matrix containing past information. As shown later, at least with the representation of time employed in this model, reference memory, like working memory, must be represented as a matrix rather than a vector. The reference memory matrix may be considered as a set of connection weights that serves as a filter that is selective for a vector representing the time that reinforcement sometimes occurs. In the simulations, at the time of reinforcement, the contents of working memory are combined with the contents of reference memory according to a linear combination rule. The parameter c is the learning rate parameter, and in the simulations it is set to .01. The use of a linear combination rule is probably not essential, but it has been used in many learning theories (see, e.g., Bush & Mosteller, 1955; Rescorla & Wagner, 1972). The same combination is sometimes expressed as the change in A being equal to $c(A_1 - A)$. In this form, the reference memory matrix is seen to approach the working memory matrix at a rate determined by the parameter c.

This completes the analysis of the storage process. The retrieval process consists of another set of oscillators and status indicators, reference memory, and a comparison between the retrieval vector and the output vector from reference memory, as described later. The oscillator and status indicator process is the same as described for storage. The purpose of separating the two in the description of the connectionist model was to make it possible to have independent variables selectively influence storage or retrieval. There is good evidence that some drugs affecting acetylcholine systems affect storage alone, but other drugs affecting dopamine systems affect retrieval (Meck, 1983).

Comparator

Because the reference memory can be regarded as a filter that has been tuned to the time of reinforcement, an input that corresponds to the time of reinforcement will be transmitted with less distortion than others. The output of reference memory, g_r, is the inner product of the memory matrix (A) and the retrieval vector, f_r. (For retrieval, the element that records whether or not a reinforcement was received was set to 0.) This output vector g_r can be directly compared to the retrieval vector f_r by the cosine of the angle between them. The similarity measure (s) is a value between -1 for an angle of 180 degrees (dissimilarity) to $+1$ for an angle of 0 degrees (similarity). The definition of the cosine is given by Equation 1.

$$S = f_r^T g_r / \sqrt{\left(f_r^T f_r\right)\left(g_r^T g_r\right)} \tag{1}$$

The numerator is the inner product of the two vectors to be compared. Positive values occur when the same sign occurs in the corresponding locations of the two vectors; negative values occur when the opposite sign occurs in corresponding locations of the two vectors. The denominator serves to scale the measure between $+1$ and -1.

The value of the similarity measure is compared with a threshold, and a response occurs if the similarity measure exceeds the threshold. In the simulations, the threshold was set at 0.5. As shown later, the use of a threshold rather than an average similarity measure may be necessary to achieve quantitative fits of data.

Some Assumptions Provide an Approximation to the Data

The first problem is to determine whether this connectionist model can fit any actual data. If it is able to do so, the second problem is to determine whether or not many variations on this connectionist model are equally able to fit the data.

A discrete-trial, fixed-interval schedule of reinforcement provides evidence for interval timing by animals. In such a procedure, the first response of an animal after a signal has been on for some fixed time, such as 50 s, may be followed by a pellet of food. As a consequence of such training, the mean response rate increases from signal onset until the delivery of food. Individual trials may be characterized by a break-run pattern—an initial low response rate followed by a high response rate (Schneider, 1969).

The peak procedure is a modification of a discrete-trial, fixed-interval schedule of reinforcement that provides further evidence for interval timing by animals (Catania, 1970; Roberts, 1981). In such a procedure, half the trials may be the same as those on a discrete-trial, fixed-interval schedule of reinforcement, but on the remaining trials there is no food and the signal remains on for a long time. As a consequence of training, on nonfood trials the mean response rate increases from signal onset until near the time of delivery of the food, and then it decreases in a slightly asymmetrical manner. Individual trials may be characterized by a break-run-break pattern—a low response rate followed by a high response rate and then another low response rate (Gibbon & Church, in press).

In one such experiment, 10 rats were trained on a 50-s peak procedure (Meck

& Church, 1984). During the final 20 three-hour sessions the mean response rate was calculated for the first 100 s after a houselight was turned on. This was expressed as a proportion of the maximum response for each rat. The median (across rats) at each 1-s interval is shown in Figure 9.2 (open circles).

The connectionist model fit to these data (the solid circles in Figure 9.2) was based on three parameters: the coefficient of variation of the oscillators, the response threshold, and a constant to translate the theoretical probability of response to the percentage of maximum response. On each trial a random normal deviate (d) was calculated with a mean of 1 and some fixed coefficient of variation. The objective time was transformed to subjective time by multiplying the objective time by 1-d. For training the system, 1,000 trials were used. This led to a fairly stable value for each connection weight in reference memory. Then 100 trials of testing were used, with the same fixed threshold. This, if the current subjective time was close enough to the output of reference memory (s ≥ .5), a response would be recorded. These calculations were made at 1-s intervals from 1 to 100 s and the results are shown in Figure 9.2.

An exhaustive search of the parameter space was made with the coefficient of variation from .10 to .35 (in steps of .05) and the threshold from .10 to .50 (in steps of .10). Some additional parameter values were examined, but the fits obtained with a coefficient of variation of .3 and a threshold of .5 were about the best observed.

The simulated result from the connectionist theory shown in Figure 9.2 approx-

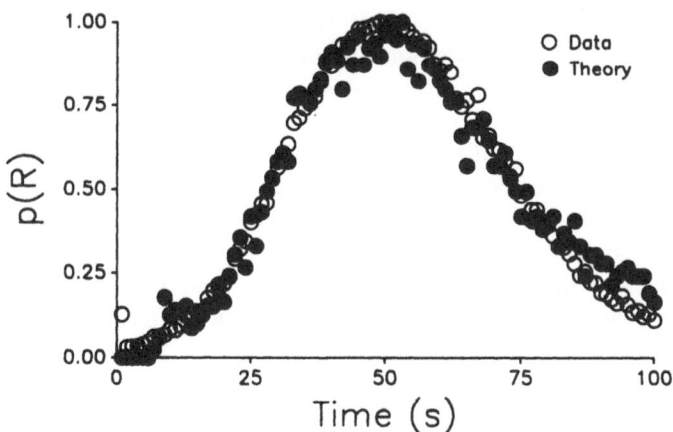

FIG. 9.2. A temporal generalization function. The response rate as a proportion of the maximum response rate (open circles) and the simulated results from the connectionist model (closed circles). The data are from Meck & Church, 1984, Figure 10. The simulations are based on 1000 trials of training and 100 trials of testing with a learning rate c = .01. The coefficient of variation was set at .3, and the threshold was set at .5.

imates the data ($\omega^2 = .988$), but there are some systematic discrepancies. In this example, and others, the theoretical curve is initially too flat and the right tail is too high. In 20 simulations the median ω^2 value was .978, and the same systematic deviations were typical.

A connectionist representation can produce results that are in qualitative agreement with three facts of timing in the peak procedure (Church & Broadbent, in press). Specifically, it produced the following results that are consistent with the data: (a) the mean probability of a response increased to a maximum near the time of reinforcement and then decreased in a slightly asymmetrical manner, (b) the probability of a response on single trials was relatively constant and low for some period, then became relatively constant and high for some period, and then again became relatively constant and low for some period, and (c) for simulations using different times of reinforcement, the mean probability of a response was about the same when plotted as a function of time relative to the time of reinforcement.

Some Assumptions Fail to Provide an Approximation to the Data

The proposed connectionist model provides a reasonable approximation to a temporal generalization function. The next problem is to determine whether or not many other connectionist models provide equally good approximations of the data.

Accurate Oscillators

If a connectionist model with an oscillator having some variability (coefficient of variation of 0.3) produces a temporal generalization gradient, one might expect that the elimination of the variability of the oscillator would produce even a sharper generalization gradient. The simulation shown in Figure 9.2 is repeated in Figure 9.3 with one exception—the coefficient of variation was 0. This means that the subjective time was equal to physical time on every trial. The results show that the simulated results always produce a response when the physical time is 50 s, but that there is no generalization function. Times near 50 s are not necessarily more similar to 50 s than times far from it. This is intuitively reasonable: a binary number is good for a precise number but it is not good as an approximation if one is counting the number of bits that are set to the same value. (For example, the binary representations of 7 and 8 have no bits in common.)

Status Indicators With +1 and 0, Rather Than +1 and −1

Each element of the status indicators (storage vector and retrieval vector) contained either a +1 or a −1. The simulation done for Figure 9.2 was redone with the values in the status indicators being recorded as +1 or 0. This change in representation may seem to be a negligible difference because two different

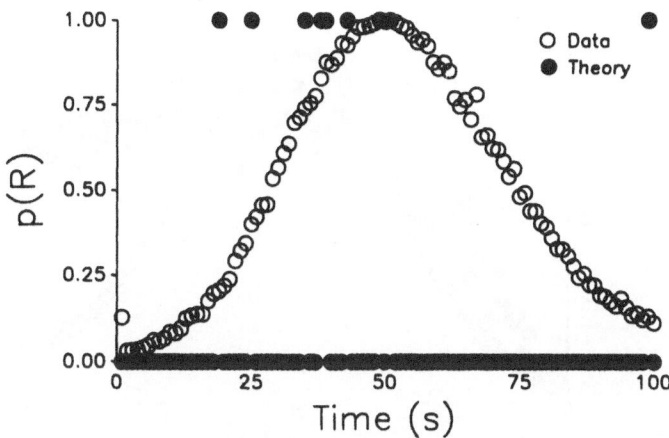

FIG. 9.3.   Accurate oscillators. The theory fails to fit the temporal generalization function when the simulation in Figure 9.2 is repeated with no
oscillator variance (coefficient of variation = 0).

values are used in each case. In fact, the difference was profound, as shown by
the simulation in Figure 9.4. With $+1$ and -1 the auto-association matrix contains
information about correlations of the sign, that is, about whether the signs of
various pairs of bits agree or not. With $+1$ and 0 the auto-association matrix
contains information about whether or not both bits in a pair were set to 1.
Thus, the auto-association matrix based on $+1$ and -1 contains "exclusive-or"
information, and the auto-association matrix based on $+1$ and 0 contains "and"
information. The correlational information in the $+1$ and -1 representation
provides a much richer and more precise representation of the times encoded in
the status indicators than does the 0 and 1 representation.

Memory as a Vector, Rather Than a Matrix

Working memory has been represented as an auto-association matrix, but all
of the information about any single time is available in the storage vector. The
simulation shown for Figure 9.2 was repeated with working memory as a vector.
This results in reference memory also being a vector. Figure 9.5 shows that this
modification failed to approximate the data. The time of the maximum response
varied from one simulation to the next when memory was a vector. Although the
response probability at long time intervals decreased, the fit was always poor.

An auto-association representation has several advantages over a vector representation. One advantage is that an auto-association matrix stores information
about the relationship between oscillators. For example, with variable oscillators,
at 50 s physical time, the oscillator for 25.6 s is set to $+1$ about half the time
(for subjective time values between 25.6 and 51.2 s), and the oscillator for 51.2

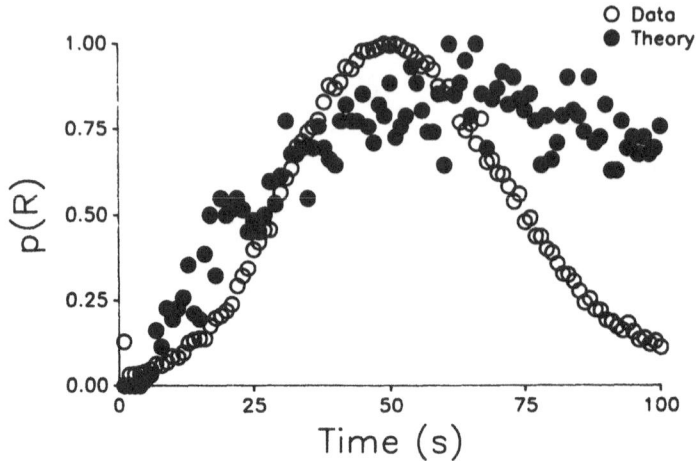

FIG. 9.4. Stimulus representation of 0 or 1. The theory fails to fit the temporal generalization function when the simulation in Figure 9.2 is repeated with status indicators containing 0 or 1, rather than − 1 or + 1.

s is also set to +1 about half the time (for subjective time values between 51.2 and 102.4 s). Because the subjective time of reinforcement is rarely less than 25.6 or greater than 102.4 s, the 25.6 and 51.2 s oscillators are seldom both set to −1 or both set to +1. Thus, the value of the element in reference memory corresponding to these two oscillators quickly approaches −1.0. Another advantage of an auto-association matrix representation is that it permits a single memory to store multiple times, although this feature is not used in the present simulations.

Comparator Decision Based Upon Average Similarity Rather Than Threshold

In all simulations thus far, the similarity of the retrieval vector (subjective time) to the output of reference memory has been compared to a threshold to determine whether or not a response occurs. Is the threshold necessary? An alternative is to average the similarity measures. Figure 9.6 shows a simulation, using the same parameters as Figure 9.2, but with mean similarity as the dependent variable rather than the mean probability that a similarity exceeds some threshold. The function rises to a maximum near to the time reinforcement is sometimes available and then it decreases. Qualitatively, it is similar to the observed temporal generalization function but quantitatively it is a poor fit. The problem is that the simulated function is highly skewed to the left and the data are slightly skewed to the right.

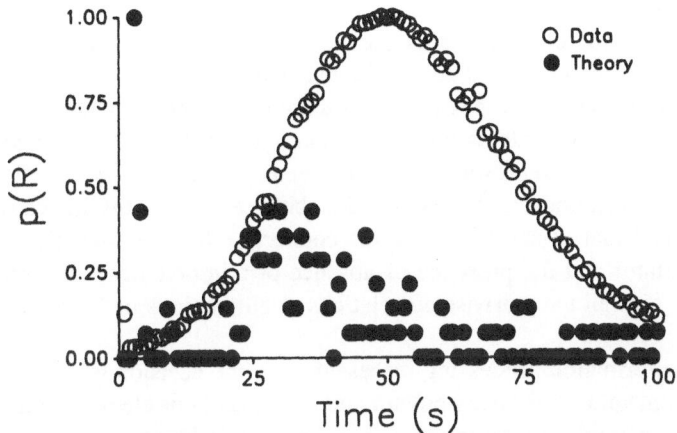

FIG. 9.5.   Memory representation as a vector. The theory fails to fit the temporal generalization function when the simulation in Figure 9.2 is repeated with working memory as a vector, rather than a matrix.

COMPARISON TO AN INFORMATION-PROCESSING MODEL OF TIMING

This connectionist model based upon oscillators can be contrasted with an information-processing model based upon an accumulator. In the information processing representation, the clock contains a pacemaker, a switch and an accumulator. Pulses from the clock may be switched into the accumulator. The reading of

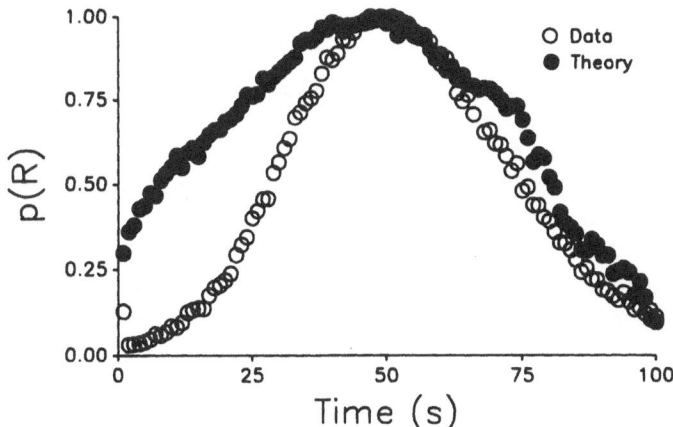

FIG. 9.6.   Decision based on average similarity. The theory fails to fit the temporal generalization function when the simulation in Figure 9.2 is repeated with decision to respond based on mean similarity, rather than on probability of similarity being greater than a threshold.

the clock is the sum of the pulses in the accumulator. In the connectionist representation, the clock contains oscillators and status indicators. The reading of the clock is the current status of the oscillators.

In the information processing representation (Gibbon, Church, & Meck, 1984), the memory contains a distribution of remembered accumulator values. In the connectionist representation, the memory is an auto-associative matrix. Each cell corresponds to the connection between the status indicator of one oscillator and that of another, or to the connection between the status indicator of an oscillator and the presence or absence of reinforcement. The matrix as a whole contains all the pairwise correlations of all the units in the status indicator vector.

In the information processing representation, the decision is based on a ratio of the current value in the accumulator and a sample of one element from memory. In the connectionist representation, the decision is based on the similarity of the retrieval status indicator to the output of the memory matrix. For both representations, the value is compared to a threshold and a response occurs if it is on the appropriate side of this threshold.

The data shown in Figure 9.2 have previously been fit by an information-processing version of scalar timing theory (Meck & Church, 1984, Fig. 10). The best fit occurred with the assumption that the probability of attention was .98, the mean threshold was .35, and the standard deviation of the threshold was .2. A fourth parameter was used to fix the response rate. The percentage of variance accounted for was higher (.998) with this information-processing model than with the present connectionist simulation, and systematic residuals were less prominent.

SUMMARY

The conclusion is that a connectionist model of timing can provide an approximation to temporal generalization functions observed in the peak procedure, and that many small variations in this model fail to do so. The major point of this chapter is that quantitative behavioral results put major constraints on the range of plausible connectionist models. To develop confidence in the model, it will be necessary to obtain quantitative fits to much more data from various timing tasks. In addition, it will be necessary to identify some unique predictions of a connectionist model and to verify them. For rapid progress, it would be particularly useful to find closed form expressions for the theory.

At present, the information processing model of scalar timing is much more advanced than the connectionist model. It accounts for more of the variance in the data and the residuals are less systematic. Closed form expressions are available and they have been successfully applied to more timing tasks. But there is good reason to pursue the analysis of connectionist models of timing because they provide a way for animals to do both interval and periodic timing with the same biological structures and functions, they provide a natural way to deal with multiple times, and the features of a connectionist model are more biologically plausible than those of the information processing model. Oscillatory processes and Hebbian synapses are biologically plausible; distributions of values and random samples from these distributions are less plausible.

ACKNOWLEDGMENTS

The authors express appreciation to Donald Blough, James Anderson, John Gibbon, and Daniel Kersten for comments and assistance in the development of ideas expressed in this article. This chapter is based on a talk given at a symposium on Neural Network Models of Conditioning and Action at Harvard University, June 3, 1989. Some of the ideas contained in this chapter were included in presentations at the Columbia University Seminar on the Psychobiology of Animal Cognition (December 8, 1988), a symposium at the meeting of the American Association for the Advancement of Science (January 19, 1989), at a symposium on cognitive aspects of stimulus control at Dalhousie University (June 13, 1989) and at the Psychonomics Society (November 19, 1989). This research was supported by a grant from the National Institute of Mental Health (RO1-MH44234).

REFERENCES

Anderson, J. A., Silverstein, J. W., Ritz, S. A., & Jones, R. S. (1977). Distinctive features, categorical perception, and probability learning: Some applications of a neural model. *Psychological Review, 84,* 413–451.

Aschoff, J. (1981). *Handbook of behavioral neurobiology: Vol. 4. Biological rhythms.* New York: Plenum.

Bush, R. R., & Mosteller, F. (1955). *Stochastic models for learning.* New York: Wiley.

Catania, A. C. (1970). Reinforcement schedules and psychophysical judgments: A study of some temporal properties of behavior. In W. N. Schoenfeld (Ed.), *The theory of reinforcement schedules* (pp. 1–42). New York: Appleton-Century-Crofts.

Carpenter, G. A., & Grossberg, S. (1987). Mammalian circadian rhythms: A neural network model. *Lectures on Mathematics in the Life Sciences, 19,* 151–203.

Church, R. M. (1989). Theories of timing behavior. In S. B. Klein & R. Mowrer (Eds.), *Contemporary learning theory,* (pp. 41–69). Hillsdale, NJ: Lawrence Erlbaum Associates.

Church, R. M. (1984). Properties of the internal clock. In J. Gibbon & L. Allan (Eds.), *Timing and time perception, Annals of the New York Academy of Sciences* (Vol. 423, pp. 566–582). New York: New York Academy of Sciences.

Church, R. M., & Broadbent, H. A. (in press). Alternative representations of time, number, and rate. *Cognition.*

Gallistel, C. R. (1980). *The organization of action: A new synthesis.* Hillsdale, NJ: Lawrence Erlbaum Associates.

Gallistel, C. R. (1990). *The organization of learning.* Cambridge, MA: MIT Press.

Gibbon, J., & Church, R. M. (in press). Representation of time. *Cognition*.

Gibbon, J., Church, R. M., & Meck, W. H. (1984). Scalar timing in memory. In J. Gibbon & L. Allan (Eds.), *Timing and time perception, Annals of the New York Academy of Sciences* (Vol. 423, pp. 52–77). New York: New York Academy of Sciences.

Grossberg, S., & Schmajuk, N. A. (1989). Neural dynamics of adaptive timing and temporal discrimination during associative learning. *Neural Networks, 2*, 151–203.

Gwinner, E. (1981). Circannual systems. In J. Aschoff (Ed.), *Handbook of behavioral neurobiology: Vol. 4. Biological rhythms* (pp. 391–410). New York: Plenum.

Kehoe, E. J. (1989). Connectionist models of conditioning: A tutorial. *Journal of the Experimental Analysis of Behavior, 52*, 427–440.

Meck, W. H. (1983). Selective adjustment of the speed of internal clock and memory storage processes. *Journal of Experimental Psychology: Animal Behavior Processes, 9*, 171–201.

Meck, W. H., & Church, R. M. (1984). Simultaneous temporal processing. *Journal of Experimental Psychology: Animal Behavior Processes, 10*, 1–29.

Miall, C. (1989). The storage of time intervals using oscillating neurons. *Neural Computation, 1*, 359–371.

Moore, D., Siegfried, D., Wilson, R., & Rankin, M. A. (1989). The influence of time of day on the foraging behavior of the honeybee, *Apis mellifera. Journal of Biological Rhythms, 4*, 305–325.

Moore-Ede, M. C., & Sulzman, F. M. (1981). Internal temporal order. In J. Aschoff (Ed.), *Handbook of behavioral neurobiology: Vol. 4. Biological rhythms* (pp. 215–242). New York: Plenum.

Moore-Ede, M. C., Sulzman, F. M., & Fuller, C. A. (1982). *The clocks that time us: Physiology of the circadian timing system*. Cambridge, MA: Harvard University Press.

Neumann, D. (1981). Tidal and lunar rhythms. In J. Aschoff (Ed.), *Handbook of behavioral neurobiology: Vol. 4. Biological rhythms* (pp. 351–380). New York: Plenum.

Rescorla, R. A., & Wagner, A. R. (1972). A theory of Pavlovian conditioning: Variations in the effectiveness of reinforcement and non-reinforcement. In A. H. Black, & W. F. Prokasy (Eds.), *Classical conditioning II: Current research and theory*. New York: Appleton-Century-Crofts.

Roberts, S. (1981). Isolation of an internal clock. *Journal of Experimental Psychology: Animal Behavior Processes, 7*, 242–268.

Schneider, B. A. (1969). A two-state analysis of fixed-interval responding in the pigeon. *Journal of the Experimental Analysis of Behavior, 12*, 677–687.

Sutton, R. S., & Barto, A. G. (in press). Time-derivative models of Pavlovian reinforcement. In J. W. Moore & M. Gabriel (Eds.), *Learning and computational neuroscience*. Cambridge, MA: MIT Press.

10 A Connectionist Approach to Conditional Discriminations: Learning, Short-term Memory, and Attention

William S. Maki
Department of Psychology
North Dakota State University

Adel M. Abunawass
Department of Computer Science
North Dakota State University

ABSTRACT

Our investigations occupy a niche situated between computational models of neurobiology of conditioning and parallel distributed processing models of human learning and cognition. We have been studying an adaptive network model of a complex discrimination (matching-to-sample, or MTS). MTS has been used for years to examine a variety of cognitive processes in animal behavior. That research could provide a valuable source of constraints on connectionist models. MTS reduces to the exclusive or problem (XOR) and, like the XOR, is learned by a multilayer network using an error back-propagation algorithm (the "generalized delta rule"). The network model exhibits some interesting effects that are shown by real organisms. For example, the model can code sample stimuli in terms of anticipated events ("prospective coding"), the network's performance of delayed MTS improves with delay training ("rehearsal"), and its matching accuracy is impaired when compound samples are presented ("shared attention"). Details of these and related simulations provide the grounds for a critical evaluation of the strengths and weaknesses of the model.

The kinds of models represented in this volume, variously referred to as neural networks, parallel distributed processing (PDP), or connectionist models, have received much attention in the past several years, but connectionism has longer history in psychology (for an entertaining summary, see Tolman 1948). The failure of simple stimulus-response (S-R) associations to account for the complexity of animal learning led to the postulation of associative mediators, coding responses, stimulus analyzers, and expectancies. All of these ideas represented attempts to make connectionist models exhibit flexible information processing.

241

Connection models of the simple S-R kind, in modern terms, are known as single-layer networks. Soon after such networks were introduced in the form of Rosenblatt's (1958) "perceptrons," it became apparent that single layer networks were limited in terms of what they could compute. In contemporary *multi* layer networks, "hidden" units intervene between S and R. As shall be explained later in this chapter, the added representational capability lets a multilayer network compute interesting functions that are analogous to complex discrimination learning problems in experimental behavioral research. Thus, enabled by advances in computer science, research on neural networks offers the real possibility of connectionist models that capture the flexibility and complexity of learning and cognition.

Special interest in connectionist models has been revived in two rather disparate areas—the neurobiology of conditioning and cognitive psychology. The first of these areas is well represented by other chapters in this volume, in which simulated neural networks have been used to model the formation of associations by classical conditioning and its neural substrates (see chapters in this volume by Moore and Alkon, Vogl, Blackwell, & Tam). In the second area, cognitive psychology, network models have become so popular that connectionism has been suggested to be a paradigm shift (Schneider, 1987). Many examples of connectionist models of cognition can be found in the popular PDP series (McClelland, Rumelhart, & the PDP research group, 1986).

Our own interests are more behavioral in orientation and our approach is more neurally inspired than neurobiologically faithful. We do not claim to model the brain but rather we aim to use a connectionist model to help organize an existing body of empirical results. We have focussed on the considerable literature on animal learning and performance of complex discriminations that has been developed in the last 20 years. These tasks are both logically and methodologically more complex than simple classical conditioning; the logical structure is conditional in nature, durations of trials are extended, and the tasks can be seen as mixture of classical and instrumental conditioning. Moreover, the level of theoretical discourse lies midway between that of classical conditioning and that of human cognition.

But that discourse is not always coherent. The class of tasks being considered have been used to investigate many kinds of cognitive processes in animals. Formation of outcome expectancies (Peterson, 1984), interference with short-term memory (Maki, Moe, & Bierley, 1977), control of rehearsal (Maki, 1981), sharing of attention among stimuli (Maki & Leuin, 1972), and use of coding strategies (Zentall, Jagielo, Jackson-Smith, & Urcuioli, 1987) have all been studied using the conditional discrimination known as matching-to-sample (MTS). Despite the common task and training procedures, there is no common theory other than some loose commitment to "animal cognition" (see chapters in Roitblat, Bever, & Terrace, 1984). Many of the experiments using MTS and related tasks can be seen as manipulations of relationships between input and output patterns designed to reveal the internal representations of those associative mappings. Simulated neural networks called pattern associators easily learn many such mappings, so it appeared to us that a connectionist model of conditional discriminations would be possible. Such a model might provide the needed theoretical integration and might even serve as a conceptual bridge between neural network models of animal conditioning (Kehoe, 1988; Klopf, 1987; Sutton & Barto, in press) and PDP models of human cognition (McClelland et al., 1986).

MATCHING-TO-SAMPLE: A BEHAVIORAL XOR

In the simplest MTS problem, the animal is presented with one of two stimuli (the "sample stimulus") and later presented with both of the stimuli (the "comparison stimuli"). The animal's task is to pick the comparison stimulus that matches the preceding sample. Sample identities, comparison positions, and trial configurations are all randomized and balanced so the problem cannot be solved in any simple way short of learning the conditional relations. A flowchart of the MTS task is contained in Figure 10.1. Note that samples and their associated comparison stimuli need not be physically identical so long as the reinforcement contingencies depicted in Figure 10.1 are followed.

MTS is particularly interesting and especially relevant to the study of adaptive

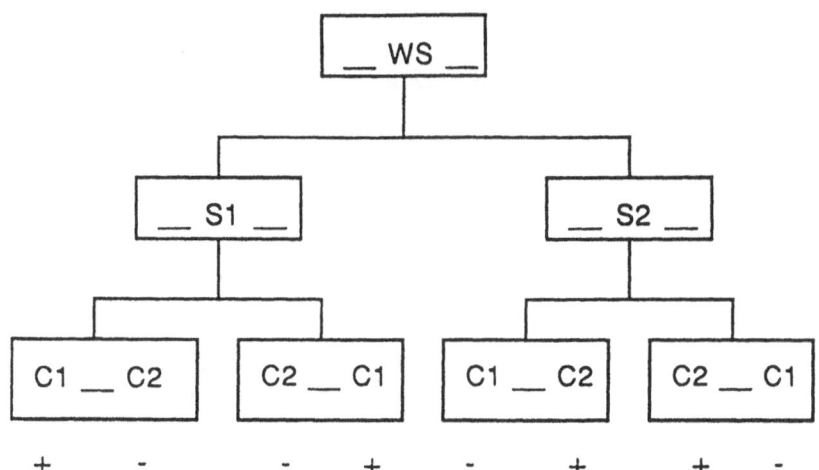

FIG. 10.1.   A flow diagram of the matching-to-sample problem. Each trial begins with a warning signal (WS). One of two sample stimuli are presented (S1 or S2). Then two comparison stimuli (C1 and C2) are presented. The plus sign marks the position of the correct (rewarded) choice. Samples and comparisons appear together when where is no delay. The four possible configurations of samples and comparisons are randomized across trials. In an example case, the stimuli 1 and 2 could be the colors red and green.

networks because the MTS task can be represented as a collection of exclusive or (XOR) functions (Maki, 1988). In the XOR, different output states are assumed depending on whether the two inputs are the same or different: 00 → 0 and 11 → 0, but 01 → 1 and 10 → 1. Table 10.1 contains logical representations of both the MTS task and the XOR. The upper part of the table describes a simple version of MTS containing two stimuli (the colors red and green) and two responses (choosing the left or the right comparison stimulus). The bottom part of the table shows only one of the XORs embedded within MTS. In general, either sample stimulus and any comparison-location combination make the legs of a gate, and the desired output from the gate is defined by one of the response positions. The result is either an XOR or an inverse XOR.

A CONNECTIONIST MODEL OF MTS

Any network that can learn the XOR problem is a candidate for a model of MTS. To build a connectionist model, we need to specify the structure of such a network and also the mathematical functions that govern learning and performance.

Representing MTS in a Network Model

The realization that MTS can be reduced to the XOR is important for selecting a network for modelling MTS. To see why, we need to consider the linear separability of classification problems and will do so by using examples from classical conditioning. Assume two conditioned stimuli (CSs), A and B, a single

TABLE 10.1
Matching-to-Sample as XOR

MTS:	Left key		Center Key		Right key		Response	
	Red	Green	Red	Green	Red	Green	Left	Right
	1	0	1	0	0	1	1	0
	0	1	1	0	1	0	0	1
	1	0	0	1	0	1	0	1
	0	1	0	1	1	0	1	0
XOR:			Input #1	Input #2				Output
			0	0				0
			0	1				1
			1	0				1
			1	1				0

Note. Presence of a stimulus or response is represented by "1" and absence of a stimulus or response is represented by "0". The matching-to-sample (MTS) problem is composed of exclusive or (XOR) and inverse XOR functions. Any center key, side key, and response define the two inputs and output of an XOR or an inverse XOR. One example is shown in the table.

unconditioned stimulus (US), and some desired conditioned reflex (R). Discrimination training of the form A+AB− makes B a conditioned inhibitor; A alone will evoke R but the compound AB will not. The left panel in Table 10.2 summarizes the relationships involved in conditioned inhibition training. Another kind of discrimination training, summarized in the middle panel of Table 10.2, is called negative patterning. A when presented alone signals the US as does B when presented alone. When presented simultaneously, however, the compound AB signals the absence of the US. (And, of course, when neither CS is presented, the US does not occur.) The traditional approach to explaining the acquisition of negative patterning has been the postulation of configural cues that emerge from the combination of elemental CSs (Rescorla & Wagner, 1972). Adding a unique cue to the compound stimulus amounts to adding another dimension (C) to the input as is done in the right panel of Table 10.2.

For a classification problem to be linearly separable, there must exist some combination of weights (coefficients for columns A and B in Table 10.2) that allows us to divide the feature space defined by dimensions A and B into two distinct regions separated by a straight line. We can do this for the conditioned inhibition problem if we assume a threshold such that R is only positive when the weighted combination of A and B is positive. In this case, the weights +1 for A and −1 for B will do. No such combination of weights exists for negative patterning, so the classification problem known as negative patterning is not linearly separable. Adding the configural cue (dimension C in the right panel of Table 10.2), however, turns negative patterning into a linearly separable problem (for example, with weights of +1, +1, and −2 for A, B, and C, respectively).

Adding extra inputs to a problem is a familiar way to achieve linear separability thus enabling nonlinear problems like the XOR to be learned by single-layer networks such as the Rescorla-Wagner model (Gluck & Bower, 1988). However, postulating cues unique to compound stimuli is not the only solution. Kehoe (1988), for example, noticed the correspondence between negative patterning

TABLE 10.2
Linear Separability in Classical Conditioning

Conditioned inhibition			Negative patterning			Negative patterning (with configure)			
A	B	R	A	B	R	A	B	C	R
0	0	0	0	0	0	0	0	0	0
0	1	0	0	1	1	0	1	0	1
1	0	1	1	0	1	1	0	0	1
1	1	0	1	1	0	1	1	1	1

Note. The binary numbers 1 and 0 represent the presence or absence (respectively) of conditioned stimuli (A and B) and a response (R). Conditioned inhibition is linearly separable, but negative patterning is not linearly separable. The addition of a configural cue (C) makes negative patterning linearly separable.

(Table 10.2) and the XOR (Table 10.1). He reported a layered network model of conditioning, in which a network learned the negative patterning problem (XOR) with the aid of hidden units. Multilayer networks learn many kinds of nonlinear problems like the XOR that single-layer networks cannot learn (Hanson & Burr, in press; cf. Minsky & Papert, 1988; Rumelhart, Hinton, & Williams, 1986). The multilayer network approach may be preferable to the unique stimulus approach for at least one reason. As Kehoe noted, increasing the number of CSs in a compound produces a combinatorial explosion of the number of unique cues that would be needed to represent the various combinations of CSs. Multilayer networks avoid this problem by learning internal representations of complex stimuli.

For present purposes, we assume a layered, feed-forward network consisting of input units, hidden units, and output units. Units in adjacent layers are assumed to be fully connected; each input unit feeds every hidden unit, and each hidden unit feeds every output unit. A minimal network is diagrammed in Figure 10.2. The functions of the input and output units should be fairly obvious. The external environment is represented by activation patterns across the input units, and actions are effected via output units. The hidden units allow for the development of internal representations that are required for coding nonlinear problems like the XOR. The specialized "teaching" inputs are used to instruct output units as to the desired response ("supervised learning").

A more complex network is presented in Figure 10.3. There we show how particular stimuli and responses in Table 10.1 get represented in the model. Two input units code the two stimuli (red and green) at each of three spatial locations (e.g., the left, center, and right keys in a pigeon operant conditioning chamber).

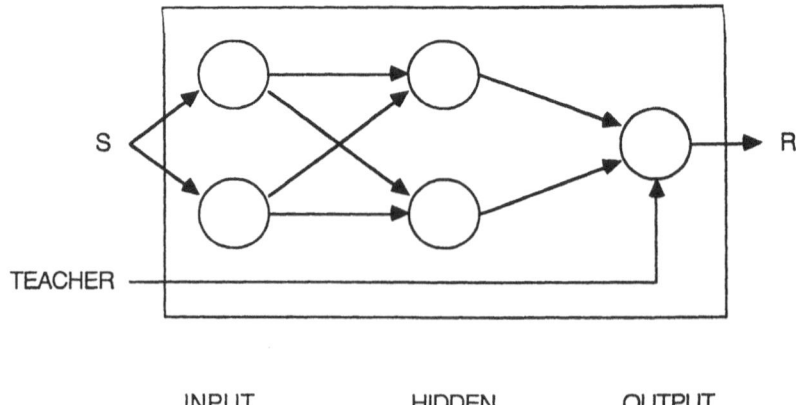

FIG. 10.2.   A multilayer network consisting of input, hidden, and output units. External environmental events (S) activate the input units and the output unit effects actions (R). The "teacher" is used during supervised learning to inform the output unit of the desired response.

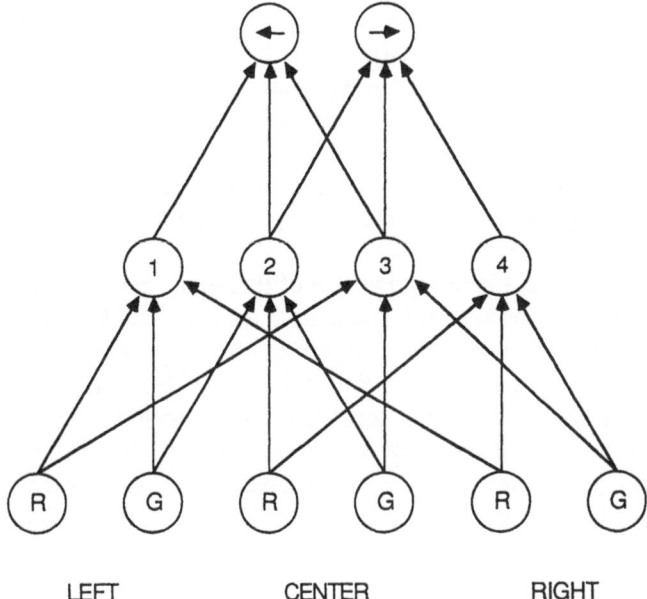

FIG. 10.3. An example of a network that learns matching-to-sample. The stimuli are the two colors red (R) and green (G). The sample stimuli are presented at the center location and the comparison stimuli are presented at the left and right locations. The network contains four hidden units (1–4) and two output units that code left (←) and right (→) responses. Not all connections are shown; in the simulations, adjacent layers were fully interconnected.

Similarly, two output units code the two responses. Each pattern in the training set thus consists of one line from Table 10.1; the first six bits set the activation levels of the input units and the last two bits serve as the teaching input indicating the correct response.

Backpropagation of Errors: The Generalized Delta Rule

Although backpropagation was contemplated by Rosenblatt (1962, p. 292) and discovered independently by other investigators (see Parker, 1987, and Werbos, 1989), its current popularity as a neural network training method is due to its derivation by Rumelhart et al., (1986). One of the attractions of the backpropagation algorithm is its conceptual simplicity. Consider again the small network diagrammed in Figure 10.2. A stimulus pattern is represented by activation of the input units. The activation values are multiplied by the weights of the connections to each hidden unit. The weighted activations are accumulated by each hidden unit, and the net input to the unit is transformed by an activation function to produce the hidden unit's activation level. The process is then repeated from the hidden units to the output units. The activation levels of output units are compared to the desired output (provided by the teacher); the resulting error signals from output units are then accumulated by hidden units. Finally, the output unit errors are used to change the weights on connections to output units from the hidden units, and the hidden unit errors are used to change the weights on connections to hidden units from input units. The same process is repeated for every input-output pattern in the training set until the output unit errors have reached acceptably small values.

The notation we use here follows that of Rumelhart et al. (1986), except that we consistently let subscripts i, j, and k index the input, hidden, and output units, respectively. For example, activation values of hidden units are symbolized by o_j and weights on connections between input and hidden units are symbolized as w_{ij}. Each simulated training trial consists of two computational "passes." In the first (forward) pass, unit activation values are computed. The activations of units in the input layer are determined environmentally (i.e., set by each input pattern in the simulation). The net input to each hidden unit j is determined by the activities of the input units i that feed unit j and the weights on each connection:

$$net_j = \sum w_{ij} o_i. \tag{1}$$

The net input is transformed by a nonlinear activation function (a sigmoid) to produce the activation value of unit j:

$$o_j = \frac{1}{1 + e^{-(net_j + \theta)}} \tag{2}$$

that takes on values between 0 and 1. A bias factor, θ, determines the operating range, over which the unit reports changes in net_j; each unit except for input units has an associated bias. Equations 1 and 2, with different subscripts, are used to compute the activation values for output units as well. Activation values are computed beginning with the input layer and ending with the output layer. This feed-forward computation is used to determine the response of the network to any arbitrary input pattern.

The second set of computations (the backward pass) performed during each training trial effects the backpropagation of error signals. In the kind of training regime considered here, each input pattern is associated with an expected output pattern. The expected output pattern is the source of "teaching inputs," t_k, presented to output units. The error signal for each output unit is simply the difference between the desired output value and the actual value computed during the forward pass, $t_k - o_k$, multiplied by the derivative of the activation function, o_j $(1 - o_j)$. So, the error signal, δ_k, for an output unit is

$$\delta_k = (t_k - o_k) o_k (1 - o_k) \tag{3}$$

Error signals for units in earlier hidden layers are computed recursively. For any unit j, the error signals of those units fed by unit j are multiplied by the weights on the appropriate connections; the sum of these products is then multiplied by the derivative of the activation function for unit j. For example, in our three-layer network, the error signal for each unit in the hidden layer is given by

$$\delta_j = (\Sigma \delta_k w_{jk}).o_j(1 - o_j) \tag{4}$$

Each weight (e.g., between units j and k) is then updated by the "delta rule,"

$$\Delta w_{jk}(t) = \eta \delta_k o_j + \alpha \Delta w_{jk}(t-1), \tag{5}$$

and

$$w_{jk}(t+1) = w_{jk}(t) + \Delta w_{jk}(t), \tag{6}$$

where η is a learning rate, α is a momentum constant that lets the current weight change be influenced by previous weight changes and, t refers to the ordinal number of the current training trial. Biases are learned like any other weights, but in the case of biases, the input activation is assumed to be on (that is, for θ_k, $o_j=1$). Equations 5 and 6, with different subscripts, are then applied to change weights on connections between units in the input and hidden layers.

Derivation of Conditioning Theories.   Sutton and Barto (1981) observed the similarities between a method of training adaptive networks known as the Widrow-Hoff rule (Widrow & Hoff, 1960) and the well known Rescorla-Wagner model of classical conditioning (Rescorla & Wagner, 1972). Perhaps not surprisingly, then, the Rescorla-Wagner model falls out as a special case from the generalized delta rule (Gluck & Bower, 1988) when simplifying assumptions are made. First assume a single layer network with CSs represented by input units and the US represented as the teaching input. Expanding Eq. 5 by substitution (from Eq. 3) and ignoring the momentum term, we have

$$\Delta w_{ik} = \eta o_i(t_k - o_k)o_k(1 - o_k). \tag{7}$$

Letting the activation function be linear rather than sigmoidal, and in particular using the identity function, $o_k = net_k$, then the derivative is 1 and Eq. 7 reduces by substitution from Eq. 1 to

$$\Delta w_{ik} = \eta o_i(t_k - \Sigma w_{ik}o_i). \tag{8}$$

Eq. 8 is formally identical to Blough's (1975) extension of the Rescorla-Wagner model,

$$\Delta V_i = \beta \gamma_i(\lambda - \Delta V_i \gamma_i). \tag{9}$$

Eliminating the generalization parameter, γ, and adding a CS salience parameter, we obtain the original Rescorla-Wagner model:

$$\Delta V_i = \alpha_i \beta(\lambda - \Delta V_i). \tag{10}$$

SIMULATION METHODS

We have used the generalized delta rule (Eq. 1–6) to teach simple, two stimulus versions of MTS to many networks similar to that diagrammed in Figure 10.3. The number of training sweeps required to learn the problem to a stringent error criterion have ranged from a couple of hundred to a couple of thousand, values that correspond well to those observed with real animals. In all our simulations, the absolute error averaged over all output units was required to be less than 0.10 (and usually less than 0.05). Throughout, the momentum coefficient (α) was set to 0.90. All the expected effects of varying the learning rate parameter (from η = 0.10 to η = 0.33) and the number of hidden units (from 4 to 12) were obtained—fewer sweeps with higher rates and larger networks.

Most simulations included 10 to 30 runs where each run began with a freshly seeded random set of weights (so each run represents a different simulated subject). At the outset of each simulation, starting weights and biases were set to small random values. The initial weights averaged 0.0 and the uniform distribution ranged between -1 and $+1$; the range was frequently smaller.

The minimal set of input/output patterns presented during training is that listed in Table 10.1. In the simulations described later in this chapter, the training set often contained a larger number of patterns needed to represent more complex experiments. The input/output patterns were presented in randomized blocks; the entire training set was scrambled and then each pattern presented one at a time. Weights were changed after each input/output pattern.

Many of the laboratory experiments of interest involved training animals on MTS and then manipulating task parameters or stimulus variables in generalization tests. We included such tests, as needed, in our stimulations. After training was complete, a series of input/output patterns were presented; for comparative purposes, the series contained the patterns from the original training set as well as the test patterns.

In most simulations, each stimulus was represented as in Table 10.1, i.e., as a single bit with 1 indicating presence and 0 indicating absence. We have done some work on representation on input patterns; representing each stimulus as a single bit or a collection of bits (features) appears to make little difference, but that variable has not been investigated thoroughly. In some simulations, sample stimulus discriminability was manipulated by reducing the difference between the codes for stimulus presence and stimulus absence. In one case that follows, we represented two samples by the pairs [0.6, 0.4] and [0.4, 0.6] instead of the usual high discriminability pairs [1,0] and [0,1]. In yet other cases described later, sample discriminability was reduced by exponential decay in an attempt to mimic the presumed effects of the passage of time. In these cases, the input bits representing the sample stimuli were multiplied by $(1 - d)^t$, where d is the decay rate ($0<d<1$) and t is the number of time units intervening between sample offset and comparison onset. In our experiments, d was set at 0.25 or 0.50.

It should be noted that we have made no attempt to make quantitative fits of the simulation results to the empirical data. We feel that such an undertaking would be premature. We judge our simulations by the goodness of fit on an ordinal scale.

OVERVIEW

Three topics of laboratory research were selected to test the model, and our work has proceeded along all three lines more or less in parallel. The three topics are coding processes in short-term memory, shared attention, and interference with short-term memory. The topics were chosen for a variety of reasons. There was no common theoretical framework between the topics. Within each topic, there is (or has been) doubt as to how the basic empirical facts should be explained, or the conditions responsible for the main phenomenon are not clear. The main findings are well documented in the experimental literature, and the research on each topic is reasonably current with published reports appearing within the last few years.

For each topic, we first report a simulation of what we regard as the main fact. We go on to demonstrate the workings of the model, for example, by analyzing patterns of activation of hidden units. Then we report the results of simulations that were motivated by more analytical laboratory experiments meant to test explanations of the phenomenon of main interest or to determine the conditions responsible for its occurrence.

MEMORY CODES

The topic of memory codes deals with the way that the sample stimulus is represented. The issue is of special importance in cases of delayed matching where the sample is not present when the choice between comparison stimuli is made. The delay could be bridged if the animal maintained a relatively faithful replica of the sample stimulus in short-term memory. But the delay could be bridged just as well if the animal transformed the representation of the sample stimulus into a representation of the comparison stimulus associated with that sample (a kind of "instruction"; Honig, 1978). A variety of experimental results now support the idea that animals can code sample stimuli in terms of associated events that occur at the end of trials (Peterson, 1984; Roitblat, 1980). The current terms used to refer to the two coding schemes are "retrospective" and "prospective" coding.

The Sample-Comparison Mapping Experiment

One source of support for the notion of prospective coding comes from experiments in which attempts are made to bias the coding process by varying the sample-comparison mapping. The logic of the mapping experiment is simple. Mapping many samples to each comparison should cause the samples to be coded in terms of their common comparison stimulus. Mapping one sample to many comparisons should cause the comparison stimuli to be coded in terms of their common sample.

Table 10.3 contains abstractions of the trial configurations in a mapping experiment performed by Zentall, Jagielo, Jackson-Smith, and Urcuioli (1987). Each of four groups of pigeons learned a different mapping. The 2S-2C condition contained two samples and two comparisons (a one-to-one mapping). In the 2S-4C condition, each sample had two associated comparison stimuli (a one-to-many mapping). In the 4S-2C condition, a pair of samples shared a common comparison stimulus (a many-to-one mapping). The 4S-4C condition contained four samples and four comparisons; each sample was associated with a different comparison (a one-to-one mapping).

Zentall et al. (1987) showed that the 2S-2C condition was learned fastest and the 4S-4C condition learned slowest. The 2S-4C and 4S-2C conditions were learned at about the same, intermediate rate. This pattern of results cannot be explained in terms of variations in the number of if-then rules (Carter & Eckerman, 1975), because the same number of rules would be needed in the 2S-4C, 4S-2C, and 4S-4C condition. Instead, it appears that the pigeons capitalized on the coding economies present in 2D-4C and 4S-2C conditions. In 2S-4C condition, each sample was mapped onto two comparisons (in Table 10.3, A → A and A → X), so the pigeons could have coded comparisons in terms of their common sample. In the 4S-2C condition, each comparison stimulus was associated with two samples (in Table 10.3, A → A, and X → A), so pigeons could have coded the

TABLE 10.3
Trial Configurations in Sample-Comparison Mapping Studies

2S–2C	2S–4C	4S–2C	4S–4C
A/A+B−	A/A+B−	A/A+B−	A/A+B−
A/B−A+	A/B−A+	A/B−A+	A/B−A+
B/A−B+	B/A−B+	B/A−B+	B/A−B+
B/B+A−	B/B+A−	B/B+A−	B/B+A−
A/A+B−	A/X+Y−	X/A+B−	X/X+Y−
A/B−A+	A/Y−X+	X/B−A+	X/Y−X+
B/A−B+	B/X−Y+	Y/A−B+	Y/X−Y+
B/B+A−	B/Y+X−	Y/B+A−	Y/Y+X−

Note. Trial configurations are presented for each of four conditions defined by the number of samples (2S or 4S) and the number of comparisions (2C or 4C). A and B represent a pair of stimuli (e.g., two orientations of a line) and X and Y represent a different pair of stimuli (e.g., two forms). Within each configuration the sample stimulus appears on the left of the slash and the comparisons appear on the right with the correct comparison flagged with a " + ". In the simulations (and experiments) an equal number of trials were presented within each session, so each configuration in the 2S–2C condition appears twice.

samples in terms of their common comparisons. Neither type of coding economy, of course, would be possible in the 4S-4C condition.

The network used in the sample-comparison mapping simulation contained 12 input units (four per key location), 12 hidden units, and two output units. Four training sets were constructed from the trial configurations in Table 10.3 coded in binary form (as in Table 10.1). Thirty replications (simulated "subjects") were conducted using each training set ("group"). Each simulated session (a sweep) consisted of one block of all trial configurations presented in a random order. The learning rate was 0.33 and the stopping criterion was an absolute average error of less than 0.10.

The results of the simulation are shown in Figure 10.4. The average number of sweeps to criterion are plotted separately for each group. The network quickly learned the 2S-2C mappings. The 2S-4C and 4C-2S mappings were learned at about the same rate. The 4S-4C mapping took the most time to learn. The results presented in Figure 10.4 thus match the empirical fact quite well (and we have replicated those results in three simulations). The network, like the pigeon, appears to have discovered the coding opportunities afforded by the 2S-4C and 4S-2C mappings.

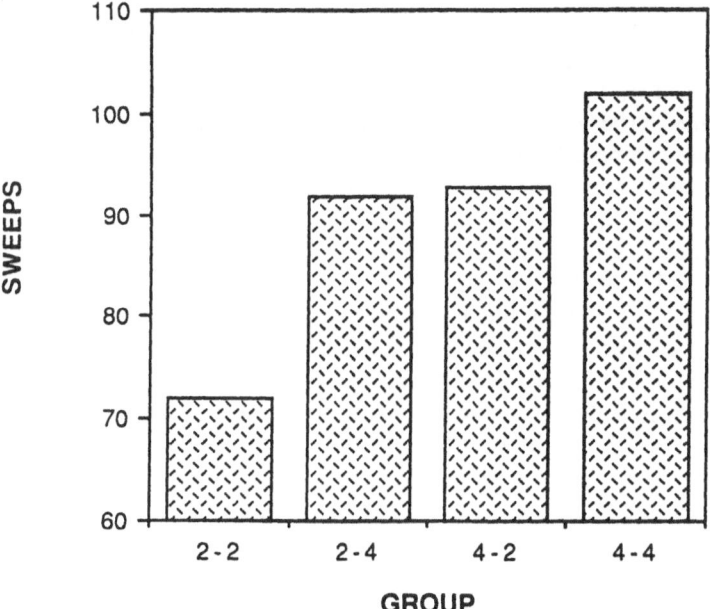

FIG. 10.4. Amount of training as a function of sample-comparison mappings. The data are average number of sweeps required to reach an error criterion.

The Representation of Mappings by Hidden Units

How did network manage to learn 2S-4C and 4S-2C conditions faster than the 4S-4C condition? Did it discover the common codings? If that were the case, we expected that the patterns of hidden unit activation should show evidence of common codes. That is, two stimuli that are treated as equivalent by the network should evoke identical responses from the hidden units. Our reasoning was based on previous studies of internal representations formed by networks after training. Rosenberg (1987), for example, taught NETtalk to translate written letters into their phonetic equivalents. He then recorded the vectors of hidden unit activities elicited by test stimuli. Pairwise correlations of these vectors indicated the degree of similarity between stimuli (as represented by NETtalk). The half matrix of correlations was the basis for further clustering analyses that revealed the structure of NETtalk's internal representations.

The sample-comparison mapping problem learned by out network is far simpler than the linguistic problem learned by NETtalk. Consequently, we devised a similar but simpler method for studying the internal representations of sample-comparison mappings. Let x and y represent the vectors of N hidden unit activation values caused by two different input patterns. The average absolute difference between the elements of x and y, $[\Sigma|(x_i-y_i)|]/N$, has some properties we would like in a cognitive distance measure. When two input patterns produce identical activations in the hidden units, the value is 0; as the activation values of hidden units caused by the input patterns become less similar, the average difference approaches 1 (because of the unit range of the sigmoidal activation function). This value can be calculated for each pair of input patterns. In our simulations, we adopted a criterion of 0.05 for judgement of similarity; if the average difference was less than 0.05, we considered the two patterns to be coded in the same way.

Our first set of test stimuli were the input patterns from the original training sets. An additional 10 simulated subjects were trained on each of the four mappings. The vector of hidden unit activations in response to each input pattern was recorded and the average difference between each pair of vectors was computed. Those differences were averaged across simulated subjects to produce the data shown in Table 10.4. Small differences (<0.05) were evident only in the data from the 2S-4C and 4S-2C conditions, and then only for pairs for trials that common codings were possible. For example, A/A+B− and A/X+Y− in 2S-4C were coded alike, as were A/A+B− and X/A+B− in 4S-2C.

The results of our second test are more revealing. In that test we presented samples and comparison stimuli separately. (Note that these patterns were novel because the network had only been trained with the full trial configurations.) Another 10 simulated subjects were trained in each of the four conditions. Two test patterns were extracted from each input pattern in the training set. One test pattern contained the binary codes for the sample stimulus, and the other test pattern contained the binary codes for both comparison stimuli; bits in unused

TABLE 10.4
Analysis of Internal Representations Produced by Different Sample-Comparison Mappings: Input Patterns in the Training Set

2 SAMPLES, 2 COMPARISONS

	A/A+B−	A/B−A+	B/A−B+	B/B+A−
A/A+B−	0.00			
A/B−A+	0.21	0.00		
B/A−B+	0.16	0.25	0.00	
B/B+A−	0.19	0.18	0.18	0.00

2 SAMPLES, 4 COMPARISONS

	A/A+B−	A/B−A+	B/A−B+	B/B+A−	A/X+Y−	A/Y−X+	B/X−Y+	B/Y+X−
A/A+B−	0.00							
A/B−A+	0.16	0.00						
B/A−B+	0.15	0.19	0.00					
B/B+A−	0.21	0.15	0.18	0.00				
A/X+Y−	0.02	0.16	0.16	0.21	0.00			
A/Y−X+	0.16	0.02	0.19	0.16	0.16	0.00		
B/X−Y+	0.16	0.19	0.02	0.18	0.15	0.19	0.00	
B/Y+X−	0.21	0.15	0.18	0.02	0.21	0.15	0.18	0.00

4 SAMPLES, 2 COMPARISONS

	A/A+B−	A/B−A+	B/A−B+	B/B+A−	X/A+B−	X/B−A+	Y/A−B+	Y/B+A−
A/A+B−	0.00							
A/B−A+	0.20	0.00						
B/A−B+	0.15	0.23	0.00					
B/B+A−	0.22	0.15	0.20	0.00				
X/A+B−	0.01	0.21	0.15	0.22	0.00			
X/B−A+	0.20	0.01	0.22	0.15	0.20	0.00		
Y/A−B+	0.15	0.23	0.02	0.21	0.15	0.23	0.00	
Y/B+A−	0.22	0.15	0.21	0.02	0.22	0.15	0.20	0.00

4 SAMPLES, 4 COMPARISONS

	A/A+B−	A/B−A+	B/A−B+	B/B+A−	X/X+Y−	X/Y−X+	Y/X−Y+	Y/Y+X−
A/A+B−	0.00							
A/B−A+	0.20	0.00						
B/A−B+	0.18	0.26	0.00					
B/B+A−	0.21	0.17	0.20	0.00				
X/X+Y−	0.12	0.18	0.20	0.15	0.00			
X/Y−X+	0.20	0.11	0.21	0.18	0.20	0.00		
Y/X−Y+	0.19	0.20	0.11	0.20	0.18	0.26	0.00	
Y/Y+X−	0.15	0.19	0.20	0.13	0.21	0.18	0.19	0.00

Note. Values in the table are the average absolute differences between vectors of hidden unit activations, $\Sigma |x_i - y_i|/N$, where x and y are the vectors produced by the input patterns identifying a row and a column within the table. In this case $N = 12$ (hidden units). Very small differences indicate common coding for pairs of input patterns and are underlined.

positions were set to zero. Otherwise, the test was conducted like that described previously. The average differences are shown in Table 10.5. In the 2S-4C condition, comparison stimuli that shared a common sample were coded alike, whereas in the 4S-2C condition, sample stimuli that shared common comparison stimuli were coded alike. In the remaining two conditions, none of the pairwise comparisons met our criterion for common coding.

These analyses of internal representations reveal what might have been responsible for the faster learning of the 2S-4C and 4S-2C conditions relative to the 4S-4C condition. After learning the 2S-4C mapping, the network commonly coded trials, in which different sets of comparison stimuli were associated with a common sample, and the different comparison stimuli were represented in the same way in hidden unit activation patterns. After learning the 4S-2C mapping, the network commonly coded trials, in which different samples were associated with common comparison stimuli, and those different samples were represented in the same way. Discovery of these common codes might reduce the number of "if-then" rules (Carter & Eckerman, 1975) to be learned.

In fact we can describe the common coding in terms of learning of disjunctive concepts in a production system framework (Langley, 1987). The general form of the rules could be

IF<SAMPLE and COMPARISON>THEN<respond LOCATION>.

In the 4S-2C condition, two explicit rules are

IF<A and AB>THEN<respond left>, and
IF<X and AB>THEN<respond left>.

The network appears to have combined the two rules into a single rule of the form

IF<(A or X) and AB>THEN<respond left>.

Similarly, two explicit rules in the 2S-4C condition are

IF<A and AB>THEN<respond left>, and
IF<A and XY>THEN<respond left>.

The network appears to have combined these two rules into

IF<A and (AB or XY)<THEN<respond left>.

Other than for descriptive purposes, however, we perceive no advantage to casting our results in terms of a production system. The connectionist model discovered the common codes quite naturally and the parallel distributed model is more easily extended to other experimental results.

TABLE 10.5
Analysis of Internal Representations Produced by Different Sample-Comparison Mappings: Tests with Sample and Comparison Stimuli

2 SAMPLES, 2 COMPARISONS

	A	B	AB	BA
A	0.00			
B	0.27	0.00		
AB	0.17	0.18	0.00	
BA	0.18	0.18	0.19	0.00

2 SAMPLES, 4 COMPARISONS

	A	B	AB	BA	XY	YX
A	0.00					
B	0.22	0.00				
AB	0.15	0.10	0.00			
BA	0.12	0.15	0.20	0.00		
XY	0.15	0.10	<u>0.02</u>	0.21	0.00	
YX	0.12	0.15	<u>0.20</u>	<u>0.02</u>	0.20	0.00

4 SAMPLES, 2 COMPARISONS

	A	B	X	Y	AB	BA
A	0.00					
B	0.22	0.00				
X	<u>0.03</u>	0.22	0.00			
Y	<u>0.21</u>	<u>0.03</u>	0.21	0.00		
AB	0.14	<u>0.20</u>	0.14	0.20	0.00	
BA	0.20	0.15	0.20	0.15	0.21	0.00

4 SAMPLES, 4 COMPARISONS

	A	B	X	Y	AB	BA	XY	YX
A	0.00							
B	0.19	0.00						
X	0.09	0.15	0.00					
Y	0.16	0.09	0.19	0.00				
AB	0.11	0.15	0.13	0.14	0.00			
BA	0.15	0.11	0.13	0.13	0.20	0.00		
XY	0.12	0.13	0.10	0.16	0.10	0.14	0.00	
YX	0.15	0.12	0.17	0.10	0.16	0.12	0.21	0.00

Note. Values in the table are the average absolute differences between vectors of hidden unit activations, $\Sigma|x_i-y_i|/N$, where x and y are the vectors produced by the sample and comparison stimuli identifying a row and a column within the table. In this case N = 12 (hidden units). Very small differences indicate common coding for pairs of stimuli and are underlined.

The Interaction of Sample Discriminability
and Sample-Comparison Mapping

Zentall, Urcuioli, Jagielo, and Jackson-Smith (1989) have discovered a variable that modulates the effect of sample-comparison mapping. They reasoned that sample stimuli that are difficult to discriminate, would bias the pigeon toward prospective processing. Four groups of pigeons (as in Table 10.3) were trained with two colors and two line orientations as samples, and pairs of colors and pairs of line orientations as comparisons. After acquisition, the pigeons were tested with delays intervening between the sample and comparisons. Performance was worse in the two conditions with four comparisons (2S-4C and 4S-4C) than in the two conditions with two comparisons (2S-2C and 4S-2C), but that difference was significant only for the less discriminable line-orientation samples.

We performed another set of simulations to check the sensitivity of the model to the interaction of the sample discriminability and sample-comparison mapping variables. In each of two simulations, the same design listed in Table 10.3 was used with 30 simulated subjects per mapping condition. In one simulation, each sample was represented by our usual binary pairs ([1,0] and [0,1]). In the second simulation, the difference between codes for stimulus presence and absence was reduced; low discriminability samples were codes as [0.6, 0.4] and [0.4, 0.6]. Following training, the networks were tested with input patterns, in which the sample stimuli had been subjected to exponential decay for 0, 1, 2, or 4 time units with a decay rate of 0.25.

The average errors of output units during the delay test are summarized in Table 10.6. The data were averaged and are presented separately for each

TABLE 10.6
Interaction of Sample Discriminability
and Sample-Comparison Mappings

Sample Discriminability	Mapping	Delay			
		0	1	2	4
High	2S–2C	0.096	0.118	0.153	0.251
	2S–4C	0.096	0.116	0.152	0.247
	4S–2C	0.096	0.123	0.164	0.263
	4S–4C	0.095	0.122	0.164	0.263
Low	2S–2C	0.096	0.142	0.206	0.334
	2S–4C	0.096	0.141	0.208	0.343
	4S–2C	0.099	0.145	0.206	0.323
	4S–4C	0.098	0.146	0.209	0.329

Note. Data are average absolute errors of ouput units. Data from each combination of sample discriminability, number of samples, and number of comparisons are based on 30 simulations. The decay rate for sample stimuli across delays was 0.25.

combination of sample discriminability, mapping, and delay. Separate analyses of variance were performed on the data from the two discriminability conditions and included number of samples, number of comparisons, and delay as factors. Although numerically small, certain key interactions are statistically significant. For the high discriminability samples, the interaction between number of samples and delay was significant; the conditions with four samples had larger errors at longer delays. The interaction between number of comparisons and delay, however, was not significant. For the low discriminability samples, a different pattern of results was obtained. The number of samples also interacted with delay, but the interaction is difficult to interpret; the conditions with four samples produced larger errors at short delays but produced smaller errors at the longest delay. The important observation is that the number of comparisons interacted with delay; at long delays, the conditions with four comparisons had larger errors than did the conditions with two comparisons.

Summary

Our attempts to simulate the sample-comparison mapping results have been reasonably successful. The simulations captured the main features of the empirical data and the analyses of internal representations show that the connectionist model indeed does form common codes for different samples associated with common comparisons in many-to-one mappings. However, more simulation work on the many-to-one mapping condition is needed and is in progress. Urcuioli, Zentall, Jackson-Smith, and Steirn (1989) recently reported additional evidence for common coding based on patterns of between-trial interference and learning of new sample-comparison associations. Intertrial (proactive) interference is a whole area in itself that has not yet been examined in our simulations of the connectionist model.

SHARED ATTENTION

Matching to Compound Samples

Influenced by developments in the areas of human perception and performance, Maki and Leuin (1972) sought evidence of a limited capacity information processing channel in animals. They devised a type of MTS known as matching to compound samples. The trial configurations are displayed in Table 10.7. In their experiment, pigeons were first trained on two independent matching problems— matching two colors and matching two achromatic line orientations. After the pigeons were performing at asymptote, a series of tests were administered. The test trials contained compound samples constructed by superimposing one of the (white) lines on one of the colors. Compound samples were followed unpredictably by either set of comparisons. The main finding was that matching to com-

TABLE 10.7
Trial Configurations in Studies of Matching to Compound Samples
(Shared Attention)

	Element samples	Compound samples	
AB	A/A+B−	AX/A+B−	AY/A+B−
relevant	A/B−A+	AX/B−A+	AY/B−A+
	B/A−B+	BX/A−B+	BY/A−B+
	B/B+A−	BX/B+A−	BY/B+A−
XY	X/X+Y−	AX/X+Y−	BX/X+Y−
relevant	X/Y−X+	AX/Y−X+	BX/Y−X+
	Y/X−Y+	AY/X−Y+	BY/X−Y+
	Y/Y+X−	AY/Y+X−	BY/Y+X−

Note. Stimuli from two dimensions, AB, and XY, are assumed. Sample stimuli are represented by a letter on the left of the slash and comparison stimuli are represented by a pair of letters on the right of the slash; the correct comparison is marked by a "+". Element samples are followed by comparisons from the corresponding dimension. Compound samples are followed by comparisons from either dimension. An additional set of compound sample trials are required to balance to irrelevant samples.

pound samples required a longer sample stimulus duration than element samples in order to maintain equal matching accuracy. (Subsequent studies showed that given a fixed sample duration, matching to compounds was less accurate than matching to elements; see Maki, Riley, & Leith, 1976). Maki and Leuin took this fact as a sign of a central processing bottleneck. They supposed that the pigeons divided their limited attentional resources between the two elements of a compound sample, had less time to process each element, and therefore performed less accurately.

We have routinely reproduced the "shared attention" effect in our simulations of matching to elements vs. matching to compounds. In general, we created a network with a number of input units sufficient to represent the two element matching problems shown in Table 10.7. After being trained to criterion concurrently on both element matching problems, generalization tests were conducted that included all the trial configurations shown in the table. The magnitude of the effect varied with the exact details of the simulation, but inevitably, the average error of output units in compound sample trials was larger than in element trials. Examples of this effect are included in the analytical studies that follow.

Competition Among Internal Representations

Here we report results of a typical simulation in which we examined the patterns of hidden unit activations produced by input patterns containing element and compound samples. The network was first trained on the two element matching problems (Table 10.7). After training, the four sample stimuli were presented alone and the activations of the four hidden units were recorded. The patterns of hidden unit activations elicited by each element sample are presented in Figure 10.5. The patterns differ between samples from the same dimension (e.g., A vs B), and the patterns are also different for samples from different dimensions (e.g., B vs X).

An example of the patterns of hidden unit activity produced by compound samples is shown in Figure 10.6. The activations produced by the B and X elements are copied from Figure 10.5 for comparison with the activations produced by the BX compound. Clearly, the internal representation of the compound is unlike the internal representation of either element. Even though the elements are familiar to the network, have well defined internal representations, and produce accurate outputs, the pattern of activation set up by the compound sample is novel and thus accuracy suffers.

Grant and MacDonald (1986) suggested that the deficit in accuracy of matching produced by compound samples could best be viewed as a coding decrement resulting from a competition between sample codes. The connectionist model is actually a realization of that theoretical idea. The pattern of inputs in a compound sample are different from any previously trained so the pattern of hidden unit activity is different. It should be noted that there is nothing special about how

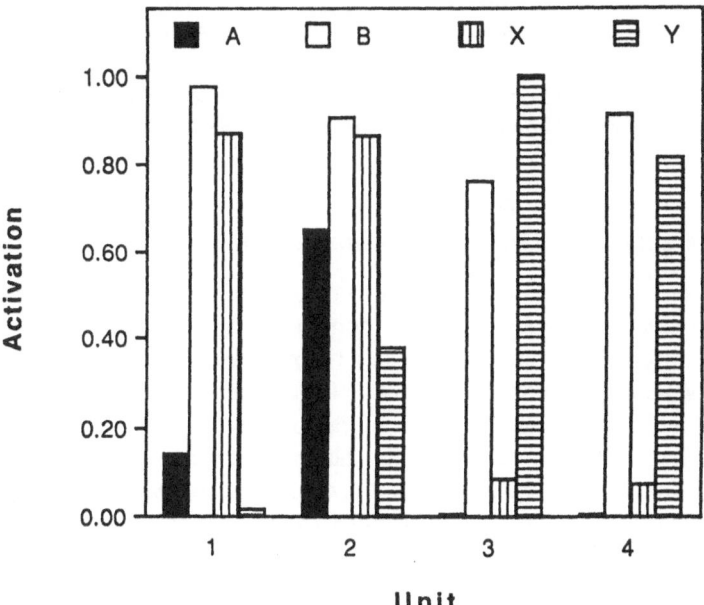

FIG. 10.5. Activation values of hidden units as a function of sample stimulus following training on element matching.

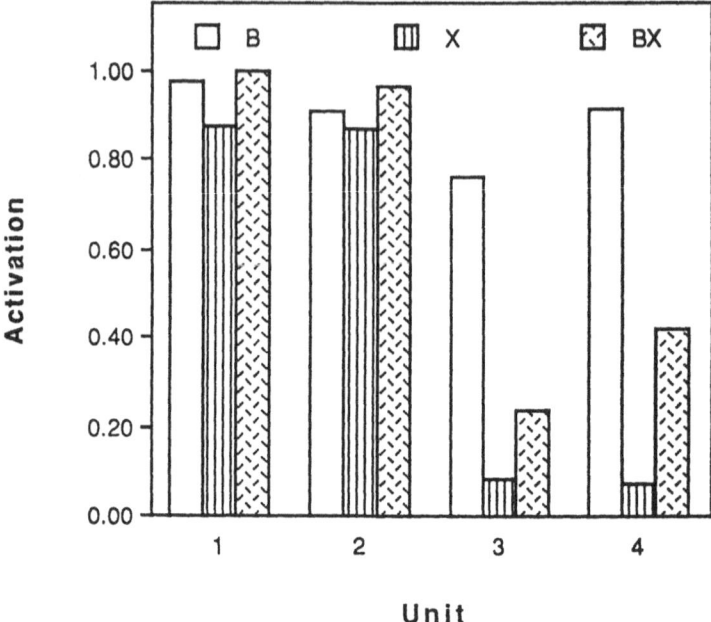

FIG. 10.6. An example of activation of hidden units during element
and compound tests following training on element matching.

the coding decrement occurs in the connectionist model because cancellation
between uncorrelated input patterns is commonplace in simulated networks.

Shared Attention vs. Generalization Decrement:
The Interaction of Sample and Comparison Stimuli

Early in the exploration of matching to compound samples, it was understood
that the reduced accuracy caused by compound samples could result from the
physical difference between a compound sample and an elemental comparison.
That is, generalization decrement and not shared attention might be responsible
for the effect of compound samples on performance.

An initial inspection of the generalization decrement hypothesis was under-
taken by Maki et al. (1976). In that study, compound comparisons were presented
along with compound or element samples. A pair of trials from Table 10.7, A/
A+B− and AX/A+B−, illustrates the point that the compound sample differs
from the element comparison. Maki et al. included conditions, in which the
generalization decrement would affect matching to element samples instead of
matching to compound samples. For example, in A/AX+BX− and AX/
AX+BS− trials, the comparisons were identical to the compound samples but

were different from the element samples. Maki et al. found that the compound comparisons produced worse performance than element comparisons, but that compound samples still were matched less accurately than element comparisons. This pattern of results did not fit the predictions they derived from the generalization decrement hypothesis.

Roberts and Grant (1978a) performed a more extensive study of matching to compound samples using both element and compound comparison stimuli. Examples of the trial configurations used in their experiments are shown in Table 10.8.

The results obtained by Roberts and Grant are summarized in Figure 10.7. The original effect is shown by the difference favoring A over AX samples when tested with elemental (A+B) comparisons. The results with compound (AX+BS− and AC+BY−) comparisons replicate closely those originally reported by Maki et al. (1976). AX samples resulted in lower accuracy than A samples with AX+BX− comparisons, but the reverse was true with AX+BY− comparisons. Based on the pattern of data presented in Figure 10.7, Roberts and Grant postulated an "elicitation" process; the presence of more than one comparison at a location was assumed to elicit a greater tendency to choose that location because all the comparison elements had extensive histories of reinforcement.

Given the success of the connectionist model with the original finding, and given our understanding its cause (the coding decrement), we elected to test the model by simulating the Roberts and Grant experiments. In one of our efforts, the network consisted of 12 input units (coding four stimuli per location), 12 hidden units, and *eight* output units. The use of so many output units represents a major departure from our other simulations. The eight outputs coded the response tendencies toward the eight comparison elements (four per location).

TABLE 10.8
Trial Types in Studies of Factors Influencing Matching to Compound Samples

Element Samples	Compound Samples
A/A+B−	AX/A+B−
A/AX+BX−	AX/AX+BX−
A/AX+BY−	AX/AX+BY−
A/AX+B−	AX/AX+B−
A/A+BY−	AX/A+BY−

Note. Stimuli from two dimenstions are represented (AB and XY). The details of the representation are the same as in Table 10.6. Only a few examples are shown here; in practice, the number of trials is much larger to accomodate the balancing for sample, location of correct comparison, and irrelevant samples and/or comparisons. See Table 10.6 for an example of such balancing.

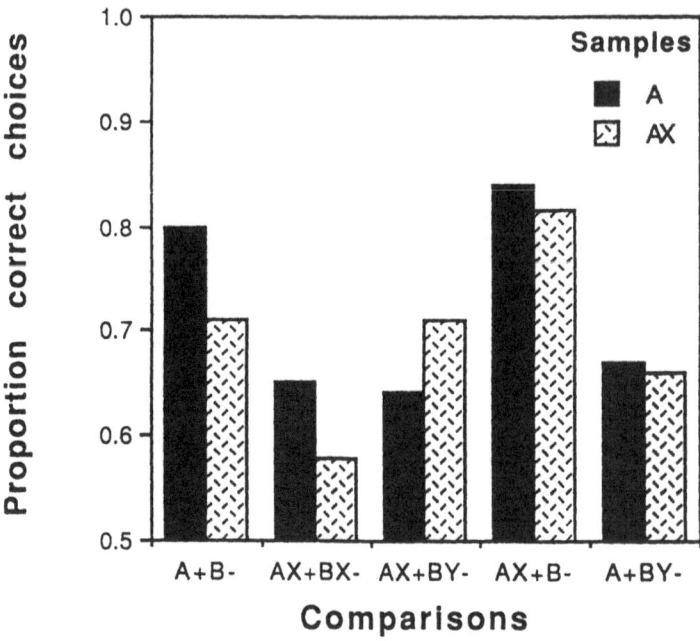

FIG. 10.7. Interaction of sample and comparison stimuli: Observed data. These data were obtained by averaging the summary data presented by Roberts and Grant (1978, Table 4) for each combination of sample and comparison stimulus conditions.

The output units were coupled to eight teaching inputs that represented the reinforced comparisons. During training with elemental comparisons, only one of the teaching inputs was active for each input pattern. Training continued until the usual error criterion was met. During testing, performance was assessed by computing the probability of choosing the correct comparison stimulus. For each input pattern, the probability was computed by summing the activations of each of the four output units responding in the location of the correct comparison and dividing by the total of the activations of all eight output units.

The average probabilities based on 30 simulated subjects are shown in Figure 10.8. In some respects, the simulation mirrors experimental findings. The basic effect is present; compound samples produce poorer matching than do element samples with element comparisons, and the same effect is shown with the AX+BS− comparisons. The effect is reversed with the redundant comparisons (AB+BY−). And, the effect disappeared with the AX+B− and A+BY− comparisons. Otherwise, however, the predictions from the simulation fit the observed

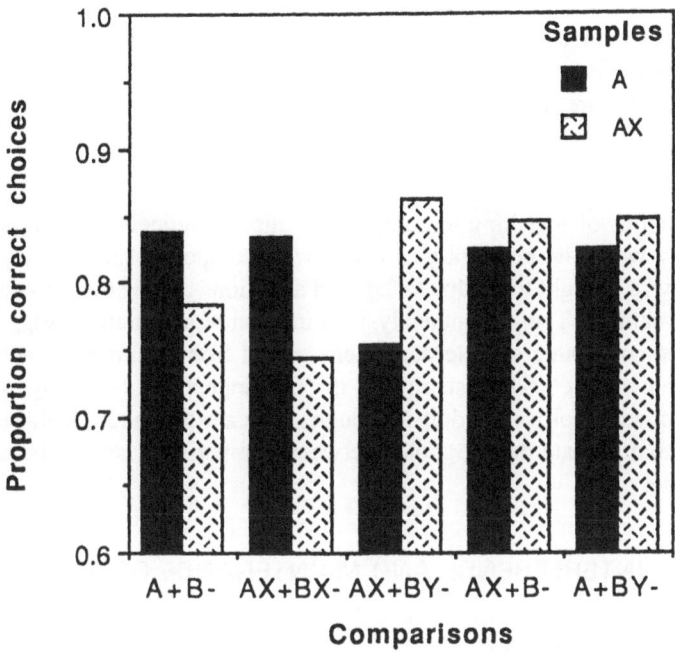

FIG. 10.8.   Interaction of sample and comparison stimuli: Simulation
results.

data poorly. The simulated AB+BX− levels are too high as are the A+BY−
levels.

We have not been able to reduce the discrepancies between the empirical
results (Figure 10.7) and simulated results (Figure 10.8) in several simulations.
We did discover that adding the elicitation process to the model improved the fit
substantially over that produced by a model with two spatial choices (cf. Figure
10.3). Otherwise, the discrepancies seem intractable. We suspect that the principle problem is the absence of sequential choice processes in the present model.
In the present version of the connectionist model, the comparison stimuli are
processed in parallel. However, both observation and theory suggests that processing of comparisons is sequential. Wright and Sands (1981) found that pigeons
performing MTS alternately looked at the two comparisons. Similarly, the
elicitation process as originally conceived by Roberts and Grant (1978a) had
the pigeon observing a single key, left or right, when the comparisons were
first presented. That key was pecked as a function of both the memory of the
sample stimulus and the number of stimuli on that key. Thus, for example,
performance was predicted to be high with the AX+B− comparisons because

of the extra stimulus at the location of the correct comparison; for the same reason, performance was predicted to be lower with the A+BY− comparisons. That difference is present in the empirical results (Figure 10.7) but not in the simulations (Figure 10.8).

Summary

Our simulations of matching to compound samples routinely have shown a loss in accuracy when the network was tested with compound samples. That effect originally was thought to be due to "shared attention" among the elements of the compound sample. The present analysis of internal representations suggests that, instead, the compound sample decrement results from an interference between uncorrelated sample codes established through independent training. Attempts to model other empirical findings, though, indicate that additional, sequential processes that operate during choice between comparisons need to be appended to the model.

INTERFERENCE AND MAINTENANCE OF STM

Learning to Remember

When a delay between sample and comparison stimuli is introduced after MTS training, dramatic drops in performance are observed. The delay gradients often take a shape suggesting exponential decay (e.g., Urcuioli et al., 1989). With practice, however, animals seem to learn to remember in the face of long delays between samples and comparisons. The finding is true of both monkeys and pigeons (D'Amato, 1973; Maki, 1979). The evidence suggests that the retention function underlying performance is displaced upwards and flattened near maximum at short delays. These observations led to the suggestion that animals learn to maintain a sample stimulus representation in an active state—to "rehearse" (Maki, 1981; Maki, 1984). We asked whether the connectionist model would show similar effects of training at long delays.

In the simulations, a delay was mimicked by systematically reducing the discriminability among input patterns. The effect of time was represented by letting reduced activation of the input units represent decay of sample-stimulus traces. The network represented in Figure 10.3 first learned MTS with no programmed delay ($t = 0$), and then training continued with the same input-output patterns but with each sample-stimulus input value reduced to $(1 - d)^t$ of its initial value (for some $t > 0$, and $0 < d < 1$). We have repeated the simulation many times with minor variations. Always the network has learned to tolerate relatively lengthy delays.

A typical result, for one simulated subject, is shown in Figure 10.8. In this simulation, d = 0.50. Following training with no delay, the network was tested on delays ranging from $t = 0$ (the training delay) to $t = 7$ time units. The resulting retention gradient resembles the common exponential decay pattern. Then the network was trained concurrently with $t = 0$ and $t = 4$. After again reaching the same error criterion as in original training, the network was retested on all eight delays. Figure 10.9 shows that the intervening delay training increased performance across the full range of delays, that the retention function was at ceiling for short delays, and that the improvement generalized to yet longer delays than those used in delay training.

Amplification of Input Activation

Our networks, of course, do not "rehearse," so some other mechanism must be responsible for the enhanced performance following delay training. The improvement appears to result from increased sensitivity of hidden units to small variations in input activation. That mechanism was revealed by an examination of connection strengths within the network. In one simulation, we calculated the root mean-

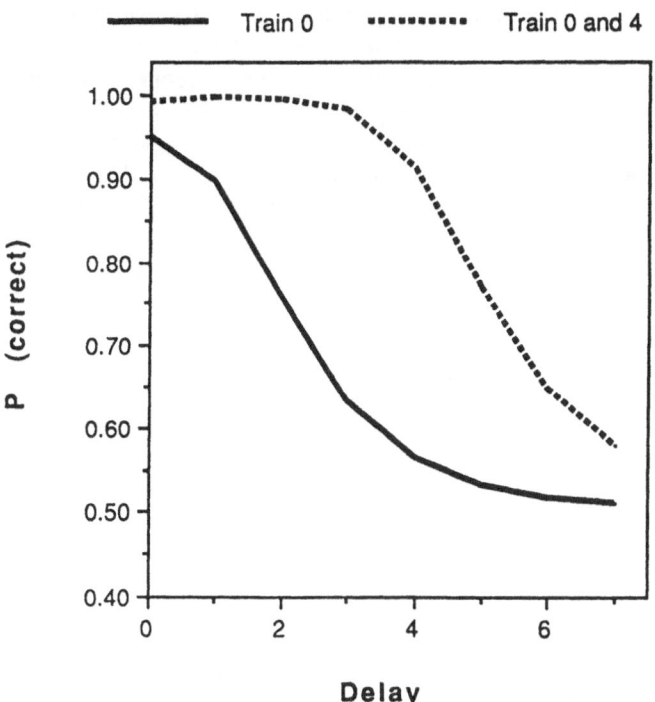

FIG. 10.9. Simulated performance of delayed MTS as a function of experience with delays. The dependent measure is the probability of a correct response; the activation of the output unit at the correct location was divided by the total activation of both output units.

squared (RMS) weights for the connections between input and hidden units. After delay training, the RMS nearly doubled. That means that small amounts of activation of input units were being amplified.

The increased sensitivity of the network to miniscule inputs has two important consequences. One is that the residual activation from prior trials should cause substantial proactive interference; we have not yet attempted connectionist simulations of data from studies of intertrial interference (e.g., Urcuioli et al., 1989). The other consequence is that delay interval stimuli that have received no particular training should cause retroactive interference. In the latter case, novel stimuli turned on during the delay would activate input units that are only weakly connected (by the initial small random weights) to hidden units. The resulting "noise" would be amplified and added to the net inputs to the hidden units. The result would be a pattern of hidden unit activity that would be novel and thus performance would suffer.

Retroactive Interference From Delay-Interval Illumination

Delayed matching performance by animals is very sensitive to delay interval stimuli. Turning on an overhead light (a "houselight") during the delay disrupts performance in monkeys (D'Amato 1973) and pigeons (Maki et al., 1977). Roberts and Grant (1978b) reported an interesting set of results dealing with the effects of the temporal point of interpolation of a brief light during the delay. The kinds of experimental manipulations they performed are diagrammed in Figure 10.10. Four conditions are represented. In the DD condition, the pigeons spent the entire delay between the offset of the sample and the onset of the comparisons in darkness. In the LD and DL conditions, a brief period of houselight illumination occurred at the beginning or end, respectively, of the delay. In the LL condition, the houselight was illuminated during the entire delay. Roberts and Grant found

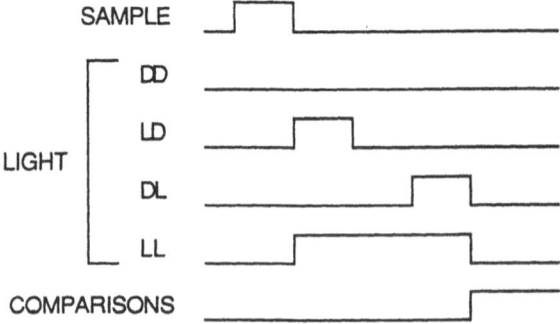

FIG. 10.10. Delay interval illumination conditions. The houselight is turned on at the beginning of the delay (LD), the end of the delay (DL), the entire delay (LL) or not at all (DD).

the usual detrimental effects of delay-interval illumination; performance in condition LL was far worse than performance in condition DD. They also showed that performance was not as poor when only part of the delay contained illumination. The important and novel finding was that performance when the houselight occurred at the end of the delay (DL) was less accurate than when the houselight occurred at the beginning of the delay (LD).

In order to simulate the "beginning-end" effect (Roberts & Grant, 1978b), the network represented in Figure 10.3 was modified so as to represent delay-interval events. The network used in this simulation contained the usual six input units needed to represent the two-stimulus MTS problem, eight hidden units, and two output units. To that network we added eight input units, all fully connected to the hidden units. Four of these extra eight units were used to represent stimuli presented by darkness; the other four represented stimuli presented by the houselight. These "delay inputs" were normally off during zero-delay training, so the weights on their connections to hidden units were left at their initial small random values. After zero-delay acquisition, delays were gradually increased until the network was performing at the criterion with $t = 6$. During the training with nonzero delays, the four "dark" input units were turned on. Weights between the dark input units and hidden units thus were modified during delay training, but weights between light input units and hidden units were left at their original values. In the final phase of the simulation, the network was tested on the four conditions diagrammed in Figure 10.10. The input units representing stimuli presented by the houselight were turned on for the first third of the delay (LD), the last third of the delay (DL), the entire delay (LL), or those units were left off (DD). As before, the passage of time was assumed to cause exponential decay ($d = 0.33$ in this simulation). The activations of input units in test patterns were multiplied by a coefficient that represented the decay; these coefficients are listed in Table 10.9.

The results of the simulation, averaged over 40 simulated subjects, are summarized in Figure 10.11, in which the average absolute error of output units is transformed (complemented) so as to represent decreased performance by downward trends. The loss in accuracy due to the light at the beginning of the delay (LD) was small, the loss due to the light at the end of the delay (DL) was larger, and the loss due to light during the whole delay (LL) was larger yet. This ordering agrees with that reported by Roberts and Grant (1978b).

The beginning-end effect was demonstrated in another way and reported by Maki (1984). When given much reinforced practice under condition of delay-interval illumination, pigeons recover from the retroactive interference caused by the houselight. The recovery occurred more rapidly for the LD than for the DL condition. In a "part-to-whole" transfer experiment, Maki also studied recovery in the LL condition as a function of prior recovery from LD or DL. Prior training on the DL condition (the hardest "part") was expected to produce the most positive transfer to the LL condition (the "whole"). Instead, recovery from LL following

TABLE 10.9
Decay Coeffiecints in Simulation of Delay-Interval
Illumination Effects

	DD	LD	DL	LL
Dark	1	1	$(1 - d)^{t/3}$	0
Light	0	$(1 - d)^{2t/3}$	1	1
Samples	$(1 - d)^t$	$(1 - d)^t$	$(1 - d)^t$	$(1 - d)^t$
Comparisons	1	1	1	1

Note. In the simulation described in the text, the delay interval was six time steps in length, so $t = 6$. The activations of input units representing darkness, the houselight, sample, and comparison stimuli were multiplied by the decay coefficients in the table. Zero coefficients indicate that the corresponding stimuli were not present. Coefficients of 1 indicate that the stimuli were available at full strength just prior to the choice. Light units were active during the first third of the delay (LD), the last third of the delay (DL), the entire delay (LL), or not at all (DD).

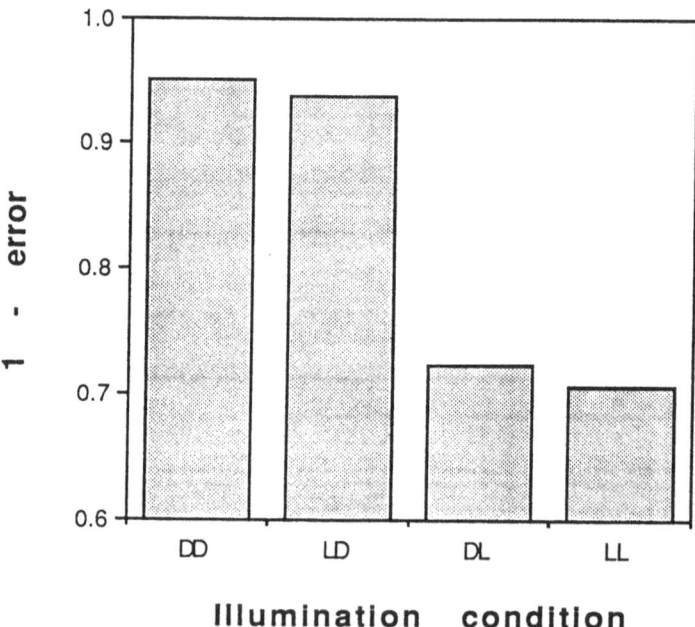

FIG. 10.11. Simulation of the effects of delay interval illumination on delayed MTS performance. The dependent measure is the complement of mean absolute error of output units.

LD training was faster than recovery from LL following DL training. Time taken to recover from the light-induced interference is summarized in Figure 10.12.

We attempted to simulate the transfer results shown in Figure 10.12. The network used in the earlier simulation of interference was taught zero-delay MTS and then trained, as before, to tolerate dark delays of length $t = 6$. Half of the simulated subjects were trained on the LD condition and the other subjects were trained on the DL condition until performance had recovered (reached original criterial levels). Then all simulated subjects were trained on the LL condition. The number of sweeps required for training are presented in Figure 10.13. In accord with the pigeon data presented in Figure 10.12, recovery from the LD condition was significantly faster than recovery from the DL condition. However, the rates of recovery from LL are reversed. In the simulation, recovery from LL following LD training was significantly slower than recovery from LL following DL training.

In retrospect, the reason for failure of the model to reproduce Maki's (1984) recovery results can be traced to the way that forgetting is represented. Consider the pattern of decay coefficients in Table 10.8. As forgetting proceeds (i.e., as the values of t and/or d grow large), the coefficients for the DL condition become similar to those of the LL condition and the coefficients for the LD condition

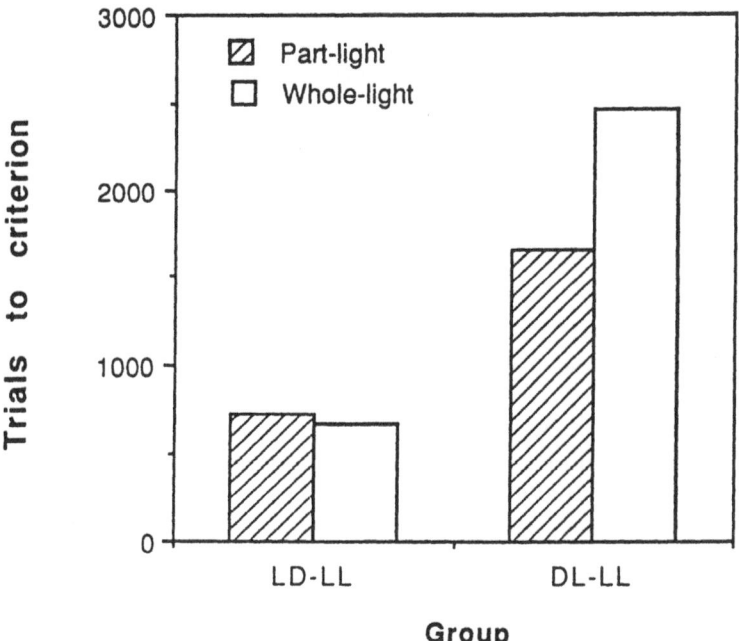

FIG. 10.12.   Effects of part-delay illumination training on reacquisition of MTS with whole-delay illumination: Observed results.

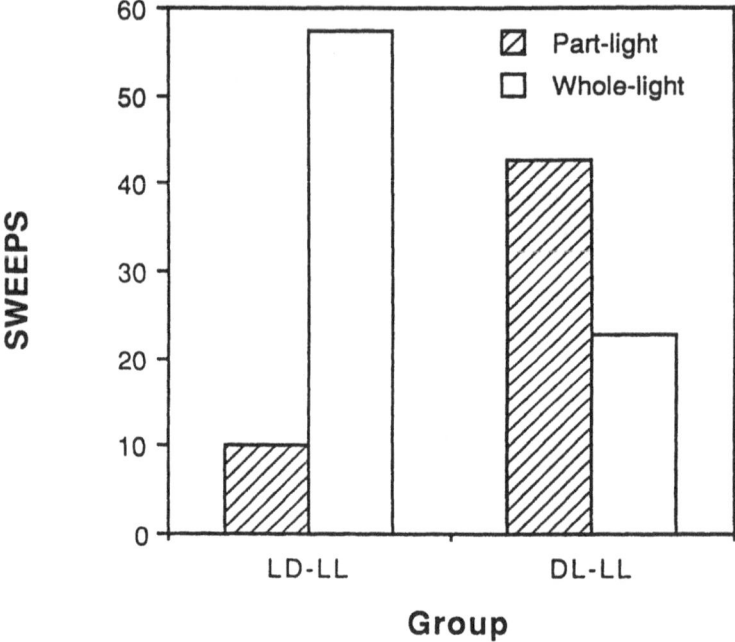

FIG. 10.13.   Effects of part-delay illumination training on reacquisition of MTS with whole-delay illumination: Simulation.

become similar to those of the DD condition. Thus recovery from the LL condition would be fastest following training on the most similar DL condition. That pattern is present in the simulation data (Figure 10.13) but it runs counter to the empirical fact (Figure 10.12).

We suspect that, as was the case with the simulations of matching to compound samples, the way we have represented time in the model is faulty. The sequential nature of an MTS trial is just ignored in the present model. A promising alternative is presented by the sequential, recurrent networks (e.g., Jordan, 1986; Rumelhart et al., 1986; Servan-Schreiber, Cleeremans, & McClelland, 1989) that are being used in studies of robotics and speech recognition. Recurrent networks also ought to prove useful for modelling other sequential problems like learning and performance of delayed matching.

Summary

Our multilayer network model has been trained to perform delayed MTS. The network learns to tolerate rather lengthy delays given delay training. The network also exhibits retroactive interference from delay interval stimuli. The amount and placement of delay-interval stimuli affect the performance of the network in much the same ways as the same variables affect the performance of real organisms in laboratory studies. However, the network model makes erroneous predictions about the effects of those variables on the amount of training required for recovery from retroactive interference. The present connectionist model needs to be modified so as to better represent the sequential nature of the events and processes involved in delayed MTS.

CONCLUSIONS

In this chapter we have reported our initial work on a connectionist model of animal learning and performance of the complex discrimination problem known as matching-to-sample (MTS). Our simulation results show that the network model exhibits some of the major properties of animal learning and performance of MTS.

• *Conditional discrimination:* The network learned the basic MTS problem by backpropagation of errors.

• *Common coding:* The network developed common internal representations of different sample stimuli that were associated with the same comparison stimulus.

• *Shared attention:* The network suffered a loss in accuracy of matching to compound samples after learning to match to the elements.

• *Rehearsal:* The network was taught to tolerate low levels of discriminability among inputs caused by trace decay during delays imposed between the sample and comparison stimuli.

• *Retroactive interference:* The network's performance decreased when confronted with noise produced by novel delay-interval stimuli; the magnitude of the loss depended on the amount and temporal locus of the interfering stimulation.

Each of these findings is well-documented in the experimental literature and is a subject of current research.

Learning of MTS by a network with hidden units was expected based on our demonstration of the logical equivalence of MTS and XOR (Table 10.1) and the fact that these kinds of networks learn the XOR (Rumelhart et al., 1986). The success of the model with the rest of the phenomena listed earlier is a reasonable return on our investment in a connectionist model of MTS. However, not all of the behavioral data on MTS can be accounted for by the relatively simple three layer network that we used in our simulations. We expected to find instances that the network model was at odds with the empirical literature, and indeed we found them. Because the discrepancies provide clues about how best to shape future versions of the model, they are important and we explore their implications in the remainder of the chapter.

About Connectionist Models

One of the virtues of connectionist models is their brain style processing. We might then ask about the degree that the mechanisms embodied in any particular model are really brain-like. Error backpropagation is a case in point. Grossberg (1987) has questioned the biological plausibility of backpropagation, arguing that backpropagation must involve a weight transport mechanism not now known to exist in the brain (see also Crick, 1989). However, several investigators are working on wiring schemes that might allow a plausible neurobiological implementation of backpropagation (Hecht-Nielsen, 1989; Stork, 1989; Zipser & Rumelhart, in press).

A more serious problem for the model considered in this chapter is its behavioral plausibility. A viable model of psychological processes should fit basic behavioral facts. It should not be surprising to learn that the backpropagation learning algorithm can be faulted on behavioral grounds. As we showed earlier, the Rescorla-Wagner model of classical conditioning can be derived from the generalized delta rule. From the time of its inception, the Rescorla-Wagner model was known to make erroneous predictions about, for example, extinction of conditioned inhibition (Zimmer-Hart & Rescorla, 1974) and the conditions required for unblocking (Mackintosh & Turner, 1971). In response, alternative theories of conditioning have been suggested, in which conditionability itself is conditioned (Pearce & Hall, 1980). We do not know how such learning schemes might fare in a connectionist model.

Backpropagation also suffers from "temporal instability" (Grossberg, 1987); new learning drastically interferes with previously learned associations. McCloskey and Cohen (1987, 1989) reported that a network trained with backpropagation on paired-associate learning tasks exhibited amounts of retroactive interference far in excess of anything observed in the human learning laboratory. We have discovered another, related problem (Abunawass & Maki, 1989); training a network on successive paired-associate lists results in cumulative negative transfer. Both of these effects make simulations of transfer tests difficult to interpret and it is no accident that for the most part we avoided such tests in the work we reported here. But simulations of transfer tests (e.g., Urcuioli et al., 1989) are important if we are to realize our goal of a connectionist model of complex discriminations. Somehow we need to provide for insulation of acquired knowledge against the interference from new learning (Sutton, 1986) and, at the same time, we need to provide the means for the new learning to occur without undue hindrance from pre-existing connections. Our present model needs to evolve toward some kind of modular structure that allows both for recognition of old patterns and for action prompted by novel inputs (such as in adaptive resonance theory; see Grossberg, chap. 4 in this volume).

About Problem Representation

The model that we've investigated is a model of some central processing that occurs during learning and performance of MTS. Like other conditioning theories much is assumed about stimulus and response processing. On the input side, a lot of preprocessing is required before incoming signals reach an "input unit." Much more work needs to be done on the kinds of inputs that get represented in the model and the codes used to represent them.

On the output side, the means by which activation of output units gets translated into behavior need to be explored. The failure of the model in the case of matching to compound samples (Roberts & Grant, 1978a) was a forceful reminder of the importance of sequential behaviors during choice (Wright & Sands, 1981). Rethinking the response processes will prompt reconsideration of the way that the consequences of responses are treated. Should the learning paradigm be changed from supervised learning to reinforcement learning?

Perhaps the most important limitation on the present version of the model is the impoverished treatment of sequential behavior. The network is temporally flat. Not only does it ignore vicarious choices, but the only process allowed during a retention interval is that of passive trace decay. The resulting limits on how delay-interval events are introduced into the stream of processing may have been responsible for the errors made by the model in predicting recovery from retroactive interference. We think that an important next step will be a sequential network model of MTS, in which the processing at each step in time is influenced by the present stimuli and the immediate past state of the network as well as the connection strengths. The research on simple recurrent networks (Servan-Schreiber et al., 1989) is serving as a guide for our work on a sequential model of MTS.

About Theoretical Integration

"Good" theories accomplish two scientific goals—explanation of existing facts and prediction of new phenomena. The first goal was the main source of motivation for our work. The controversy over the explanations of the facts (and even the labels we have given them) was partly responsible for the work we reported here. Our announced aim was to use a connectionist model that would provide a coherent framework within which we could integrate the known facts about learning and performance of conditional discriminations. We made some good progress in the direction of interpreting various empirical phenomena within the common framework of the internal representations acquired by a multilayer adaptive network. The second goal so far has proved elusive. There is no doubt that connectionist models can generate interesting predictions (Sutton & Barto, in press), but, perhaps because of our preoccupation with simulations of existing data, we have not yet coaxed a novel prediction from the network model about learning or performance of MTS. Future work will determine whether a connectionist model of conditional discriminations can generate testable predictions and thus serve the heuristic function traditionally sought by experimental scientists.

REFERENCES

Abunawass, A. M., & Maki, W. S. (1989). Cumulative negative transfer during successive training: Analysis of a second sequential learning problem. *Proceedings of the 1989 International Joint Conference on Neural Networks*, Vol. I, 623 [abstract].

Blough, D. S. (1975). Steady-state data and a quantitative model of operant generalization and discrimination. *Journal of Experimental Psychology: Animal Behavior Processes, 1*, 3–21.

Carter, D. E., & Eckerman, D. A. (1975). Symbolic matching by pigeons: Rate of learning complex discriminations predicted from simple discriminations. *Science, 187*, 662–664.

Crick, F. C. (1989). The recent excitement about neural networks. *Nature, 337*, 129–132.

D'Amato, M. R. (1973). Delayed matching and short-term memory in monkeys. In G. H. Bower (Ed.), *The psychology of learning and motivation: Advances in research and theory* (Vol. 7, pp. 227–269). New York: Academic Press.

Gluck, M. A., & Bower, G. H. (1988). Evaluating an adaptive network model of human learning. *Journal of Memory and Language, 27*, 166–195.

Grant, D. S., & MacDonald, S. (1986). Matching to element and compound samples in pigeons: The role of sample coding. *Journal of Experimental Psychology: Animal Behavior Processes, 12*, 160–171.

Grossberg, S. (1987). Competitive learning: From interactive activation to adaptive resonance. *Cognitive Science, 11*, 23–63. (Reprinted in S. Grossberg (Ed.), *Neural networks and natural intelligence*. Cambridge, MA: MIT Press, 1988).

Hanson, S. J., & Burr, D. J. (in press). What connectionist models learn: Learning and representation in connectionist networks. *The Behavioral and Brain Sciences*.

Hecht-Nielsen, R. M. (1989). Theory of the backpropagation neural network. *Proceedings of the 1989 International Joint Conference on Neural Networks*, Vol. I, 593–605.

Honig, W. K. (1978). Studies of working memory in the pigeon. In S. H. Hulse, H. Fowler, & W. K. Honig (Eds.), *Cognitive processes in animal behavior* (pp. 211–248). Hillsdale, NJ: Lawrence Erlbaum Associates.

Jordan, M. I. (1986). Attractor dynamics and parallelism in a connectionist sequential machine. *Proceedings of the Eighth Annual Conference of the Cognitive Science Society* (pp. 531–546). Hillsdale, NJ: Lawrence Erlbaum Associates.

Kehoe, E. J. (1988). A layered network model of associative learning: Learning-to-learn and configuration. *Psychological Review, 95*, 411–433.

Klopf, A. H. (1987). A neuronal model of classical conditioning. *Animal Learning & Behavior, 16*, 85–125.

Langley, P. (1987). A general theory of discrimination learning. In D. Klahr, P., Langley, & R. Neches, R. (Eds.), *Production system models of learning and development* (pp. 99–161). Cambridge, MA: MIT Press.

Mackintosh, N. J., & Turner, C. (1971). Blocking as a function of novelty of CS and predictability of US. *Quarterly Journal of Experimental Psychology, 23*, 359–366.

Maki, W. S. (1979). Pigeons' short-term memories for surprising vs. expected reinforcement and nonreinforcement. *Animal Learning & Behavior, 7,* 31–37.

Maki, W. S. (1981). Directed forgetting in pigeons. In N. E. Spear & R. R. Miller (Eds.), *Information processing in animals: Memory mechanisms.* Hillsdale, NJ: Lawrence Erlbaum Associates.

Maki, W. S. (1984). Some problems for a theory of working memory. In H. Roitblat, T. Bever, & H. Terrace (Eds.), *Animal Cognition* (pp. 117–133). Hillsdale, NJ: Lawrence Erlbaum Associates.

Maki, W. S. (1988, November). *Extending connectionist models to animal cognition.* Paper presented at the meeting of the Psychonomic Society, Chicago.

Maki, W. S., & Leuin, T. C. (1972). Information processing in pigeons. *Science, 176,* 535–536.

Maki, W. S., Moe, J. C., & Bierley, C. M. (1977). Short-term memory for stimuli, responses, and reinforcers. *Journal of Experimental Psychology: Animal Behavior Processes, 3,* 156–177.

Maki, W. S., Riley, D. A., & Leith, C. R. (1976). The role of test stimuli in matching to compound samples by pigeons. *Animal Learning & Behavior, 4,* 13–21.

McClelland, J. L., Rumelhart, D. E., & the PDP Research Group. (1986). *Parallel distributed processing: Explorations in the microstructure of cognition. Volume 2: Psychological and biological models.* Cambridge, MA: MIT Press.

McCloskey, M., & Cohen, N. J. (1987, November). *The sequential learning problem in connectionist modeling.* Paper presented at the meeting of the Psychonomic Society, Seattle.

McCloskey, M., & Cohen, N. J. (1989). Catastrophic interference in connectionist networks: The sequential learning problem. In G. Bower (Ed.), *The psychology of learning and motivation: Advances in research and theory* (Vol. 24, pp. 109–165). New York: Academic Press.

Minsky, M. L., & Papert, S. A. (1988). *Perceptrons: An introduction to computational geometry (Expanded edition).* Cambridge, MA: MIT Press.

Parker, D. B. (1987). Optimal algorithms for adaptive networks: Second order backpropagation, second order direct propagation, and second order Hebbian learning. *Proceedings of the International Joint Conference on Neural Networks,* Vol. II, 593–600.

Pearce, J. M., & Hall, G. (1980). A model for Pavlovian learning: Variations in the effectiveness of conditioned but not unconditioned stimuli. *Psychological Review, 87,* 532–552.

Peterson, G. B. (1984). How expectancies guide behavior. In H. Roitblat, T. Bever, & H. Terrace (Eds.), *Animal cognition* (pp. 135–148). Hillsdale, NJ: Lawrence Erlbaum Associates.

Rescorla, R. A., & Wagner, A. R. (1972). A theory of Pavlovian conditioning: Variations in the effectiveness of reinforcement and nonreinforcement. In A. H. Black & W. F. Prokasy (Eds.), *Classical conditioning II: Current theory and research* (pp. 64–99). New York: Appleton-Century-Crofts.

Roberts, W. A., & Grant, D. S. (1978a). Interaction of sample and comparison stimuli in delayed matching to sample with the pigeon. *Journal of Experimental Psychology: Animal Behavior Processes, 4,* 68–82.

Roberts, W. A., & Grant, D. S. (1978b). An analysis of light induced retroactive inhibition in pigeon short-term memory. *Journal of Experimental Psychology: Animal Behavior Processes, 4,* 219–236.

Roitblat, H. (1980). Codes and coding processes in pigeon short-term memory. *Animal Learning & Behavior, 8,* 341–351.

Roitblat, H., Bever, T., & Terrace, H. (Eds.). (1984). *Animal cognition.* Hillsdale, NJ: Lawrence Erlbaum Associates.

Rosenberg, C. R. (1987). Revealing the structure of NETtalk's internal representations. *Proceedings of the Ninth Annual Meetings of the Cognitive Science Society.* (pp. 537–554). Hillsdale, NJ: Lawrence Erlbaum Associates.

Rosenblatt, F. (1958). The perceptron: A probabilistic model for information storage and organization in the brain. *Psychological Review, 65,* 386–408.

Rosenblatt, F. (1962). *Principles of neurodynamics: Perceptrons and the theory of brain mechanisms.* Washington, DC: Spartan.

Rumelhart, D. E., Hinton, G. E., & Williams, R. J. (1986). Learning internal representations by error propagation. In D. E. Rumelhart, J. L. McClelland, & the PDP Research Group (Eds.), *Parallel distributed processing: Explorations in the microstructure of cognition. Volume 1: Foundations* (pp. 318–362). Cambridge, MA: MIT Press.

Schneider, W. (1987). Connectionism: Is it a paradigm shift for psychology? *Behavior Research Methods, Instruments, & Computers, 19,* 73–83.

Servan-Schreiber, D., Cleeremans, A., & McClelland, J. L. (1989). Learning sequential structure in simple recurrent networks. In D. S. Touretzky (Ed.), *Advances in neural information processing systems 1* (pp. 643–652). San Mateo, CA: Morgan Kaufman.

Stork, D. G. (1989). Is backpropagation biologically plausible? *Proceedings of the 1989 International Joint Conference on Neural Networks,* Vol II, 241–246.

Sutton, R. S. (1986). Two problems with backpropagation and other steepest-descent learning procedures for networks. In Charles Clifton (Ed.), *Proceedings of the eighth annual conference of the Cognitive Science Society* (pp. 823–831). Hillsdale, NJ: Lawrence Erlbaum Associates.

Sutton, R. S., & Barto, A. G. (1981). Toward a modern theory of adaptive networks: Expectation and prediction. *Psychological Review, 88,* 135–170.

Sutton, R. S., & Barto, A. G. (in press). Time-derivative models of Pavlovian reinforcement. In J. W. Moore & M. Gabriel (Eds.), *Learning and computational neuroscience.*

Tolman, E. C. (1948). Cognitive maps in rats and man. *Psychological Review, 55,* 189–208.

Urcuioli, P. J., Zentall, T. R., Jackson-Smith, P., & Steirn, J. N. (1989). Evidence for common coding in many-to-one matching: Retention, intertrial interference, and transfer. *Journal of Experimental Psychology: Animal Behavior Processes, 15,* 264–273.

Werbos, P. J. (1989). Backpropagation and neurocontrol: A review and prospectus. *Proceedings of the International Joint Conference on Neural Networks,* Vol I, 209–216.

Widrow, B. E., & Hoff, M. E. (1960). Adaptive switching circuits. *IRE WESCON convention record, 4,* 96–104.

Wright, A. A., & Sands, S. F. (1981). A model of detection processes during matching to sample by pigeon: Performance with 88 different wavelengths in delayed and simultaneous matching tasks. *Journal of Experimental Psychology: Animal Behavior Processes, 7,* 191–216.

Zentall, T. P., Jagielo, J., Jackson-Smith, P., & Urcuioli, P. (1987). Memory codes in pigeon short-term memory: Effects of varying the number of sample and comparison stimuli. *Learning and Motivation, 18,* 21–33.

Zentall, T. P., Urcuioli, P. J., Jagielo, J. A., & Jackson-Smith, P. (1989). Interaction of sample dimension and sample-comparison mapping on pigeons' performance of delayed conditional discriminations. *Animal Learning & Behavior, 17,* 172–178.

Zimmer-Hart, C. L., & Rescorla, R. A. (1974). Extinction of Pavlovian conditioned inhibition. *Journal of Comparative and Physiological Psychology, 86,* 837–845.

Zipser, D., & Rumelhart, D. E. (in press). Neurobiological significance of new learning models. In E. Schwartz (Ed.), *Computational neuroscience.* Cambridge, MA: MIT Press.

11 On the Assignment-of-Credit Problem in Operant Learning

J. E. R. Staddon
Y. Zhang
Duke University

ABSTRACT

Operant conditioning is the selection of particular activities by the activity-dependent occurrence of pleasant or unpleasant events (*reinforcers*). Reinforcers are most effective if they closely follow the target activity (contiguity), but it is also important that the reinforcer and activity be correlated (contingency). No formal model for the action of response-reinforcer contiguity exists and several anomalies—cases where the behavior contiguous with the reinforcer is not strengthened, or an activity is strengthened by response-independent reinforcement—indicate flaws in the informal contiguity account. We show that an exceedingly simple, nonassociative (context-free) process provides a qualitative account for both anomalous and nonanomalous properties of operant conditioning. The process can be extended to permit associative effects, including classical conditioning and stimulus control; it may therefore represent the initial processing stage for all conditioning in higher vertebrates.

Simple learning—operant and classical conditioning—is thought to be the key to the evolution of human intelligence, hence has been the focus of much neurobiological research with vertebrate and invertebrate animals (e.g., Weinberger, McGaugh, & Lynch, 1985). Most emphasis has been placed on classical conditioning, in which a neutral stimulus, the conditioned stimulus (CS), acquires the capacity to elicit a conditioned response (CR) because of a training history, in which the CS has been a reliable predictor of an unconditioned stimulus, such as food delivery, that elicits its own, unconditioned, response. Nevertheless operant conditioning—the modification of freely occurring ("emitted") behavior by re-

279

warding or punishing consequences (*reinforcement*)—is probably the more primitive function.

The simplest form of adaptive behavior is what Fraenkel and Gunn (1961) call kinesis—a much-studied example is bacterial chemotaxis (e.g., Berg, 1983; Koshland, 1980). Kineses fit the definition of operant behavior—behavior guided by its consequences; they are not examples of classical conditioning or related processes such as habituation or sensitization (cf. Staddon, 1983, chap. 4).

Note that there is a distinction between operant *behavior* and operant *conditioning*. Chemotaxis involves the modification of behavior by its consequences, but does not involve the kind of context dependency seen in the operant conditioning of higher vertebrates and some large-brained invertebrates: A rat learning the location of food may at first behave in a random-search fashion reminiscent of chemotaxis; but, unlike a food-seeking bacterium, the rat will remember the location of food when replaced in the same environment a day later. The rat, but not the bacterium, has some internal representation of the context in which it found food the first time and can use that representation to shortcut random search when replaced in the same situation after a delay. Our model deals with the first stage, finding food, but not the second, recalling its location.

The two conditioning *procedures,* operant and classical, are no longer thought to tap separate underlying *processes* (cf. Mackintosh, 1983; Staddon, 1983). If they share underlying mechanisms, and if operant conditioning is the more primitive function, then any reasonable learning model should naturally be capable of showing both types of conditioning. Yet most models are designed to account for the results of classical conditioning experiments, and have no explicit provision for operant conditioning, with its strong dependence on response-reinforce relations and endogenous variability.

The way that operant reinforcement selects one activity over others—the *assignment-of-credit* problem—is taken for granted in most learning models, which are examples of what has come to be known as *supervised* learning. For example, in the old Bush–Mosteller (1951) stochastic learning model and its many descendants (e.g., Horner & Staddon, 1987; Sternberg, 1963) reinforcement is assumed to selectively strengthen the "reinforced response" without any explicit discussion of how the organism knows what that response is.

The term assignment of credit arose from early discussions in artificial intelligence (AI) of how a learning system can decide what event—or activity, or change in its own internal state—is the actual cause of something good (or bad) that has happened to it (Minsky, 1961). Several accounts of operant and classical conditioning (Dickinson, 1980; Revusky, 1977; Staddon, 1983) in effect argue that the so-called laws of conditioning have evolved to optimize credit assignment in natural environments—although these accounts do not use the AI term.

In supervised learning, a "teacher" (reinforcement) is assumed to present explicit error information to the learning system in a way that has no obvious biological parallel (see for example the recent comments of Crick, 1989, on the biological implausibility of backwards error propagation in supervised learning). With a few exceptions (e.g., Sutton, 1984) assignment of credit is satisfactorily solved for most learning theorists by the intuitively appealing but unformalized mechanism of temporal contiguity. It is extraordinary that there is no developed formal theory or model of the temporal-contiguity account of response selection in operant conditioning. The informal theory amounts to little more than the assertion that activities are "strengthened" by contiguity with reinforcement, and the more closely the reinforcement follows the activity, the greater the strengthening.

Apparent anomalies always pose problems for an informal theory, and the prevailing vague view of response-reinforcer contiguity learning is no exception. The furor aroused in the late 1960s and early 1970s by so-called "biological constraints on learning," evidenced by anomalous phenomena such a autoshaping, superstition and instinctive drift (cf. Breland & Breland, 1961; Seligman & Hager, 1972; Staddon & Simmelhag, 1971), dealt "general learning theory" a blow, from which it has yet to recover. These phenomena are anomalies either because the activity contiguous with reinforcement is not the one actually strengthened or because activities continue to occur despite response-independent reinforcement. They have never really been reconciled with the standard contiguity account of operant reinforcement (Gardner & Gardner, 1988).

We present a model that reflects these three concerns: (a) The need for a well defined process to solve the assignment-of-credit problem in operant conditioning; (b) the need to reconcile supposed exceptions to a contiguity account; and (c) the requirement that any valid learning model be as capable of operant as classical conditioning. We show that the simplest possible explicit model for the action of contiguity immediately suggests an explanation for apparent exceptions and can be extended to incorporate the associative properties of classical and operant conditioning.

A SIMPLE PARALLEL MODEL: STANDARD EFFECTS

Reinforcement is the defining property of operant conditioning: certain kinds of events—food, access to a mate, freedom from restraint, electric shock—have selective effects if their occurrence (or nonoccurrence, in the case of electric shock) is made to depend on the occurrence of some identifiable activity. Under most conditions—most responses, most kinds of response-reinforcer dependency—the effective activity is facilitated relative to others. Many years of study of this phenomenon have revealed a number of what might be called *standard properties* of reinforcement, as well as a number of *anomalous properties*.

Positive reinforcement is defined by its *selective* effect: when the occurrence of a reinforcer is made to depend on the occurrence of a particular response, the response probability should increase, and when the reinforcer is no longer presented, or is presented independently of responding, response probability should decline. Selection must be reversible between at least one pair of activities: capable of strengthening reinforced activity A at the expense of activity B that is not reinforced, and vice versa. In addition to this property of *selection*, reinforcement is also sensitive to *delay*, and *contingency*. We first define our model and show how it provides a parameter-free account of all these standard properties, then give an account of the anomalous properties.

Biologically valuable events, such as the occurrence of food or the availability of a mate, have two general effects on organisms: they arouse the animal in various ways; and the arousal dissipates with the lapse of time (Killeen, Hanson, & Osborne, 1978). Most theories of adaptive behavior also recognize some notion of response strength and competition among incompatible activities. Variability, the ability to generate a repertoire of different activities, is an essential ingredient of operant behavior. We embody these five properties—arousal, adaptation, strength, competition and variability—in two linear discrete-time equations: We define for each behavior a variable, V_i, its strength. The competition rule is winner-take-all: the activity with highest V value is the one that occurs. This is the only non-linearity in the model. The equations describe the changes in V values from one discrete-time instant to the next in the absence of reinforcement, or following reinforcement:

$$\text{After nonreinforcement: } V_i(t + 1) = a_i V_i(t) + \varepsilon(1 - a_i), 0 < a_i < 1, \quad (1)$$

$$\text{After reinforcement: } V_i(t + 1) = a_i V_i(t) + \varepsilon(1 - a_i) + b_i V_i(t), \quad (2)$$

where $V_i(t)$ is the strength of the ith activity in discrete-time instant t, a_i and b_i are parameters that depend on *both* the activity *and* the reinforcer and ε is a random variable sampled independently (iid) for each activity in each time instant. Equation 1 describes a filtered noise process, since $V_i(t + 1)$ is just a weighted average of noise plus its previous value. Term $a_i V_i(t)$ represents adaptation: because $a_i < 1$, this term reduces to zero with repeated iterations. a_i can also be thought of as a measure of short-term memory (STM): the larger a_i, the more persistent the effects of any forced change in V_i. The effect of a on the response of Equation 1 to an impulse input (i.e., a single unit increment) is illustrated in Figure 11.1. The top curve shows the highest a value: the effect of the increment persists with little decrement over time. The lowest curve shows the smallest a value: the brief increment is almost at once swamped by the noise term. The other two curves show intermediate effects.

Term $b_i V_i(t)$ in Equation 2 represents the arousal effect of a hedonic event, which we assume acts on *all* activities. If $b_i > 0$, the effect is to increase V_i (a positive reinforcer); if $b_i < 0$, the effect is to reduce V_i (a punisher).

Note that the relation $a_i + b_i < 1$ must hold if V_i is not to rise without limit in repeated iterations of Equation 2.

The way this process works is best seen by example. Consider first the case of

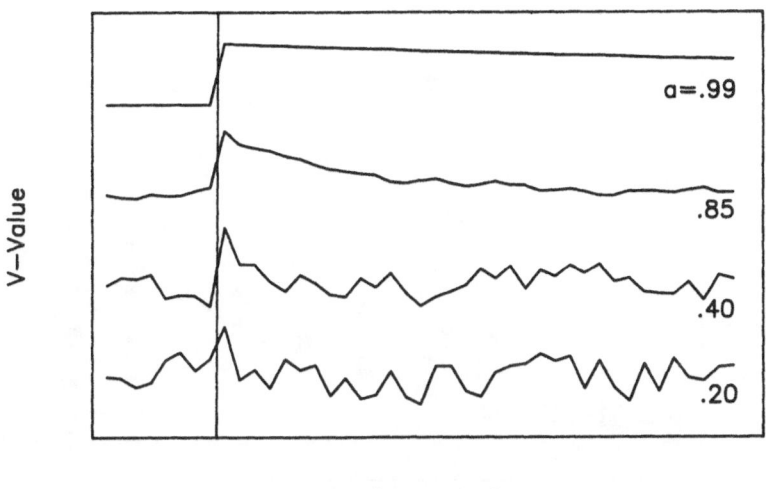

Time(Iterations)

FIG. 11.1 Effect of the STM parameter a on the impulse response of Equation 1 in the test: $V_i(t+1) = a_iV_i(t) + \in(1-a_i) + X$, where $X = 1$ during one time instant (vertical line) and is 0 otherwise. When a is high, an impulse is remembered for a long time; when a is low the effect is swamped by noise after only a few iterations.

two or more identical activities (i.e., $a_1 = a_2$; $b_1 = b_2$), which permits derivation of all the standard properties of operant reinforcement. In the absence of reinforcement, because the two parameters are the same for both activities, each activity will occur equally often, on average. If positive reinforcement is delivered for each occurrence of Activity 1, then at that instant by the highest-wins competition rule $V_1 > V_2$, hence the increment to 1 must be greater than the increment to 2: $bV_1 > bV_2$. If the reinforcement occurs frequently enough that the increment in V_i does not decay to zero by the next reinforcement, V_1 will be steadily incremented relative to V_2, so that Activity 1 will come to dominate. This conclusion holds whichever activity is reinforced; thus the process satisfies the reversibility condition.

The essential feature of the reinforcement mechanism is that immediate, response-contingent, reinforcement adds *some* increment to *all* activities, but the *largest* increment goes to the higher-V activity. Making each increment proportional to the V value is the simplest way to accomplish this, but other methods would probably give similar results. Adding a constant increment to all does *not* work.

Figure 11.2 shows the asymptotic effects of reinforcing every occurrence of one behavior, in a set of four, for a wide range of parameter pairs. The reinforced behavior is always facilitated, and the proportion of time taken up increases with increases in either parameter. We have obtained similar results for ensembles of two, four, and eight identical activities.

Figure 11.3 shows the effects of reinforcement delay. The same increasing pat-

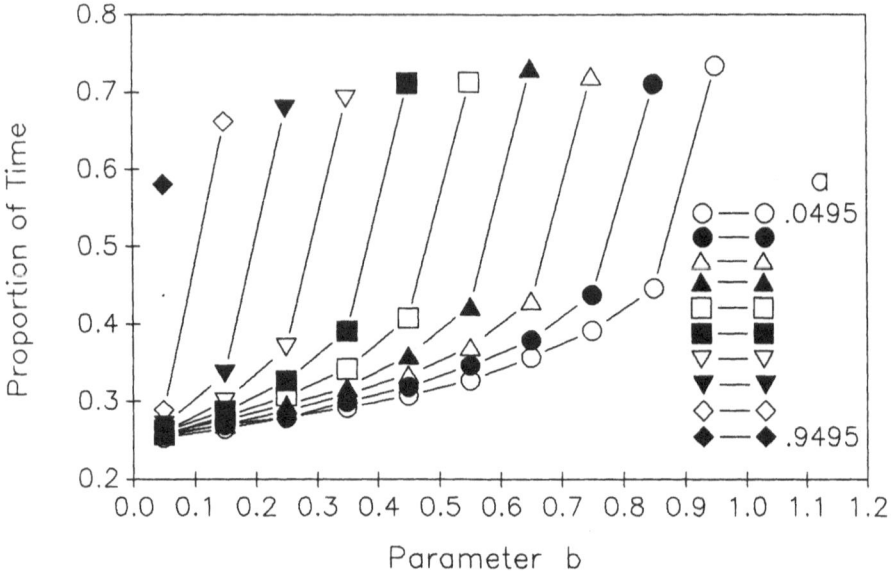

FIG. 11.2. Parameter Space. Simulation of the effects of 100% reinforcement on one of four identical (same a_i and b_i) activities for a range
of a and b values (0.0495–0.9495 in increments of 0.1). Ordinate shows
the proportion of time (iterations) taken up by the reinforced activity.
Note that in the absence of any reinforcing effect each activity should
take up about 25% of the time. Each point in this and the next two
figures is the average of 2×10^5 iterations. The random variable $\in$ in
Equations 1 & 2 had a rectangular distribution over the interval 0 –1.

tern is seen in the parameter-space plot (Top), but now STM-parameter a has the
greatest effect (compare the highest point on each curve: the profile is decreasing in
Figure 11.3, but increasing Figure 11.2). The higher the a value, the more likely
the effective response will persist across the delay, and the less likely, therefore,
that the delayed reinforcer, when it actually occurs, will strengthen some other
behavior. The Bottom panel of Figure 11.3 shows selected data from the Top panel
plotted in standard delay-of-reinforcement-gradient form. The declining pattern is
relatively independent of the values of a and b so long as they sum to a constant.

Note that if two identical activities with high a values are both reinforced, but
with a small difference in delay, then the activity that is more contiguous with
reinforcement will eventually predominate. The system is potentially very sensitive to small contiguity differences. Animals also may be similarly sensitive
(Killeen, 1978), which is paradoxical given the existence of so-called "superstitious" behavior, which has sometimes been interpreted as a failure to discriminate
much larger contiguity differences (see discussion in Staddon & Simmelhag,
1971). Our model is consistent with both "superstition" and high sensitivity to
contiguity, hence can resolve this paradox.

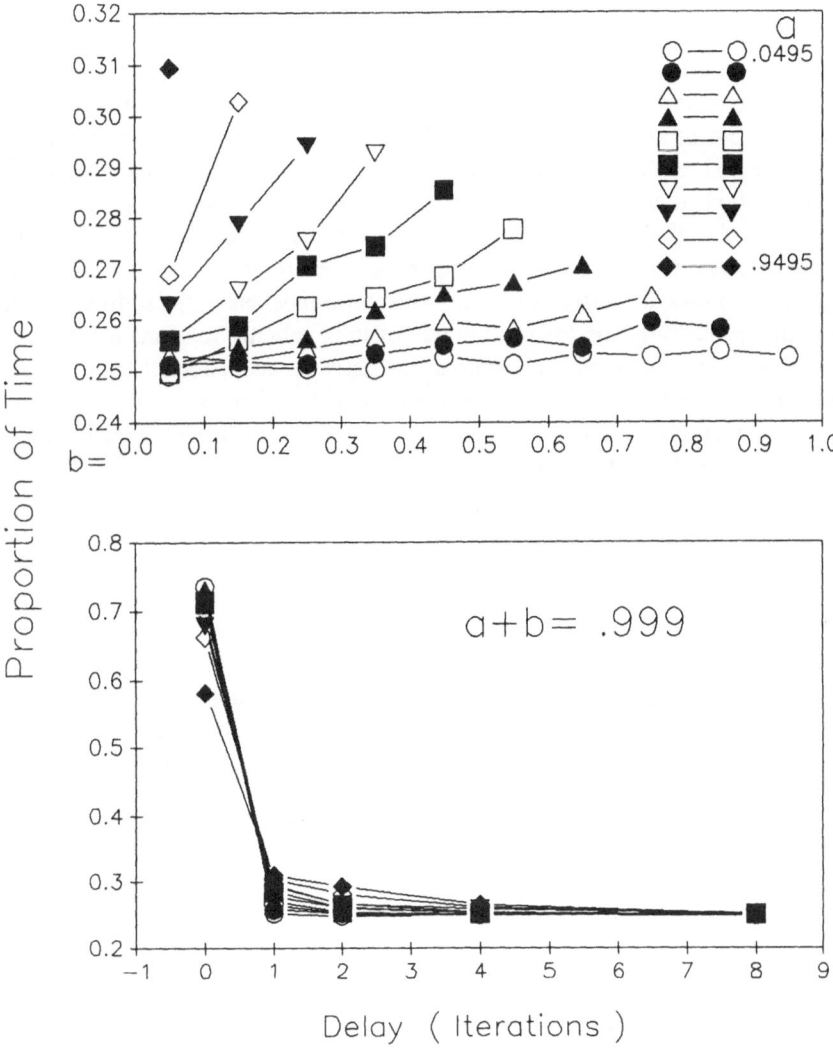

FIG. 11.3. Effect of reinforcement delay. Top panel: Simulation of the effects of 100% reinforcement on one of four identical activities for a range of a and b values when the reinforcement is delayed one iteration. Bottom panel: Delay-of-reinforcement gradient (1–8 iterations) for a range of parameter pairs, for which $a + b = 0.999$. In all these simulations the delay procedure used only a single delay timer; thus, a delayed reinforcer had to be collected before the timer restarted. This procedure forces a reduction in the maximum possible reinforcement frequency as the prescribed delay increases. If this constraint is eliminated by using multiple delay timers, the delay gradient is considerably shallower but retains the same declining form.

Contingency is the fact that the strengthening effect of reinforcement depends on its correlation with the reinforced behavior, not just on contiguity. Thus, if the target behavior is reinforced intermittently, or if it is reinforced every time but reinforcement also occurs at other times, the behavior will be strengthened less than if it is reinforced invariably or exclusively. Figure 11.4 shows that our model has both these properties: Vertical comparison shows the effect of the probability of response-dependent reinforcement: the higher the probability, the larger the proportion of time taken up by the reinforced activity. Horizontal comparison shows the effect of "free" (response-independent) reinforcers: the higher this probability the lower the level of the explicitly reinforced act.

Thus, this system provides a simple qualitative account for all the standard steady-state properties of operant reinforcement. The account is essentially parameter free, because the pattern of results is the same for all parameter pairs, so long as all activities are identical.

Anomalous Effects

Most anomalies arise when activities have *different* parameter values. We discuss three anomalies: superstitious behavior, typologies, such as the distinction between emitted and elicited behavior, and instinctive drift.

"Superstitious" behavior (Skinner, 1948; Staddon & Simmelhag, 1971; Timberlake & Lucas, 1985) is typically produced when an animal such as a pigeon

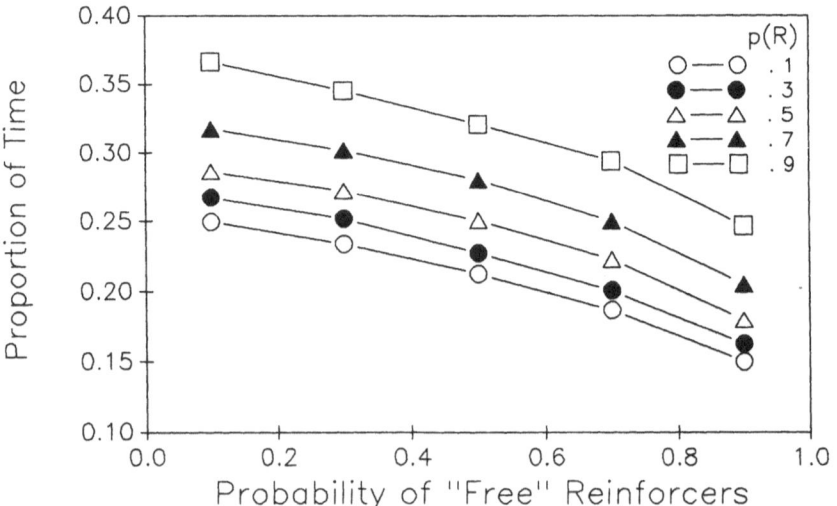

FIG. 11.4. Effects of contingency. Each curve shows the effect on the level of the reinforced behavior of variation in the probability of response-independent reinforcement, for a given probability of response-dependent reinforcement. Parameter values: $a = b = 0.4495$.

is given periodic response-independent food reinforcement. If food delivery is frequent enough, the animal develops various kinds of vigorous stereotypies, despite the fact that food delivery is independent of its behavior. Our model is too simple to account for the temporal properties of this phenomenon (see Staddon & Simmelhag, 1971, for details), but the model can, under restricted conditions, produce apparently stereotyped behavior, even with a set of identical activities. An example is shown as the four cumulative records on the Left in Figure 11.5. The curves show the levels of four identical activities, with high-*a* and low-*b* values, under continuous reinforcement. It is clear from the figure that over any given epoch on the order of 100–1,000 iterations the distribution of activities will be highly skewed, with one activity tending to predominate. Examples are indicated by the vertical lines: over the epoch labeled "A", for example, Activity 1 predominates; whereas during epoch B, Activity 2 is the dominant one. Eventually, every activity will have its place in the sun; but over a limited epoch, one activity will seem to have been preferentially strengthened.

The Right panel in Figure 11.5 shows the effect of reducing reinforcement probability: the epochs, over which behavior appear stereotyped, are now much shorter.

The stereotypy illustrated in the Left panel is the outcome of a positive-feedback process that resembles a suggestion of Skinner. Skinner's (1948) explanation for all superstitious behavior is that when a (response-independent) reinforcer is delivered some behavior will be occurring and will be automatically strengthened; if the next reinforcer follows soon enough, the same behavior will

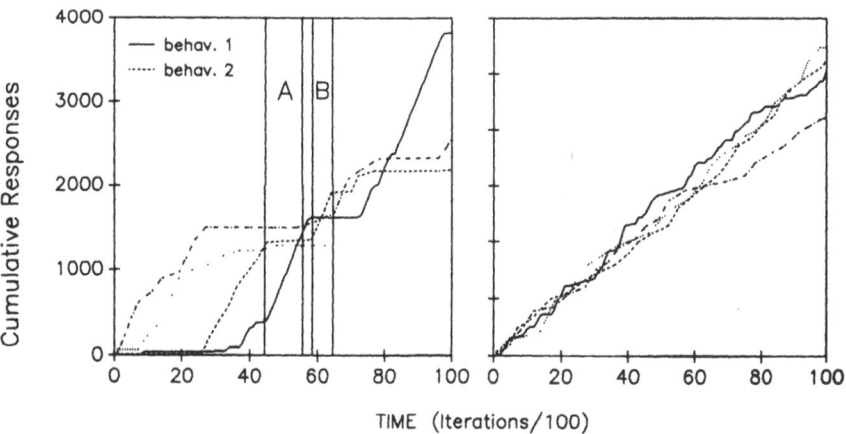

FIG. 11.5.   Effects of absolute probability of "free" reinforcement ("superstitious" behavior). Left panel: Simulated cumulative records of four identical activities (a = 0.9495, b = 0.0495) when reinforcement occurs during each iteration. Right panel: Cumulative records when reinforcement occurs with p(R) = 0.5.

still be occurring and will be further strengthened, and so on, until the behavior appears to dominate.

Two ingredients are lacking in Skinner's account: an explicit statement about the temporal properties of behavior that are necessary for the process to work (the STM parameter, a, in the present model); and recognition of the importance of the time window over which observations are carried out. Our analysis shows that (a) the STM parameter must be very high for the process to work at all, even with reinforcers occurring at the maximum rate. (b) The time window is critical: if observations are continued for long enough, no activity will predominate. Thus, the effect of "free" reinforcers is simply to increase the autocorrelation of activities, rather than to single out one over others permanently (see Staddon & Horner, 1989, for a theoretical discussion of this issue in the context of stochastic learning models). And (c) even a modest reduction in reinforcement probability causes a sharp reduction in stereotypy. Since reinforcement is in fact intermittent in all published demonstrations of superstition, Skinner's process, at least in our version of it, is unlikely to be responsible. We suggest an alternative later.

Nevertheless, this analysis of superstition shows how our model accounts for the reliable finding of *hysteresis* in operant conditioning: the fact that once a response has been strengthened via response-contingent reinforcement, it may persist even when reinforcers are delivered independently of responding (Herrnstein, 1966). Given a high enough a value, even intermittent "free" reinforcers may be enough to maintain an already strengthened activity, at least for a while, and for longer than if reinforcement were withdrawn entirely.

The distinction between *emitted* and *elicited* behavior parallels the procedural difference between operant and classical conditioning (Skinner, 1938): elicited behavior (salivation is an example) is behavior elicited by a reinforcer but not modifiable by operant conditioning; emitted behavior (lever pressing is an example) is not usually elicited by the reinforcer, but is modifiable by operant conditioning (the rat learns to press the lever to get food).

The existence of a typology is usually a sign that we lack understanding of the underlying process[1]. It is interesting, therefore, that the distinction between these two types of behavior emerges in a natural way from our model. Recall that the

---

[1]We cannot resist the temptation to give a favorite quote from the great Francis Bacon, just to dispel any impression that this insight is at all novel: "The axioms now in use, having been suggested by a scanty and manipular experience and a few particulars of most general occurrence, are made for the most part just large enough to fit and take these in: and therefore it is no wonder if they do not lead to new particulars. And if some opposite instance, not observed or not known before, chance to come in the way, *the axiom is rescued and preserved by some frivolous distinction;* whereas the truer course would be to correct the axiom itself (*Novum Organum* Aphorism XXV, Bk. I, our italics)" The emitted/elicited distinction may perhaps be termed "frivolous" without intending any disrespect to its generations of adherents. What we propose is not so much a change in the "axiom," as Bacon calls it, but simply an attempt to specify, in a true dynamic model, what the axiom—the principle of response selection in operant conditioning—actually is.

two parameters a, the STM parameter, and b, the arousal parameter, must add to less than one if V is to be bounded. Moreover, as Figure 11.1 illustrates, parameter a strongly determines the effect of operant (consequential) reinforcement: when a is small, even large b values have little effect. The importance of a is exaggerated by delay or intermittency of reinforcement: when a is low, even the highest b value is insufficient to permit much strengthening of the behavior by operant reinforcement (Figure 11.3). Moreover, because $a + b < 1$, a highly elicitable (high-b) activity must have a low a value, hence will be only weakly susceptible to strengthening by operant reinforcement. Conversely, activities with high a values will be readily conditionable (we show an example shortly), but will not generally be elicitable, because they must also have low b values. Thus, the dichotomy between emitted and elicited behavior may be a consequence of the complementary relation between the arousal and STM parameters that is forced by stability considerations.

Instinctive drift is perhaps the most striking exception to a contiguity account of reinforcement. Breland and Breland (1961) reported several instances that conform to the following pattern: Behavior A (e.g., a raccoon putting a wooden egg into a chute) is successfully "shaped" by response-contingent food reinforcement; but after a while Behavior A is supplanted by Behavior B ("washing" the egg between the animal's forepaws), which is part of the raccoon's natural foraging repertoire. Behavior B is inappropriate in this context because it competes with Behavior A, on which food delivery actually depends. Behavior A is contiguous with the reinforcer, but Behavior B is ultimately the one strengthened.

A variety of naturalistic interpretations have been offered for these effects (and for the related phenomenon of "superstition"), but they follow rather simply from our model on the assumption that "instinctive" behaviors are characterized by very high a values (STM persistence). Given a set of activities with moderate a and b values, and one activity with a very high a value (and nonzero b), reinforcement of one of the moderate-a activities will cause it to predominate initially (because it has a higher b value than the "instinctive" activity). But, because increments to the high-a activity cumulate more effectively than the (larger) increments to the reinforced activity, it may predominate eventually, even if it is never contiguous with reinforcement.

These effects are illustrated in the cumulative records in the top panel of Figure 11.6. Each set of four records shows four activities, three with moderate a and b values, one with a high a value and low b value. The Left records show the free-operant (unreinforced) levels of the four activities; notice the low frequency of the high-a activity. (Although the form of Equation 1 ensures that all activities will have the same average V-value [this assumption could obviously be relaxed], the variance of the V-values will be inversely related to a. Thus, a single high-a activity will occur less often on average than an activity with lower a-value when there is more than one low-a activity.) The Center and Right panels show the effect of reinforcing one of the low-a activities with a probability of 0.25 or

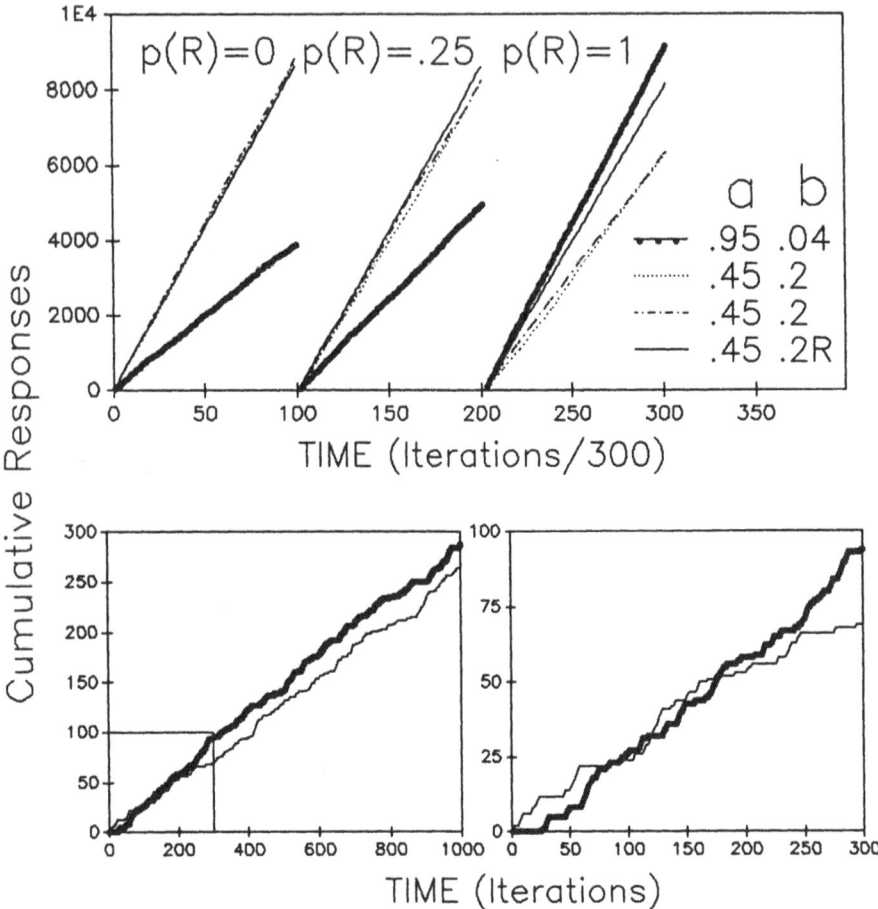

FIG. 11.6   Instinctive drift. Top panel, Left: Cumulative records of four activities in the absence of reinforcement (operant levels): high-*a*, low-*b* activity is the least frequent (see parameter values in the table). Center: The effect of reinforcing one of the three intermediate-*a* activities (*a* = 0.45, *b* = 0.2R) with probability 0.25. Right: Effect of reinforcing every occurrence of the intermediate activity. Bottom panel: Magnified pictures of the beginning of the record in the Right panel at the top ("instinctive" and reinforced activities only).

1.00. The increasing reinforcement probability has two effects: it causes the level of the reinforced activity ("R") to increase above the level of the other two low-*a* activities; but it causes a disproportionate increase in level of the high-*a* activity, which becomes predominant when p(R) = 1 (Right). The two bottom panels in Figure 11.6 show a magnified picture of initial acquisition in the p(R) = 1 condition, illustrating the transition from dominance by the reinforced, low-*b*, activity to predominance of the "instinctive", high-*a,* activity. If a low-*a*-value activity is reinforced, it may predominate initially, but will be supplanted by the high-*a* value activity, even if it has a lower "arousability" (*b* value).

Unsignaled Shock-Avoidance

Avoidance behavior in the absence of a "safety" signal has always posed a problem for the contiguity view of reinforcement because there is nothing for the avoidance response to be contiguous with. An explicit reinforcing event is lacking in what is termed a shock- rate-reduction schedule, for example (Herrnstein & Hineline, 1966). In such a schedule brief shocks are delivered with probability $p(Sh) = x_s$ during each sampling interval, δt (δt is usually on the order of a second or two). If the animal makes the designated avoidance response, $p(Sh)$ is reduced to a value $x_R < x_s$ until the next shock occurs, when it returns to x_s. Rats learn (with some difficulty) to make the avoidance response on such procedures, despite the fact that no tangible event occurs contiguous with the avoidance response. Our model can accommodate this behavior if it is assumed that electric shock and other aversive events reduce, rather than increase, V values (i.e., $b < 0$). In such a case, electric shocks reduce all V values, but the V value of the avoidance response will be reduced less than others because on average it is followed by shock after a longer delay. Hence, it will be favored in the competition and, in an ensemble of identical activities, will come to predominate.

CONCLUSION

The standard steady-state, context-free properties of operant conditioning—selection, delay-of-reward, contingency and unsignaled avoidance—are all consistent with a simple assignment-of-credit mechanism that assumes each activity can be represented by a "strength" variable, V, that corresponds to filtered noise. Each V value is transiently augmented or decremented by pleasant or unpleasant events in an amount proportional to its value. The activity with the highest V value is the one that actually occurs. These qualitative predictions are parameter free.

By allowing different activities to have different time constants and arousal effects, the major anomalies can also be explained.

The present model is nonassociative because the properties of operant conditioning that it explains have no necessary relation to context dependency. It can be extended so as to take on associative properties, however, although we can only sketch the possibilities here. The model can be implemented by a one-layer "n-flop" (mutually inhibitory) network, with global, nonmodifiable inputs for "noise" (term ε) and reinforcement (bV_i). An additional layer of nodes representing stimuli (including the reinforcement event), fully connected to the first with modifiable weights, allows for the formation of the modifiable long-term stimulus-response and reinforcer-response connections involved in classical conditioning and stimulus control (i.e., the context sensitivity of operantly conditioned responses). Models of this general type should be capable of, and should show at one and the same time, both classical and operant conditioning, hence can potentially accommodate the rich set of data now available on modification of the conditioned response with continued training, CS-sensitivity of the CR, stimulus-stimulus modulation of various sorts, and different kinds of operant-classical interaction (e.g., Holland, 1979; Mackintosh, 1983; Rescorla, 1985).

Because the operant behavior of vertebrates is always context dependent, our model is obviously only part of the whole story, so it is inappropriate to look for quantitative agreement between facts and predictions. The model also does not deal with anything that depends on long-term memory, such as extinction and reconditioning, stimulus effects, or timing processes. Because the model does not pretend to encompass the whole process of conditioning, it is hard to know what to make of a number of partial exceptions (e.g., the pecking response is both elicited and operantly conditionable, intermittent reinforcement sometimes sustains more behavior than continuous, etc.)—since they may reflect later stages in the process. What we have shown is that there is a surprising correspondence between the qualitative properties of this very simple response-selection process and the steady-state properties of operant behavior—suggesting that such a process may form the "front end" for the complex neural processing involved in the operant and classical conditioning of higher animals.

ACKNOWLEDGMENTS

Preliminary versions of this work appeared in Staddon (1988) and Staddon & Zhang (1989). We thank Jennifer Higa, Peter Holland, Nancy Innis, Peter Killeen, and Alliston Reid for comments on earlier versions. Research supported by grants from NSF.

REFERENCES

Bacon, F. (1825/1620). Novum organum. In *The works of Francis Bacon, Lord Chancellor of England* (B. Montagu, Ed.) London: William Pickering.

Berg, H. C. (1983). *Random walks in biology.* Princeton: Princeton University Press.

Breland, K., & Breland, M. (1961). The misbehavior of organisms. *American Psychologist, 16,* 681–664.

Bush, R. R., & Mosteller, F. (1951). A mathematical model for simple learning. *Psychological Review, 68,* 313–23.

Crick, F. (1989). The recent excitement about neural networks. *Nature, 337,* 129–32.

Dickinson, A. (1980). *Contemporary animal learning theory.* New York: Cambridge University Press.

Fraenkel, G. S. & Gunn, D. L. (1961). *The orientation of animals*. New York: Dover.

Gardner, R. A., & Gardner, B. T. (1988). Feedforward versus feedbackward: An ethological alternative to the law of effect. *Behavioral and Brain Sciences, 11*, 429–493.

Herrnstein, R. J. (1966). Superstition: A corollary of the principles of operant conditioning. In W. K. Honig (Ed.), *Operant behavior* (pp. 33–51). New York: Appleton-Century-Crofts.

Herrnstein, R. J., & Hineline, P. N. (1966). Negative reinforcement as shock-frequency reduction. *Journal of the Experimental Analysis of Behavior, 9*, 421–30.

Holland, P. C. (1979). Differential effects of omission contingencies on various components of Pavlovian appetitive conditioned responding in rats. *Journal of Experimental Psychology: Animal Behavior Process, 5*, 178–193.

Horner, J. M., & Staddon, J. E. R. (1987). Probabilistic choice: A simple invariance. *Behavioural Processes, 15*, 59–92.

Killeen, P. R. (1978). Superstition: A matter of bias, not detectability. *Science, 199*, 88–90.

Killeen, P. R., Hanson, S. J., & Osborne, S. R. (1978). Arousal: Its genesis and manifestation as response rate. *Psychological Review, 85*, 571–81.

Koshland, D. E. (1980). *Bacterial Chemotaxis as a Model Behavioral System*. New York: Raven Press.

Mackintosh, N. J. (1983). *Conditioning and associative learning*. New York: Oxford University Press.

Minsky, M. (1961). Steps towards artificial intelligence. *Proceedings of the Institutes of Radio Engineers, 49*, 10–30.

Rescorla, R. A. (1985). Associative learning: Some consequences of contiguity. In N. M. Weinberger, J. L. McGaugh, & G. Lynch (Eds.), *Memory systems of the brain* (pp. 211–230). New York: Guilford Press.

Revusky, S. H. (1977). Learning as a general process with an emphasis on data from feeding experiments. In N. W. Milgram, L. Krames, & T. M. Alloway (Eds.) *Food Aversion Learning*. (pp. 1–71). New York: Plenum.

Seligman, M. E. P., & Hager, J. L. (1972). (Eds.), *Biological boundaries of learning*. New York: Appleton-Century-Crofts.

Skinner, B. F. (1938). *The behavior of organisms*. New York: Appleton-Century-Crofts.

Skinner, B. F. (1948). "Superstition" in the pigeon. *Journal of Experimental Psychology, 38*, 168–72.

Staddon, J. E. R. (1983). Adaptive behavior and learning, New York: Cambridge University Press.

Staddon, J. E. R. (1988). On the process of reinforcement. *Behavioral and Brain Sciences, 11*, 467–469.

Staddon, J. E. R., & Horner, J. M. (1989). Stochastic choice models: A comparison between Bush–Mosteller and a source-independent reward-following model. *Journal of the Experimental Analysis of Behavior, 52*, 57–64.

Staddon, J. E. R., & Simmelhag, V. L. (1971). The "superstition" experiment; A reexamination of its implications for the principles of adaptive behavior. *Psychological Review, 78*, 3–43.

Staddon, J. E. R., & Zhang, Y. (1989). Response selection in operant learning. *Behavioural Processes, 20*, 189–197.

Sternberg, S. (1963). Stochastic learning theory. In R. D. Luce, R. R. Bush, & E. Galanter (Eds.), *Handbook of mathematical psychology* (Vol. 2, pp. 1–120). New York: Wiley.

Sutton, R. S. (1984). *Temporal credit assignment in reinforcement learning*. Unpublished PhD dissertation, Department of Computer and Information Science, University of Massachusetts.

Timberlake, W., & Lucas, G. A. (1985). The basis of superstitious behavior: Chance contingency, stimulus substitution, or appetitive behavior? *Journal of the Experimental Analysis of Behavior, 44*, 279–99.

Weinberger, N. M., McGaugh, J. L., & Lynch, G. (1985). *Memory systems of the brain*. New York: Guilford Press.

12

Behavioral Diversity, Search and Stochastic Connectionist Systems

Stephen José Hanson
*Learning and Knowledge
Acquisition Research Group
Siemens
Princeton, New Jersey*
*Cognitive Science Laboratory
Princeton University*

ABSTRACT

Biological systems engaged in "strategic" type behavior—foraging, avoiding threats, sexual selection—tend to become "activated" or aroused. Concomitant with the observable consequences of arousal in terms of the rate and vigor of activity (Killeen, 1975) are changes in behavioral diversity (Staddon & Simmelhag, 1971; Hanson, 1980). The dynamics of variability in behavior during challenge can serve at least three purposes: (1) it allows the organism to entertain multiple hypotheses in a static environment, (2) maintains a prediction history (in terms of activity variability) which is local, recent and cheap, and (3) allows the organism to revoke or revise strategies which may not lead to globally optimal outcomes. Problem solving and learning in many different contexts can be seen as a tension between directed search and random variation and combination of previously attempted strategies. In environments with a large number of constraints and variables to negotiate, directed search is intractable and must be amended with more local or more heuristic strategies that may be reflected in the judicious introduction of noise inherent in biological or neural systems.

"Plus ca change, plus c'est la meme chose"

CONNECTIONISM AND ANIMAL LEARNING THEORY: A NEXUS

Recent years have seen an explosion of research in neural networks, or what has become more generally known as *connectionism*. The roots of connectionism can be found in many disciplines, giving the area a strength due to the many fields that it intersects and a weakness in that the connectionism seems diffuse and ill-defined. Much of what amounts to the central issues in connectionist research (cf. Hanson & Burr, 1990) finds a comfortable home in the cognitive science field, which is itself a mixture of computer science, psychology, philosophy, linguistics and other fields. Nonetheless, despite this apparent diversity, the roots of connectionism are deeply and uniquely entwined with psychology and in particular animal learning theory.

Some common features between connectionism and animal learning theory include general learning mechanisms, interaction of learning and perception, associationism, computational analogues of classical and operant conditioning, and a focus on behavior or performance of the system in some closed environment. Notwithstanding these similarities, connectionism is not as some might believe "born-again" behaviorism. Rather there are crucial differences that may allow both connectionism and animal learning theory to interconnect in productive ways. For example, animal learning theory has tended to focus on problems that are related to relatively simple organisms in impoverished environments. Obviously, such focus is an advantage for quantitative analysis of behavior and modeling, however, this sort of simplification leads to a "scale" problem that has caused a schism between behavioral psychologists and cognitive psychologists for some years now. It has never really been clear how associationism, or reinforcement processes would "scale up" to phenomena like problem solving, language acquisition, and reasoning. Connectionism, however, can seemingly move between simple organisms in impoverished environments (Barto, 1985; Gluck & Thompson, 1987; Kehoe, 1989; Tesauro, 1990) to complex issues involving language acquisition and human categorization phenomena (Elman, 1988; Gluck & Bower, 1988; Hanson & Kegl, 1987; Rumelhart & McClelland, 1986).

Animal learning theory has also eschewed representational problems. For various historical reasons behavioral psychologists have studied temporal and feedback relations between behavior and environment without speculating about the sorts of representations that might support the behavior-environment interaction. Consequently, key problems related to memory retrieval, perceptual encoding and learning have been left relatively unexplored. Connectionism provides an interface between the performance/environment learning paradigms and a general representation language, in which hypotheses about the nature of the representation and the dynamic interaction with learning processes can be entertained.

The present chapter is an attempt to open up at least one channel between connectionism and animal learning theory, having to do with the variability or diversity of behavior and the nature of search in complex high-dimensional parameter spaces. There are roughly two parts to the chapter. In the first part of the chapter, behavioral data will be discussed that establishes some aspects of behavioral diversity in biological systems. The second part of the chapter discusses some aspects of connectionism and its relation to search and stochastic learning algorithms. There are three themes that will recur throughout the chapter:

- Changes in feedback and information (reinforcement success or failure) can produce increases in diversity.
- "Reinforcement" as presently used in many connectionist accounts is vastly over-simplified especially with multi-response domains and high-(feature space)-dimensionality.
- Diversity under constraint (e.g., goals) can lead to efficient rapid search.

First, we consider the historical context that behavioral variability was established as a potential determinant in the reinforcement process and the potential nature of variational principles and their function.

SUPERSTITIOUS PIGEONS

The diversity of behavior at any one moment makes it unlikely during even very simple schedules of reinforcement that only a single response (unit) is modified. It is unclear how extensive in time the reinforcement process is and therefore how many responses are being affected by any single reinforcing event. The problem is demonstrated by simply arranging a schedule of reinforcement totally independent of the organism's behavior.

Two outcomes are possible from this type of scheduling arrangement: Behavior may drift and become even more diverse due to the explicit lack of a response contingency, or a stereotypic response or set of responses will begin to emerge due to the efficacy of the conditioning process.

To show that such response independent schedules would cause conditioning, Skinner (1948) presented food to hungry pigeons every 15 seconds (fixed time 15 seconds) regardless of their behavior. After a few sessions (about 100 food presentations) it was clear that a stable stereotypic response emerged (one that at least two observers could agree on in counting instances). The apparently random acquisition of these stereotypic responses prompted Skinner to dub the process "superstitious conditioning".

Indeed, the nature of the responses that resulted from the response independent schedule had many earmarks of a superstitious kind of process, that is, these responses seemed to be the product of some sort of random selection process. Only six out of the eight subjects adopted clear stereotypic responses and of those six subjects there were five qualitatively different responses shared among the subjects. The responses could not be considered simple topographic variants of the same response (e.g., there was counter-clockwise turning, pendulum type motion of head and body and incomplete pecking or brushing directed at but not touching the floor). These responses also appeared and developed at different rates and for some of the subjects the predominant behavior would drift (become less dominant and another behavior would take its place) especially after extinction.

Thus, according to Skinner, conditioning was not simply a function of the experimenter's prior experimental intentions (response-reinforcer dependencies) nor the subject's a priori organization of behavior (phylogeny) but occurred because a reinforcement had some critical contiguity with a response. He presumed that subjects have some difficulty distinguishing a true response-reinforcer dependency from accidental correlations of a response and reinforcer. If this is the case then two questions arise: (a) Is superstitious conditioning simply a discrimination failure?, that is, do organisms have a great difficulty sorting those behaviors that cause incentives from those that do not? and (b) to what extent are other behaviors that arise form periodic incentives related to the predominant (superstitious response) behaviors generated? This second question is somewhat dependent on the first, for if organisms have no difficulty discriminating behaviors that cause events from those that do not, then an explanation of the process that accounts for Skinner's superstitious behaviors will need to be generated.

CAN ANIMALS DISCRIMINATE CAUSALITY?

The first question was addressed more recently and elegantly by Killeen (1978). In a modified "yes-no" signal detection task, Killeen "asked" pigeons whether their last response, which preceded a stimulus change, caused that stimulus change or not. Noise was introduced by a computer that concurrently pecked with the pigeon and had a similar rate and pattern. Could the pigeon discriminate those responses it made to produce reinforcement from those made by the computer while the pigeon was responding?

Within a few sessions typical detection performance (hits + CR/noise trials) of a pigeon in this procedure is well above 90%. Even given the topographic and force variation of each pigeon's own response, they still had no difficulty sorting caused from uncaused stimulus change. The pigeons, however, could be biased to respond "uncaused" when in fact they had just caused the stimulus change. This biasing was accomplished by simply paying off the pigeons more (through longer duration of food presentation) for making "uncaused" responses independent of the actual outcome (see Figure 12.1). This manipulation produced typical ROC curves suggesting that detection was independent of the bias or more simply the causality of the situation could be subverted for behavior that was more valuable. Contrary to Skinner's conclusions, then, superstitious behaviors do not result from a simple discrimination failure.

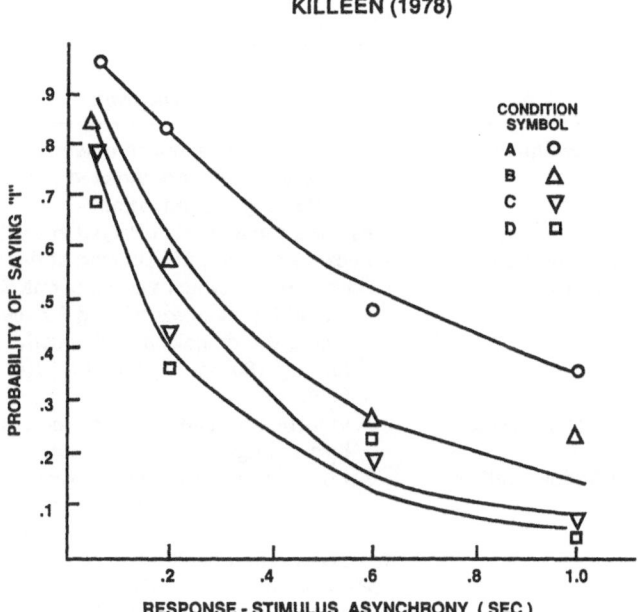

FIG. 12.1. Probability of saying "I" caused the stimulus change as a function of the delay between a response the stimulus change. The conditions A-D refer to decreasing payoff for saying "I".

THEN WHAT'S ALL THE BEHAVIORAL DIVERSITY GOOD FOR?

Staddon and Simmelhag (1971), perhaps anticipating the Killeen study, decided to reexamine the superstition experiment by Skinner and observe the structural aspects of the behaviors more carefully. After Staddon and Simmelhag determined various discrete categories of behavior for the pigeon (approximately 16; see Table 12.1), they observed the frequency with which each category occurred when food was presented independently of (FT 12'', Skinner's manipulation) and dependently on (FI 12'') the pigeons behavior.

The results were similar to Skinner's; all pigeons exhibited several stable behaviors that emerged after several sessions during both schedules. Skinner's observations, however, were apparently confounded with time because he reported the diversity of the behaviors across pigeons without regard to the time in the interval that the behavior occurred. Although Staddon and Simmelhag had noted similar behaviors to those Skinner noted, behaviors like counter-clockwise turns and general ambulation were observed only early in the interval (e.g., within the first 3–5 seconds of the last feeding). However, other behaviors like

Response no.	Name	Description
R_1	Magazine wall	An orientation response in which the bird's head and body are directed toward the wall containing the magazine.
R_2	Pecking key	Pecking movements directed at the key.
R_3	Pecking floor	Pecking movements directed at the floor.
R_4	¼circle	A response in which a count of one ¼ circle would be given for turning 90° away from facing the magazine wall, a count of two for turning 180° away, three for 270°, and four for 360°.
R_5	Flapping wings	A vigorous up and down movement of the bird's wings.
R_6	Window wall	An orientation response in which the bird's head and body are directed toward the door of the experimental chamber containing the observation window.
R_7	Pecking	Pecking movements directed toward some point on the magazine wall. This point generally varied between birds and sometimes within the same bird at different times.
R_8	Moving along magazine wall	A side-stepping motion with breastbone close to the magazine wall, a few steps to the left followed by a few steps to the right, etc. Sometimes accompanied by (a) beak pointed up to ceiling, (b) hopping, (c) flapping wings.
R_9	Preening	Any movement in which the beak comes into contact with the feathers on the bird's body.
R_{10}	Beak to ceiling	The bird moves around the chamber in no particular direction with its beak directed upward touching the ceiling.
R_{11}	Head in magazine	A response in which at least the beak or more of the bird's head is inserted into the magazine opening.
R_{12}	Head movements along magazine wall	The bird faces the magazine wall and moves its head from left to right and/or up and down.
R_{13}	Dizzy motion	A response peculiar to Bird 49 in which the head vibrates rapidly from side to side. It was apparently related to, and alternated with, Pecking (R_7).
R_{14}	Pecking window wall	Pecking movements directed at the door with the observation window in it.
R_{15}	Head to magazine	The bird turns its head toward the magazine.
R_{16}	Locomotion	The bird walks about in no particular direction.

pecking and pacing tended to cluster towards the end of the interval (about 3–4 seconds before the feeding; see Figure 12.2).

These observations led Staddon and Simmelhag to hypothesize that at least two kinds of behavior between successive feedings could be distinguished (see Staddon, 1977 for a proposal of a third kind of behavior). The early behaviors were termed interim, presumably because of their position between a reinforcement and subsequent terminal behavior, whereas the late behaviors were called terminal. Functionally, interim behaviors seem to provide a diversity for behavior that could anticipate potential contingencies (e.g., turn counter-clockwise for food). Terminal behaviors seem to be simply those behaviors most likely elicited by the incentive event. If, for example, a mate was presented every 12 seconds behind a plexiglass wall, bowing and cooing, in addition to other sexual displays, might be expected as terminal behaviors (cf. signtracking literature, Hearst & Jenkins, 1974).

To account for these and other observations, Staddon and Simmelhag proposed a theory based on an analogy to Darwin's natural selection theory of evolution. Their theory states that interim behaviors provide for the "units" of selection similar to genes. Interim behaviors also represent the variation of behavior in a particular context. The functional nature of terminal behavior is less clear, although Staddon and Simmelhag assume they may be akin to a selected phenotype. These behaviors also apparently compete with interim behaviors (Hinson & Staddon, 1979) and appear more appropriate for the incentive event than for the response contingency (Williams & Williams, 1969; negative auto-maintenance).

The results obtained by Killeen and by Staddon and Simmelhag cast serious doubt on Skinner's account of superstitious conditioning. Although there may be some sort of selection process produced due to periodic food presentation, this process does not result from a breakdown of casuality, nor from a random selection of bits of behavior chosen irrespective of their nature nor does the selection seem to affect only a small subset of behaviors that are locally contiguous with the incentive.

DOES DIVERSITY CHANGE AS FUNCTION OF REINFORCEMENT?

Hanson (1980) developed a specific time allocation model to measure how diversity expands and contracts in time. The time allocation theory was predicated on two assumptions: First, that the arousal from periodic feedings increases asymptotically to some constant level and is shared among all the behaviors elicited by the particular context, and second that under any constant conditions (e.g., continued exposure to a chamber or schedule of food presentation) behavioral diversity will tend to increase to some relative maximum (relative entropy principle). Given that time may be most easily allocated to those behaviors most

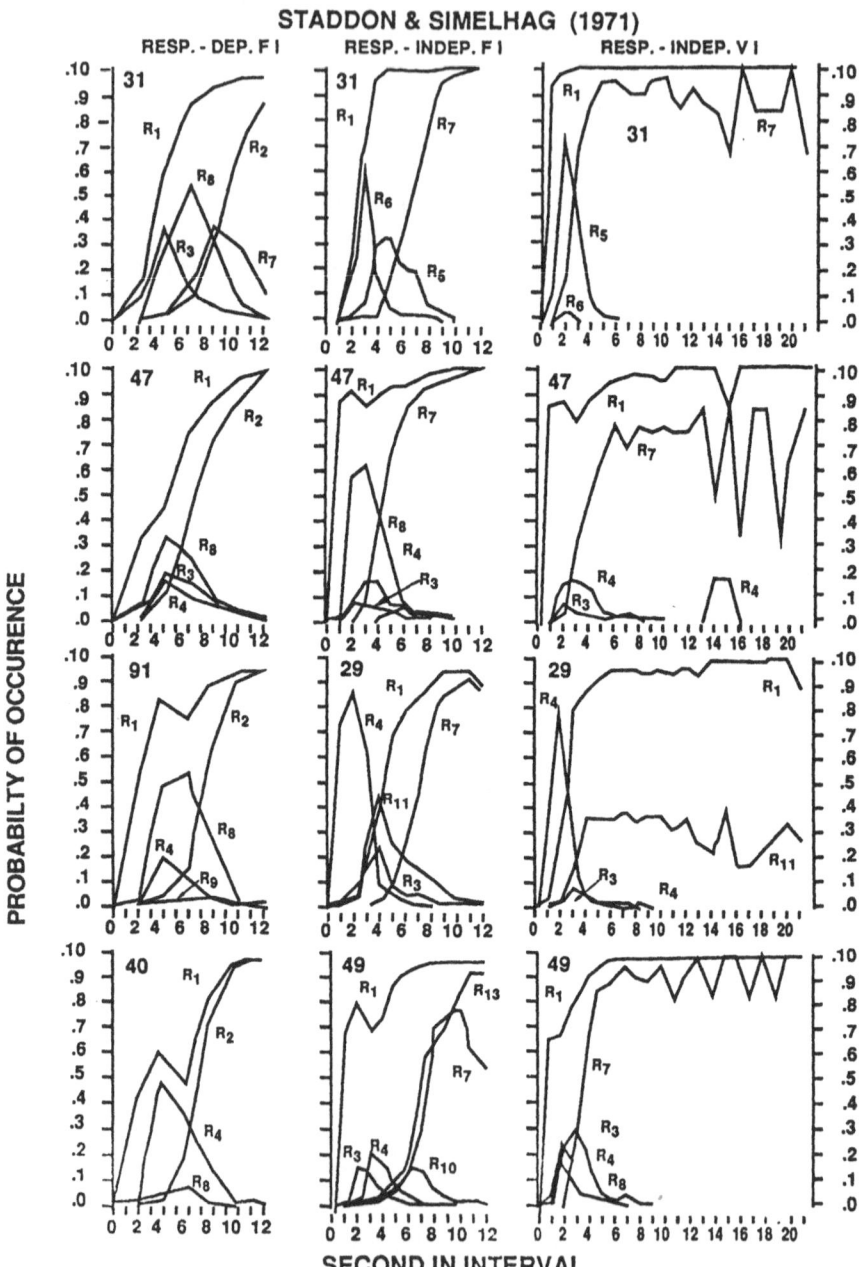

FIG. 12.2. Likelihood of responding for one or 16 different possible behaviors (see Table 12.1) in different points in an interval. Recording was done during feeding conditions (fixed time or variable time food presentation) for six individual pigeons.

likely activated by the incentive event then the relative entropy principle provides a specific prediction of the distribution of activities in any fixed time period. This sort of modelling is an attempt at quantitatively elucidating the "principles of variation" and "principles of selection," to which Staddon and Simmelhag alluded. In what follows the time allocation model is introduced and validated with a number of simple habituation and conditioning experiments.

The Relative Entropy Principle

Fill a cigar box with 1,000 coins and vigorously shake it. Suppose you started shaking when all coins showed heads. If, after a few minutes of shaking, you examined the contents of the box you might find that the distribution of tails and heads showing had changed considerably, say now there are about 300 tails showing and 700 heads. Continued shaking will lead to more tails and less head showing. Tails will not, however, predominate if the shaking continues indefinitely. As you have probably guessed, the final outcome of all this shaking is an equal proportion of tails and heads.

When given a random jolt coins tend to a state where the possible outcomes (heads or tails) of the sample space are uniformly and randomly distributed. It appears to the naive observer that coins "like" to be in an equal proportion to one another.

There are two constraints of these systems that make them easier to analyze. First these systems may be thought of as existing in one of many mutually exclusive states. For example, the box of coins may all show heads, the air in the room may aggregate in a layer close to the floor, and evolution may have produced brain architecture similar to a Von-Neumann architecture; these are all single states of each respective system. The system may find itself in any one of its many states at any moment in time. The illusion of purpose is created because there are many more ways for some states to happen over all others.

Second, to ensure that the number of ways a state can occur remains unique, each category must be stochastically independent of any other. That is, the occurrence of one category in the last instant of time should have no effect on the occurrence of any other category in the next instant of time. Clearly, if coin tosses began to affect one another, say a tail induced a head in the next throw, the number of ways an all head distribution could occur would change and this bias would indeed represent a purpose (that of a dishonest gambler, perhaps).

Given these two conditions, consider a simpler system where the states may be easily enumerated. If there are only four (n) coins in the cigar box then the sample space, the number of possible outcomes (states *with regard* to sequence, (that is, HTTH is different from HHTT) is 16 (2^n). The number of unique outcomes *without regard* to order is much smaller, in this case only five ($n + 1$) unique number of heads and tails. Clearly, the two heads and two tails distribution has the largest number of ways to occur at six.

Notice that the number of ways an independent state can occur divided by the total number of outcomes is equivalent to its binomial probability: $\left(\begin{smallmatrix} 4 \\ 2 \end{smallmatrix}(\frac{1}{2})^2(\frac{1}{2})^2 = 6/16\right)$, 6 out of 16 possible ways to occur is certainly not an overwhelmingly large probability, but as the number of coins increases, however, to say 1,000, then the likelihood that the predominant distribution or state is the 2 heads-2 tails state approaches .90. Of the $2^{1,000}$ outcomes of this system more than 10^{300} will be found in the 2 heads-2 tails state. As the number of throws increases indefinitely, then the system given any time sample will be found "purposefully" approaching the 2 tails- 2 heads distribution. This fact, of course, is equivalent to stating as the number of throws of a balanced coin increases the proportion of heads and tails appearing approaches $\frac{1}{2}$; an observation that would have avoided the apparent tortuous language of the present analysis of coin tosses, but would have also not allowed the derivation of a general form for the most likely distribution (the one that has the most ways of occurring) of any type of categorical system.

An important component of the derivation of a most likely distribution is a mathematical expression for the calculation of the number of ways a distribution of state of a system can occur. For the two category coin system the binomial coefficient is handy. In the previous example for the 1,000 coins, the exact number of outcomes of the 2 head and 2 tail state can be found by calculating its binomial coefficient, call it W.

$$W = \frac{N!}{H! \, (N-H)!} = \frac{1,000!}{500! \, 500!} = 10^{300}$$

It may be easily shown that any other distribution of heads and tails like 300:700 or 499:510 occurs with less ways than the 500:500 distribution. In general, when no constraint occurs on the occupancy of the categories of a multicategory system, the uniform (equal probability) distribution will be the outcome of random perturbations to the system.

What is particularly nice about the binomial coefficient is that it is easily generalized to multicategory systems with categories greater than two (Feller, 1968). The analog of the binomial for multicategory systems is the multinomial coefficient. For the multinomial only the denominator changes, instead of the number of heads and tails, the number of each category is represented, thus:

$$W = \frac{N!}{n_1! n_2! \ldots n_k!} \tag{1}$$

Where k is the number of categories of the system and n_1 is the number of events of the first category and n_2 is the number of events of the second category . . . and n_k is the number of events of the kth category, given the constraint

$\sum\limits^{k} n_i = N$. It is easy to show that the binomial is a special case of the multinomial (let $k = 2$) and that the multinomial also takes on a maximum when the distribution over categories is uniform.

This analysis of the simple coin example should point to the answer of our earlier queries. That a system under random influence has a "preference" for a particular state has little to do with purpose. Rather, the state a random system settles to after a time is simply a stationary (one that can occur the most ways) distribution of a completely random process. Flips of coins, motions of air molecules and the pathways of evolution all share this tendency of random systems to occupy the state with the largest number of ways to occur, albeit these systems do so under varying kinds and complexes of constraints. Thus, the equal proportion of heads and tails state occurs because there are more ways for it to occur, air distributes uniformly through a fixed space because there is more ways for it to do so, and evolution has produced a trait like a brain because of all the other evolutionary pathways under similar constraints (and there are many constraints in this case) this pathway had the most of occurring. This is the principle of relative entropy.

Behavior as Time Allocation

An organism's behavior may be framed in the previous terms. Consider, for example, a pigeon, foraging for grain and food bits in some finite resource area. Suppose we could catalog basically five different behaviors the pigeon engages in; namely, pecking, perambulation, observe ground, wingflap (or even flying), and beak digging. For the previously described model, it must be assumed that these behaviors may be treated as discrete mutually exclusive categories. Of course, it would be ideal to have some nonarbitrary way of cutting up an organism's behavior; it has been found, however, that independent observers will generally describe a "continuous" stream of behavior in discrete categories and in counting instances generally agree on what these categories are. To meet the first assumption of the model, then, the behavior of the foraging pigeon may be seen as an outcome of a "5-sided" random process.

The second assumption of the model to be met by behavior is that the mutually exclusive categories are independent in time. At this moment, digging in the ground should have no effect on the probability in the next moment of pecking. As unlikely as it may seem that behavior satisfies this condition, the few *complete* behavioral transition matrices that have been constructed and correctly analyzed typically show no 2-tuple dependence. For example, one very extensive study of the rat under fixed stimulus conditions was done by Timberlake (1969). In this study nine categories of behavior were constructed from five categories of exploratory behavior and four categories of grooming. Over a 10-minute period of constant conditions, a rat's probability of engaging in one behavior category and then changing to another behavior category was recorded over all behaviors. This procedure resulted in a behavioral transition matrix. Calculating the marginals of such a matrix and forming an expected transition probability given stochastic independence provides a chi-square test. Timberlake found that there were no significant first-order (two-tuple) transitions that deviated from the independence hypothesis. Behavior occurred randomly in time.

Another extensive study of behavioral categories was done by Staddon and Simmelhag (1971). They reported sequential patterns of the pigeon during periodic food presentation (see Figure 12.3). Unfortunately the patterns they showed represented only *partial* transition matrices and although many of the transition probabilities were relatively high, it is unclear whether a chi-square test would show that these probabilities were different from chance under the assumption of independence because this depends on the estimated independent probability of a behavior holding all others constant.

It is true, however, that under many circumstances where a sequence may impart some information affecting the fitness of the individual (e.g., courtship, waggle dances) stereotypic *k*-tuple sequential dependencies exist. When this occurs the model proposed can handle it in two ways. First, the nature of the categories might be changed, so that the sequential dependencies are considered

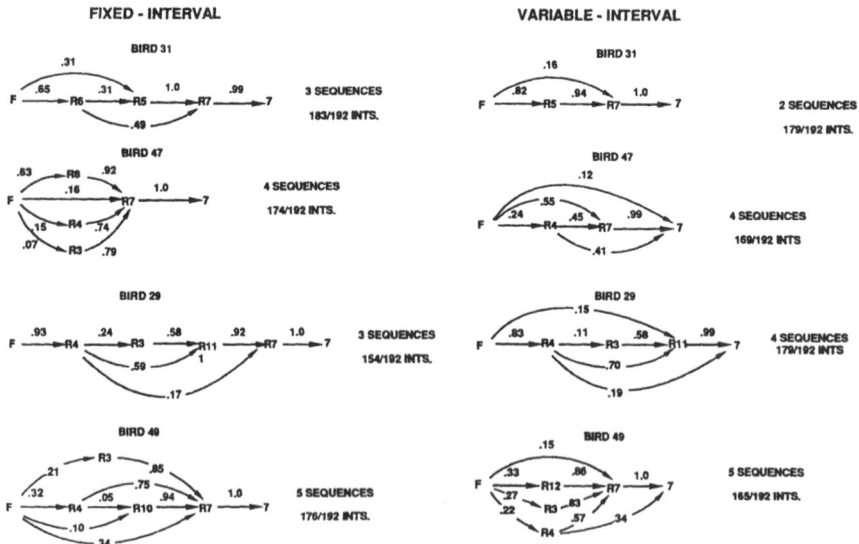

STADDON & SIMELHAG (1971)

FIG. 12.3. Sequential dependencies of the 16 behaviors in the Staddon & Simmelhag (1971) behavior catalog.

units of behavior comprising new categories; second, the probability model might be adjusted from an independence to any number of dependence type structures.

Measuring Behavioral Diversity: Brillouin Entropy Index

Over a 5-minute period how might we characterize the diversity of our foraging pigeon? Intuitively, it seems W may provide an index of the diversity. Recall that for any multicategory system W takes on a maximum value when the distribution over categories is uniform. So for a heavy loading of certain categories, like 4 minutes beak digging and 1 minute pecking, W would be low indicating that the diversity is low, that the behavior in the 5-minute period at any moment is very predictable, and that this particular distribution is very unlikely. Whereas, a distribution of activities where one minute is devoted to each behavior, W is at a maximum indicating that the organism is conforming to the relative entropy principle, very chaotic at any moment, and exhibits the most likely distribution from chance alone.

In fact, Brillouin (1962) adopted a form of W to describe the entropy of physical systems that have k distinct mutually exclusive categories. For various reasons, Brillouin adopted the form:

$$Entropy = S = 1/N \log (W). \tag{2}$$

One interesting aspect of this form is that the smallest n_i gets large (i.e., they all get large) the Brillouin entropy index approaches the information index of Shannon and Wiener.

$$S \rightarrow \rightarrow information = \sum^{k} \left(\frac{n_i}{N}\right) \log \left(\frac{n_i}{N}\right). \tag{3}$$
$$\min (n_i) \rightarrow \infty$$

This relation caused Brilloin to speculate about the nature of entropy, information, and a notion he introduced called neg-entropy (Brillouin, 1962). These particular connections are of less interest (see Fast, 1962, for a comment on Brillouin's Negative Entropy) than the fact that the Brillouin index or the multinomial coefficient provides a way to summarize the diversity of an organism in a single number (see also Krebs, 1978, and Pielou, 1977, for variance estimates).

Applying the Brillouin index to the foraging pigeon is relatively straightforward. Consider the first distribution of activities mentioned, 4 minutes of beak digging and 1 minute of pecking (here 1 minutes will serve as the basic unit of time allocation although any unit would work) and 0 minutes in all other behavior categories. Thus we have:

$$W = \frac{5!}{4!1!0!0!0!} = 5$$

ways this distribution may occur. This measure may be normalized with the maximum behavior entropy that can occur in this time period, 1 minute in each behavior:

$$W_{max} = \frac{5!}{1!1!1!1!1!} = 120$$

The ratio of the raw entropy to the maximum entropy provides a relative entropy measure that varies from zero to one, and in this case is .04.

The Brillouin index is more convenient when working with a large number of units per category and requires simply logging the W index and dividing by the total number of units.

$$S = 1/5 \log (5) = .32$$

Normalizing with the Brillouin form of the W_{max} we have

$$S_{max} = 1/5 \log (120) = .95$$

which gives a large relative diversity than the W index alone of .33.

The Brillouin index will be preferred to the multinomial because of its convenience both computationally and statistically (there are at least variance formulas for the Brillouin index) its theoretical connection to information theory (although it will not be investigated here) and its relation to the following derivation of the most likely (temporal) distribution of activities, the Boltzmann distribution.

Stationary Distribution of Behavioral Diversity: The Boltzmann Distribution

Before deriving the most likely distribution of activities, there are two constraints on activation that must be considered. Indeed, without these following constraints the most likely distribution of activities would always be uniform. The first constraint is a key assumption of the present diversity theory; it is assumed that the ease of activation of behaviors could be scaled on an equal interval scale (in fact a logarithmic-interval scale, see Stevens, 1959). And, second, time to engage in all activities is finite, that is, behavior is a closed system under time.

There is an abundance of behavior in any context. Ethologists have long noted that organisms will engage in other behaviors during activating events (aggressors, mates) that seem to displace the appropriate behavior for the context. Thus, Herring Gulls in the heat of an aggressive encounter over territorial boundaries may begin to tug or pull at grass on the edge of a boundary (Tinbergen, 1960), a raccoon may interrupt a well learned routine of crawfish-token exchange to "wash" the token prior to placing it in a slot for the fish (Breland & Breland, 1961) and a rat previously sated for water during a schedule of food presentation will (given the opportunity) drink so much water as to border on water intoxication

(Falk, 1966). In all these cases another ongoing system of incentive motivation that supports an appropriate set of behaviors (e.g., aggression in the presence of a border threat) also elicits other behaviors seemingly unrelated to the nature of the motivation of the context (e.g., grass pulling in response to a border threat).

Given a context, in which more than one behavior occurs (there is a clear abundance of behavior) categories of behaviors are not obviously orderable. At best a nominal scale would seem to be extractable from such a measure of behavior. On the other hand, a general account of the reinforcement process has been predicated on the ordering of behavioral categories. Premack (1965) proposed a theory of reinforcement, in which the ordering of categories of behavior was based on the likelihood of each behavior's occurrence. Premack argued that in a context where several behaviors will occur those that occur at the highest frequency will cause a frequency increase of others with lower frequency when the low frequency response must occur for some limited access to the high frequency behavior. Thus, consummatory responses are not special with regard to the reinforcement process, consummatory responses are responses that are simply the most convenient ones for the experimenter to deprive and therefore cause them to occur at a higher frequency.

Implicit in Premack's theory is that behaviors are orderable in some sort of preference hierarchy. If behavior A serves as a reinforcer for behavior C and behavior C serves as a reinforcer for etc. . . . then preference as an intrinsic property of behavior results in an ordinal scale (at least).

Because scale types are hierarchical, higher order scale types (e.g., metric scales) must always contain lower order scale properties (e.g., ordinal). It is surprising to find the converse, that phenomena that have been assumed only measurable (the measurement achieved at that point; cf. Stevens, 1951, 1959) on a lower order scale type to contain higher order scale type information.

Because we have no scale theory for arousal or activation states for simplicity then, each mutually exclusive category of behavior will be assumed to lie equidistantly from one another on an arousal scale. Behavior further down the scale require more overall arousal to occur given a particular context. These levels and their nature will be dependent on the context and the incentive event. For example, organisms in the presence of a receptive mate would be most likely to engage in sexual displays with the highest frequency, whereas food incentives might cause exploration and foraging activities to be the most dominant behaviors and thus occupy the lowest activation or arousal levels.

Over these hypothesized equidistant levels a form for the distribution of activities will be derived using the relative entropy principle. This theoretical distribution will be compared to data. If this analysis clarifies the nature of the various behaviors in the context and generates further experiments for stronger tests it will be retained as a reasonable working hypothesis.

The second constraint is trivial and results from the definition of W; there is a finite amount of time to engage in any activity and engaging in one activity deletes potential time from the rest of the behavioral ensemble. Time is the currency of the behavior system. This constraint has found use in the recent "economic-optimal" models and is usually referred to as the linear constraint (cf. Staddon, 1979).

With these two constraints and the relative entropy principle a most likely distribution may now be derived. This distribution, of course, is flat or uniform when no constraints exist on the occupancy of categories but it will be shown that constraints can drastically alter this form. If an organism has a fixed amount of time and a fixed number of behaviors (each one successively more difficult to activate) and these behaviors are also emitted randomly in time, then what *distribution* of time in each behavior might be expected?

Mathematically this question is easily framed and solved. This particular derivation is commonly found in most introductory statistical thermodynamics texts and uses methods of the calculus of variations. One such method is the method of Lagrange or undetermined multipliers. In this method each constraint is introduced with a dummy multiplier with the quantity to be maximized. This leads to several simultaneous differential equations all set to zero (homogeneous) and once solved reveals the most likely distribution of activities.

If behavior tends to settle to that distribution with the greatest ways of occurrence than we might consider maximizing W given the activation and time constraints. Recall that the number of ways a particular distribution of activities can occur is given by:

$$W = \frac{T!}{\sum_{i=1}^{k} t_i!} \qquad (4)$$

and the two constraints may be written as:

$$T = \sum_{i}^{k} t_i, \ A = \sum_{i}^{k} t_i a_i \qquad (5)$$

where a_i are the assumed levels of arousal associated with some mutually exclusive class of behaviors, b_i, and A is the constant amount of arousal in the context.

Two transformations of W will make it simpler to analyze. First consider:

$$\log (W) = \log T! - \sum^{k} \log t! \qquad (6)$$

This transformation, of course, makes W (for all practical purposes) the Brillouin diversity index. The second transformation depends on having enough units per category (over a 100 units) such that an approximation of the factorial can be used, Sterlings' approximation:

$$\log t! = t \log t - t \qquad (7)$$

which simplifies log (W) considerably,

$$\log (W) = T \log T - T - \sum_{i}^{k}(t_i \log t_i - t_i) \tag{8}$$

or equivalently,

$$= T \log T - \sum_{i}^{k}(t_i \log t_i) \tag{9}$$

The derivative of this function is now quite simple to find. Lagrange's method requires that the previously listed function include the two constraints,

$$L = \log (W) + \lambda T + \beta A \tag{10}$$

This linear function of log (W) and its constraints, called a Lagrangean (L), may be rewritten using equations 7 and 4,

$$L = T \log T - T - \sum_{i}^{k}(t_i \log t_i) + \lambda \sum_{i}^{k} t_i + \beta \sum t_i a_i \tag{11}$$

It is a necessary condition and can be shown in this case to be a sufficient condition to have all the partial derivatives of the Lagrangean vanish (with respect to the time allocated to each behavior) for a maximum to exist. Thus, for the most likely distribution of activities we require:

$$\frac{\partial L}{\partial t_i} = 0 \tag{12}$$

or equivalently,

$$\frac{\partial L}{\partial t_i} = \sum(\log t_i + 1 + \lambda + \beta a_i) = 0 \tag{13}$$

which vanishes when each and every term of the series vanishes, that is when,

$$\log t_i + 1 + \lambda + \beta a_i = 0. \tag{14}$$

Exponentiating both sides

$$t_i \, e^{(1 + \lambda + \beta a_i)} = 1 \tag{15}$$

and solving for the distribution (t_i)

$$t_i = B \, e^{-\beta} \, a_i, \text{ where } B = e^{-(1 + \lambda)}, \tag{16}$$

which is the most likely distribution given a fixed amount of time and behaviors associated with equidistant arousal or activation levels.

This function is well known in thermodynamics, it is the Boltzmann distribution. In the present case it relates the time spent in any activity to the arousal level that any behavior occupies (a_i). The Boltzmann distribution makes the prediction that in a fixed period of time (T), the time spent in arousal levels requiring more activation will decrease exponentially under constant stimulus conditions (*A*).

It is notable that this type of plot has been used before in a linguistics context. These type of frequency-rank curves have been called Zipf curves and were named for the linguist George Kingsly Zipf, who originally plotted word frequencies against their ranks. There are actually two kinds of Zipf curves. A Zipf curve is either (a) a relation between the frequency of occurrence of an event and the number of events occurring with that frequency, or (b) a relation between the frequency of occurrence of an event and its rank when the events are ordered according to frequency of occurrence. The second type of Zipf curve is of the same kind proposed here. Zipf, however, had no deductive account of his functions and plotted them on double log coordinates apparently because of the range of values he typically used and because these coordinates tended to linearize the data suggesting a power function to be a good description of the Zipf curve.

To illustrate a Zipf curve and a clear *failure* of the Boltzmann Distribution prediction, a classic case is reproduced in Figure 12.4. Here is a sample of the 29,899 different words from the 260,430 total words in James Joyce's novel, *Ulysses*. Although Zipf's (1949) original plot shows all 29,899 different words, the gist of the relation can be gathered from the first 1000 words plotted at varying intervals. The plot is obtained in the same manner proposed for a Boltzmann curve, the frequency of a word is plotted against its rank according to its frequency. It can be seen that the plot is clearly curvilinear in the semilog coordinates of Figure 12.4 and is well described as Zipf noted by a power function.

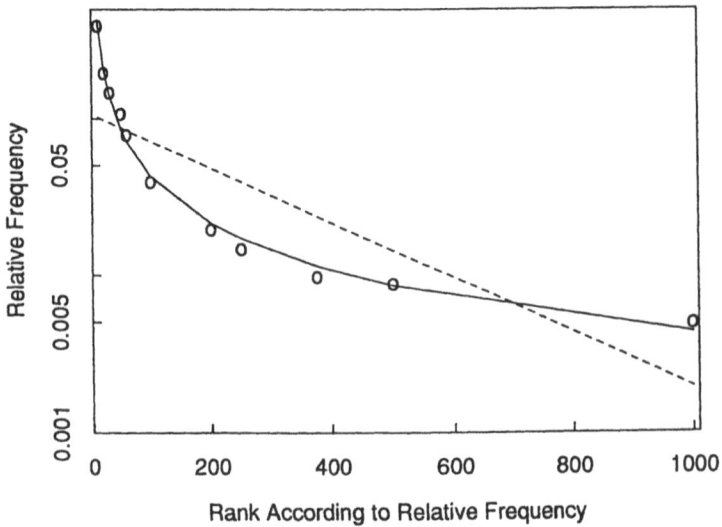

FIG. 12.4.   "Zipf Curve": Word frequencies from James Joyce's *Ulysses* plotted as a function of rank of relative frequency.

The dashed line in the figure is the prediction of the present theory. Apparently, when sequential dependencies are many, such as in written language, a deviation from the Boltzmann curve is quite likely.

In the next section the previous measures and models are used to analyze data collected from animals behaving in simple learning experiments.

EXPERIMENTS

General Methods

The same subjects and following general procedure was used throughout all of the experiments. Only new methodological details will precede results sections, subjects and apparatus sections will be omitted.

Subjects.   Subject consisted of three experimentally naive pigeons of various breeds. Subjects throughout all experiments were maintained at 85% of free feeding weight.

Apparatus.   A standard Lehigh Valley pigeon box was used with the front wall replaced by a clear piece of plexiglass. The left side key was not used and covered with a neutral grey tape. The hopper aperture, located centrally 7.7 cm above the floor, was illuminated with 28 volt dc light bulb and a photocell for detecting changes in hopper illumination was placed in the hopper aperture. This photocell was located approximately .3 cm from the front edge of the hopper aperture and the interior of the hopper was painted black to reduce reflection to the photocell. Feeding duration was timed from the bird's first entry into the hopper.

Above and to the right of the hopper was a response key approximately 3 cm in diameter that required about .13 N to operate. The response key was capable of being transilluminated red by a filter located directly behind the response key and in front of a 28 volt dc light bulb.

To record general activity, a Lafayette activity platform (model A501) was used and was located centrally beneath the chamber. The transducer of this activity monitor consisted of a permanent magnet suspended within a coil. This kind of system eliminated the need for any nonrepeatable mechanical adjustment and provided for equal sensitivity across a wide range of animal weights. Pilot studies indicated that a setting of the monitor close to three or four for weights between 250–400 grams seemed to be primarily sensitive to shifts in weight due to foot movements.

To monitor and record all behaviors a Sony videotape recorder, a Panasonic monitor and a Sanyo camera was used. Once a session was taped, observers used a button box to "decode" the continuous record of behavior into various mutually exclusive predetermined categories. The button box consisted of 12 buttons in a

3×4 array mounted on a metal panel with appropriate labels under each button corresponding to each category of behavior.

Extraneous noise was masked by 86 db of white noise, and internal chamber ventilation was provided by a fan outside the box. Stimulus programming and response collection was done with a minicomputer.

General Procedure

Observer Procedure.   Categories of behaviors were determined by correlating several (at least four) observer categorizations to discover a minimum number of common categories. This resulted in 17 behaviors (cf. Staddon & Simmelhag, 1971) and are described in Table 12.2. These categories were labeled on the button box and all observers were given several sessions of practice until at least 50% of the variance of each record were equivalent (.707 correlation of approximately 400 observations/record). A program was developed to analyze the data that calculated the correlation between two observer records, collapsed the records if the correlation was high enough and then calculated the entropy per trial, over a trial and over the entire session. Depressing any button scored a 100 msec count for that particular category of behavior. Thus, at the end of the session a category counter represented both the frequency (in 100 msec units) with which this behavior occurred and the time spent engaging in this activity. Releasing the button automatically caused a "null" category to score a 100 msec count, and continued to score 100 msec counts until any button is pushed.

Subject Procedure.   The following experiments provided a continuous record of behavioral change for each subject. General activity was collected throughout all conditions and experiments whereas videotapes were collected either in the first or last sessions (typically two) of a condition. Tapes were decoded within two days of taping by at least two observers, both within and across subject comparisons were made in all conditions. General activity was correlated against observation measures in most of the following experiments.

EXPERIMENT 1: THE EFFECTS OF HABITUATION ON BEHAVIORAL DIVERSITY

Behaviorally, habituation is the decline in activity of a response system exposed to a fixed set of conditions. Usually, experiments investigating habituation use a relatively isolated response system (a single neuron, for example) and a well defined stimulus. In the present investigation the whole organism (its response repertoire) will serve as the response system and the background environment will be the source of the habituation. To further complicate matters, the subjects in the present study will be deprived of food. The deprivation manipulation

TABLE 12.2
Hanson, 1980: Behavior Categories for the Pigeon

General Movement (GM)	Any sort of random locomotion not clearly directed at a wall or stimulus. This could include circling and other stereotypic actions.
Preening (P)	Any action directed at the body of the pigeon with another part of its body. This can include 'dog scratching' type responses.
Head Movement (HM)	Any movement of the head in any direction which is not part of an orienting or pecking response.
Floor Peck (FP)	Pecking at the floor.
Ceiling Peck (CP)	Pecking at the ceiling. (This almost never occurred).
Wing Movement (WM)	This includes active behaviors like flying and quiescent ones like stretching which are not involved in preening activities.
Peck Intelligence Panel (PKIP)	Pecking anywhere on the wall where the food is located. (See Peck Food).
Peck Food Aperture (PFA)	Pecking in or around within an inch of food Aperture.
Peck Plexiglas (PKP)	Pecking anywhere on the plexiglass wall.
Peck Other Walls (PKOW)	Pecking anywhere on any of the other walls.
Observe Food Aperture (OFA)	An orientation and directed head movement in the area of the food Aperture.
Observe Plexiglas (OP)	An orientation and directed head movement in front of the plexiglass wall. No locomotion (see Pace).
Observe Other Walls (OOW)	An orientation and directed head movement in front of any other wall (2 others) besides the food wall and the plexiglass wall.
Observe Intelligence Panel (OIP)	An orientation and directed head movement in front of intelligence panel.
Pace Intelligence Panel (PIP)	Walk back and forth in an almost stereotypic manner in front of the intelligence panel. Head movement always occurred simultaneously with this behavior.
Pace Plexiglas (PP)	Walk back and forth in an almost stereotypic manner in front of the plexiglass wall. Head movement always occurred simultaneously with this behavior.
Pace Other Walls (POW)	Walk back and forth in an almost stereotypic manner in front of the other walls. Head movement always occurred simultaneously with this behavior.
No Decision (ND)	Any time a button was not pushed.

ensures a reasonably high level of general activity (Bolles, 1967; Shettleworth, 1975).

Although these conditions are not suited to a detailed and controlled study of the various parameters of habituation, they do provide an ideal set of conditions to examine the diversity of an organism under relatively simple fixed stimulus conditions. This will allow an initial assessment of the present diversity theory. Exposure to the simple conditions during habituation should result in a stationary distribution of activities over the season; according to the theory this distribution should be Boltzmann (exponential).

Method

All subjects were exposed to constant illumination and white noise in the experimental chamber, no other stimuli occurred. There were three such habituation sessions each 20 minutes in length. All three sessions were videotaped and decoded (see general methods). After each session subjects were returned to their home cage and given enough grain to maintain their deprivation weights.

Results and Discussion

As expected, habituation sessions were characterized by a decline in exploration and general movement. By the second session, general movement had practically ceased and by the third, except for W46 who moved his head occasionally, all subjects sat stationary. The first habituation session, therefore, was used for the following analyses.

On the left of Figure 12.5 (a), are shown Boltzmann curves for habituation, that is, the relative time spent (the time in activity "j" divided by the total time) is plotted against its *rank* in terms of time spent in each activity. For example, a Boltzmann curve for W37 was constructed in the following manner. First, all of the relative times spent in each activity were rank ordered. Second, the rank number associated with the rank of that relative time was assigned to the behavior. For instance, for W37, head movement was the most frequent behavior and would be assigned a rank of one; preening was next in rank and would be assigned a rank of two, etc. Third, the time spent for head movement was plotted against one, the time spent for preening against two, etc. and a Boltzmann curve was produced.

If a relative time dropped below .001 it was not plotted, in fact in all of the following experiments all Boltzmann curves are truncated at .001 since the reliability of the observer as well as the behavior would not seem assured below this point, although this criterion was arbitrarily chosen.

In the semilog coordinates, that is, coordinates where just the y axis is logged, the data (the unconnected squares) linearize. Linearalization due to the log y transformation indicates that an exponential function (the straight line in Figure

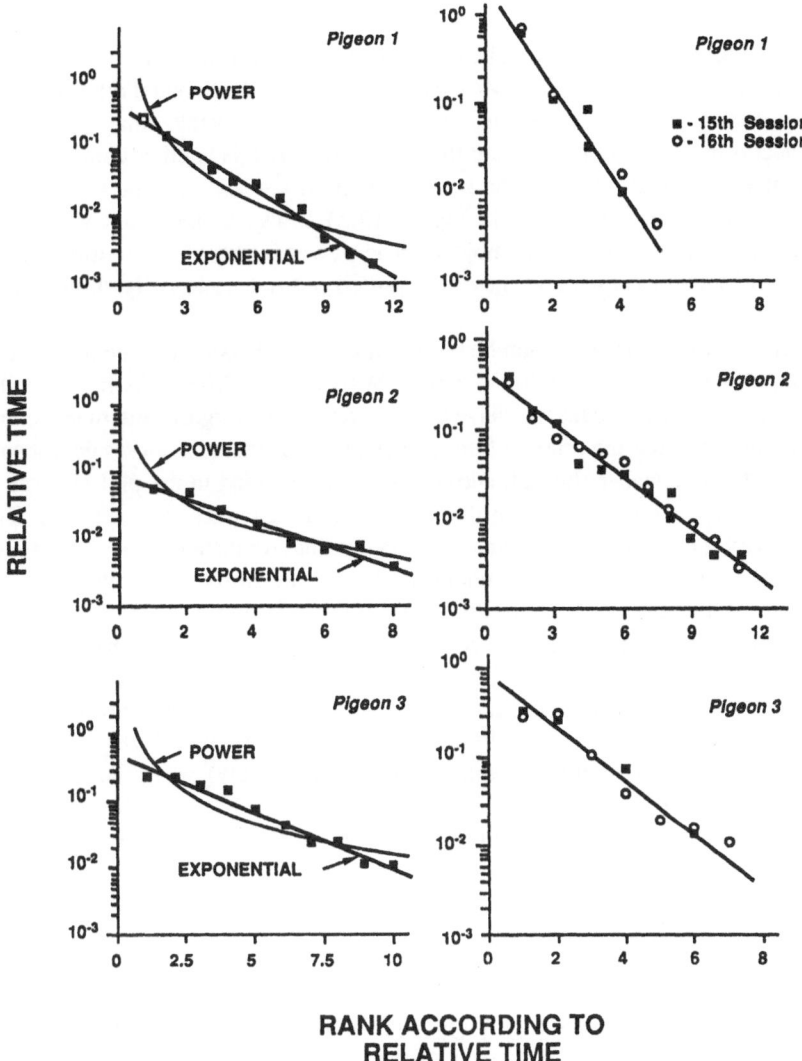

FIG. 12.5. (a) left column are "Boltzmann curves (straight lines)" for three pigeons during habituation conditions. Points represent relative frequency of an activity, curved lines represent best fit power functions. (b) right column are Boltzmann curves for same three pigeons in a later fixed time feeding condition. Notice how all distributions sharpen, i.e. diversity decreases overall.

12.5) is a good description of the data over x (this is clear since when $y = \exp(-bx)$ is logged on both sides of the equation $\log y$ is now linear both in x and the parameter b: $\log y = -bx$). Moreover, because the Boltzmann distribution is an exponential function, the semi-log coordinates have revealed that the data are well accommodated by the present theory. Also shown is another common two-parameter nonlinear function, the power function, $y = ax^b$. The power function in this figure (the curved line in Figure 12.5) shows a significant systematic deviation from the data for all subjects. The power function was routinely fit to the all Boltzmann curves and may be reported but not necessarily shown in all figures.

Across subjects both the number of reliable ($>.001$) occurring behaviors and the slope of the Boltzmann curve varied. W46 and W37 have the most similar slopes and W47 seemed least diverse because he had the largest Boltzmann slope.

Finally, note that the relative frequencies in the Boltzmann curves do not sum to one. This is because the null activity was not included in the plot but was in the total time (excluding it from the total time just changes the intercept). The null activity included various forms of doing nothing; for instance, the "dead time" between responses, ambiguous response forms, and truly just doing nothing. Although, there are cases where doing nothing has important behavioral significance, in the present case it was considered more of a junk category due to its heterogeneous nature, but will be reported for each experiment and discussed later. In the present experiment the percent of time in null category was .82, .02 and .3 for W46 and W47, respectively. The null category did not vary systematically with diversity of the subjects but (not surprisingly) was related to their relative activation.

In the next experiment an attempt to provide evidence for one of the assumptions of the present entropy theory was made. If under constant conditions entropy tends toward a relative maximum, then a single stimulus change should produce a momentary increase in entropy.

EXPERIMENT 2: DOES DIVERSITY INCREASE AFTER A SINGLE REINFORCEMENT?

In the entropy theory outlined earlier an important assumption, that behavioral entropy increases to some relative maximum under fixed conditions, was made. Habituation conditions do not provide controlled or salient enough stimuli for careful measurement of any possible entropy increase. Even though entropy seems relatively high and constant during habituation, the variability that is evident might be due to transitory changes incurred by various stimuli impinging randomly on the subject. To increase the likelihood of finding a specific entropy increase a "one trial a day" methodology will be used (cf. Bolles, 1967; Killeen, Hanson, & Osborne, 1978). In this type of experiment behavior after the stimulus event is isolated from any other contaminating or confounding stimuli, that is, any behavioral change should be caused primarily by the present stimulus event.

Method

After Experiment 1, all subjects were hopper trained. Hopper training consisted of raising the hopper until the subject ate (interrupted light received by the photocell in the hopper). The hopper then remained raised for 6 seconds. At the offset of food subjects were returned to their home cage until the next session when they were exposed to the hopper once again. Hopper training was continued for all subjects for six more sessions.

After hopper training, all subjects were exposed to a one trial a day methodology. Subjects received one 2 second feeding that occurred randomly with a mean of 10 minutes but never less than 2 minutes from entry into the experimental chamber. There was also a probability ($p = .14$) that no feeding would occur at all. After feeding, behavior was time sampled in 120 second bins for 20 minutes. Both general activity and the categories of behavior in Table 12.1 were collected before and after food presentation. These conditions were maintained for 16 trials. Because of the response cost of videotaping (I had only two tapes) and decoding, only 5 sessions of the 16 were randomly chosen for videotaping.

Results and Discussion

Prior to feeding, activity averaged over 14 sessions (2 sessions per subject did not include food) was low and decreasing for all three subjects. Immediately following food offset, activity rose to a maximum and decreased exponentially thereafter. These trends can be seen in Figure 12.6. Exponential functions (solid lines) were fit to the post food activity and were found to describe the data well (with a few deviations 84% (median) variance was accounted for). Time constants (time where 60% of activity has declined) were estimated from each subject and found to be 588, 454, and 296 seconds, averaging to 411 seconds (harmonic mean). This average value is quite similar to those values collected by Killeen, Hanson, and Osborne (1978) in a similar one trial a day experiment (harmonic mean = 370 seconds).

Entropy patterns (as calculated by Equation 2 and normalized by maximum entropy) showed trends similar to the activity patterns. Shown in Figure 12.7 is the entropy for the 18 behaviors over the five sessions (including null category) measured before and after behavior. Prior to feeding, behavior entropy is low and characterized by only two or three behaviors (preening, head movement, and null). After feeding, entropy immediately surpasses prefeeding levels and then decays in an exponential or linear manner (for W37 the increase from before feeding to just after feeding is significant, $t = 3.32$, $p < .025$). Behavior just after feeding tended to be more active type behaviors like wingflap, general movement and wall pace. These behaviors gave way (300–400 seconds after

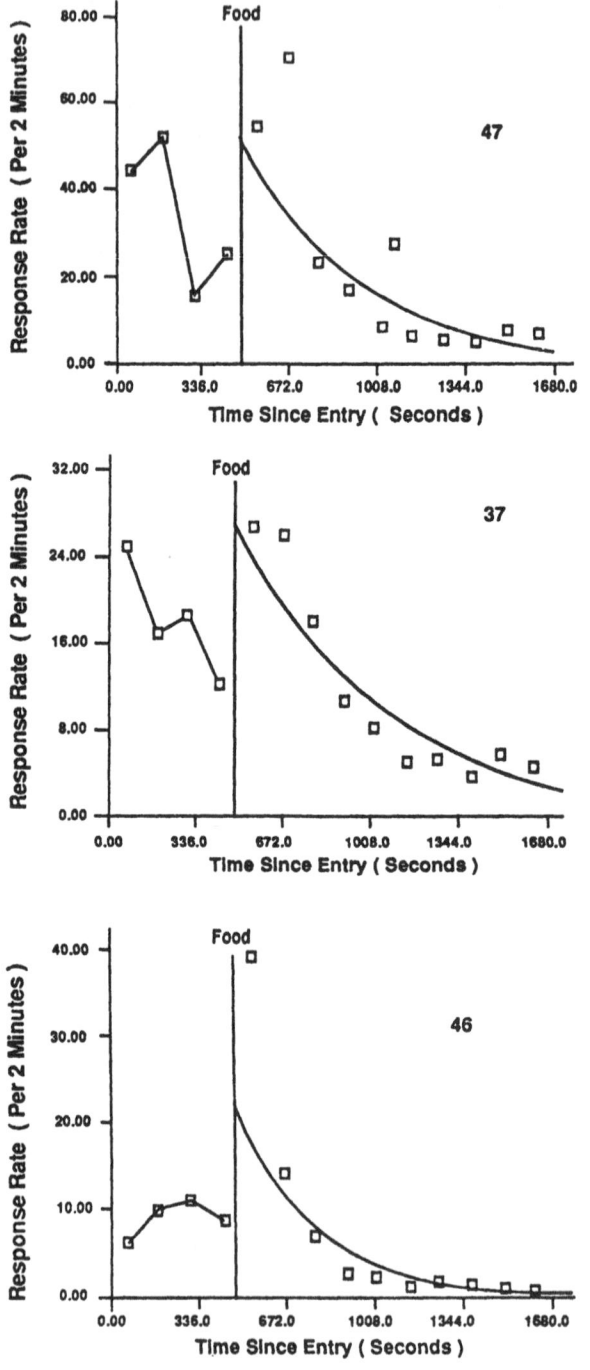

FIG. 12.6.   Activity patterns for the one-trial-a-day feeding experiment. Shown prior to vertical line in the graph is the pre-feeding activity that is generally decreasing. At food presentation (to the right of the vertical line) entropy rises and then decreases in relatively linear fashion.

feeding) to more quiescent behaviors that had also occurred before feeding. The activation of behavior after feeding is also accompanied by a general increase in diversity. Comparing the activity and diversity patterns indicates again that the intensity and diversity of behavior can be disassociated. Although, activity is low, and general movement has almost ceased, the diversity of behavior may still be high due to the variety of behaviors occurring that do not include intense behaviors.

EXPERIMENT 3: THE EFFECTS OF PERIODIC REINFORCEMENT ON DIVERSITY

In the last experiment it was shown that behavioral entropy increases with incentive presentation. Boltzmann curves were argued to result when constant conditions are maintained, causing entropy to tend toward a relative maximum and remain there. It was also shown in the last experiment that if constant conditions are not maintained, behavioral entropy decreases after stimulus offset. Does entropy increase to a relative maximum when constant conditions are maintained? Does this stability of entropy result in stable Boltzmann distributions as predicted?

Method

The same subjects used in Experiment 2 were shifted to a periodic schedule of food presentation, a fixed time 30 second schedule. A session consisted of 50 2-second food presentations 30 seconds apart and independent of the pigeon's behavior. All pigeons were exposed to 15 sessions of the fixed time 30 food schedule, 750 food presentations. General activity was collected from all sessions although only the first and last two sessions of the periodic food presentation were videotaped.

Results and Discussion

In the first session of periodic food presentation general activity was quite variable across trials for all subjects. In the first few trials for two of the subjects, activity rose rapidly to a peak and then declined to a lower relatively constant level. The initial peak is characterized by high intensity behaviors like wingflap and flying into the ceiling. These behaviors tended to cause the activity monitor to "flutter" momentarily and then return to a more normal operation rate. Without the high peaks the remaining pattern over trials is more of a negative acceleration than bitonic. By the 14th and 15th session, activity has stabilized and is a relatively constant function over trials for all subjects. Behavioral entropy was similar over trials to general activity but much less variable.

Boltzmann curves for the last two sessions for each subject are shown in the right half of Figure 12.5 (b). As usual the Boltzmann curves were truncated at

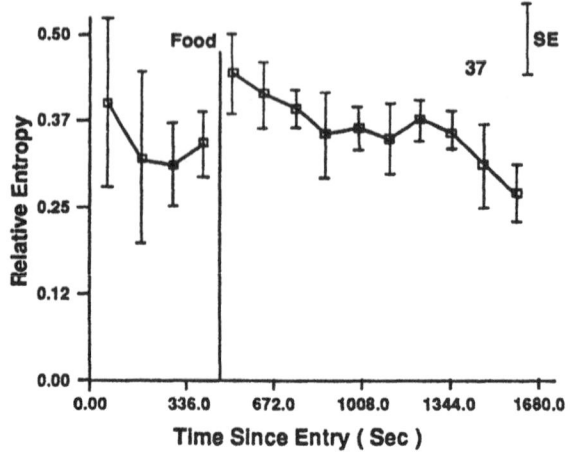

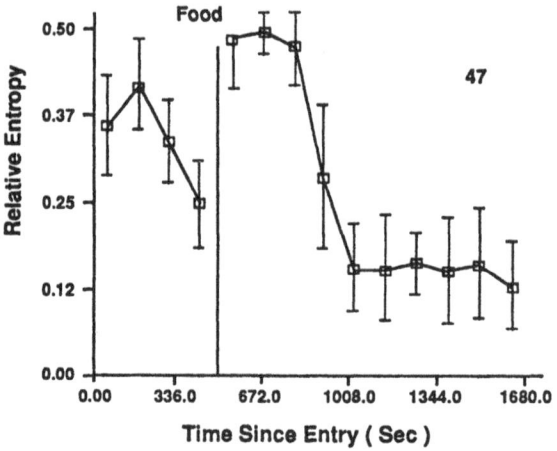

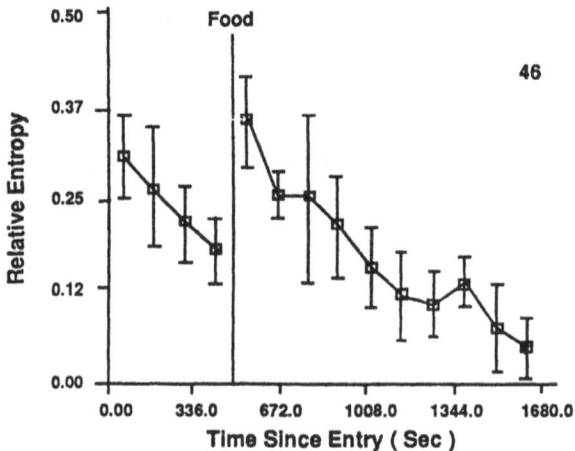

.001. Over the remaining three log cycles the curves for all three birds and both sessions linearize with 93% to 98% of the data variance accounted for. Null behavior for all three birds occupied about 17% of the total session time.

Between bird slopes were significantly different whereas between session (within bird variable) were not significantly different. The most diverse bird (smallest slope) was W37 with slopes of $-.456$ (14) and $-.429$ (15). These slopes for all birds were different from their habituation slopes. The slope change from habituation to periodic conditioning was consistent within each bird; the Boltzmann curve sharpens (less diversity) from habituation to a rate of food presentation of two per minute. This loss of diversity is due to a reallocation of time spent in activation levels lower in rank to those higher in rank. This also causes the number of behaviors to be truncated by the criterion sooner, reducing the overall number of behaviors. It has been shown (not surprisingly) that diversity increases when an organism is deprived of food (Bolles, 1967; Shettleworth, 1975). Continuous food presentation tends to lower diversity. In the present case it has also been shown that diversity decreases if food is presented periodically. Whether diversity increases or decreases, however, it appears to be well described by a Boltzmann distribution.

Not only does time spent in each category change from habituation to periodic food conditions but the behaviors tend to reorder and new ones appear to move down from lower ranks. This observation was confirmed by performing rank order correlations on the behavior ranks for the last two sessions of periodic conditioning and the first session of habituation. The results of this analysis is shown in Table 12.3 with the respective behaviors form each condition for each subject. The results of the correlations are clear: For each subject the last two conditions of periodic feeding were highly correlated, whereas the behavior ranks during the habituation condition were quite different in order and content during periodic food presentation. The behaviors that contribute the most to this reordering from habituation to periodic feeding are general movement and pacing behaviors that tend to move up in rank during feedings and behaviors like preening and observing, tend to move up in rank during habituation, sometimes to the exclusion of that behavior. One account of the reordering of behavior categories from habituation to periodic feeding is in terms of the underlying arousal levels that might be associated with each behavior category. Recall from habituation to periodic food conditions the diversity for all subjects decreased although the general activation increased. This type of relation between these behavior properties has two clear functions: First, limiting diversity makes it more likely that a

---

FIG. 12.7. Brillouin entropy measures during the one-trial-a-day feeding experiment. Shown prior to the vertical line in the graph is the pre-food event diversity. At food presentation (to the right of the vertical line) entropy rises and then decreases in relatively linear fashion.

TABLE 12.3
Hanson 1980: Comparison of Behavior Ranks for the Last Two
Feeding Sessions (14, 15) with Habituation (HA)

Subject	Condition		
	14	**15**	**Hab**
W47	PP	PP	OP
	GM	GM	P
$r_s\ 14\ vs\ ^{15}_{<}=_{.01}.88,$	PIP	OP	PP
	OP	PIP	OFA
	WM	OFA	OIP
	POW	WM	GM
$r_s\ Hab\ vs_{.035}\ ^{14\ \&\ 15\ =}_{,\ p>.1}$	OIP	P	OOW
W46			PKIP
	PIP	PIP	OIP
	POW	POW	PIP
	GM	GM	POW
	OIP	OIP	OP
$r_s\ 14\ vs_{p\ <\ .01}\ ^{15}.89,$	OOW	OP	OP
	OP	PP	PP
$r_s\ Hab\ vs_{.035}\ ^{14\ \&\ 15\ =}_{,\ p>.1}$	PP	OOW	FP
W37	OIP	OIP	OOW
	OP	OP	GM
$r_s\ 14\ vs_{p\ <}\ ^{15}=_{.01}.93,$	GM	PP	OFA
	POW	GM	HM
	PP	PIP	P
	PIP	POW	OIP
$r_s\ Hab\ vs_{.035}\ ^{14\ \&\ 15\ =}_{,\ p>.1}$	OOW	P	OP
	P	WM	GM
	PKIP	OOW	WM
	WM	PKIP	OOW
	HM	HM	
	OFA	OFA	

smaller set of behaviors will come into contact with the prevailing environment conditions and second, increasing arousal makes it more likely that this smaller set of behaviors are those of greatest intensity. This behavioral response of the system represents a "focusing" of behavior in two senses both that result in an overall increase in *behavioral density* per unit time.

This density increase must also be appropriate for the given environmental conditions. In the present case the increase in general movement, observation and pacing maximizes the probability of finding the source of food in the experimental context, a set of behaviors common to foraging strategies. A sexual partner might be expected to increase the intensity and decrease the diversity of a different set of behaviors (i.e., courtship behaviors).

TEMPORAL PATTERNS

Temporal patterns of activity serve to reinforce the previous conclusions concerning the relationship of behavioral entropy and arousal. In Figure 12.8 the activity patterns for all three birds in various sessions are shown. On the ordinate is the response rate per 3 second bin. On the abscissa is the time between 30 second feedings sampled at every 3 seconds. During the last two conditions of the periodic feeding condition, general activity increased dramatically from the first session of periodic feeding. And generally the first session of periodic feeding was higher than the habituation (no food) condition. Clearly, as noted by many others (see Staddon, 1977, for a review) food has a moment by moment activating effect on the organism.

The activity pattern during the last two sessions does not rise from either the habituation pattern nor from the first session pattern uniformly. The activity pattern of these last two sessions are bitonic and skewed. This kind of effect has usually been interpreted to be an indication of temporal control. The concentration of activity (mode) in the early portion of the interval makes it difficult to support an anticipation of food hypothesis. Theorists have, however, invoked expectancy, elicitation and response competition arguments to account for the varied forms of these temporal patterns (see Gibbon, 1977; Killeen, 1979). And although the mathematical description of these patterns have generated robust and sometimes complementary temporal control models the general activity data, has neither been comprehensive nor "sharp" enough to eliminate various competing mathematical forms (see particularly Killeen, 1979, pp. 49–51).

This difficulty may be due to the relation between diversity and activation. In some of the previous experiments activity and diversity have been correlated. Because activity as measured, is comprised of many different kinds of behaviors, it is possible that the temporal control activity exhibits is a byproduct of the temporal control of diversity.

In Figure 12.9 the temporal patterns of diversity in the 30 seconds between

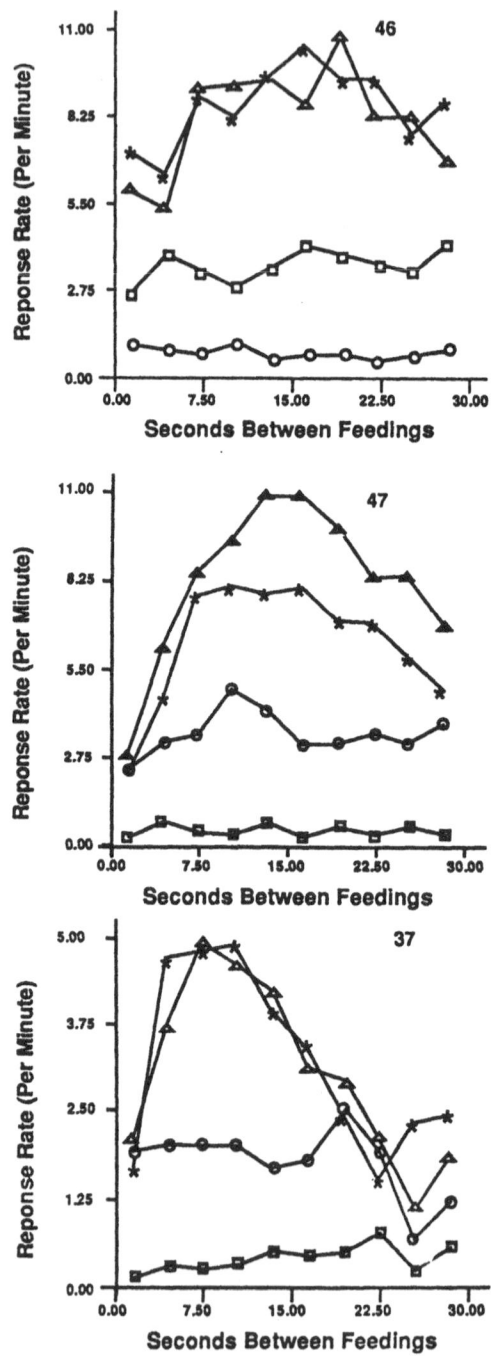

FIG. 12.8. Temporal activity patterns averaged over a single session during fixed time feeding conditions. Shown are the habituation patterns, first feeding session and last two feeding sessions (14th and 15th).

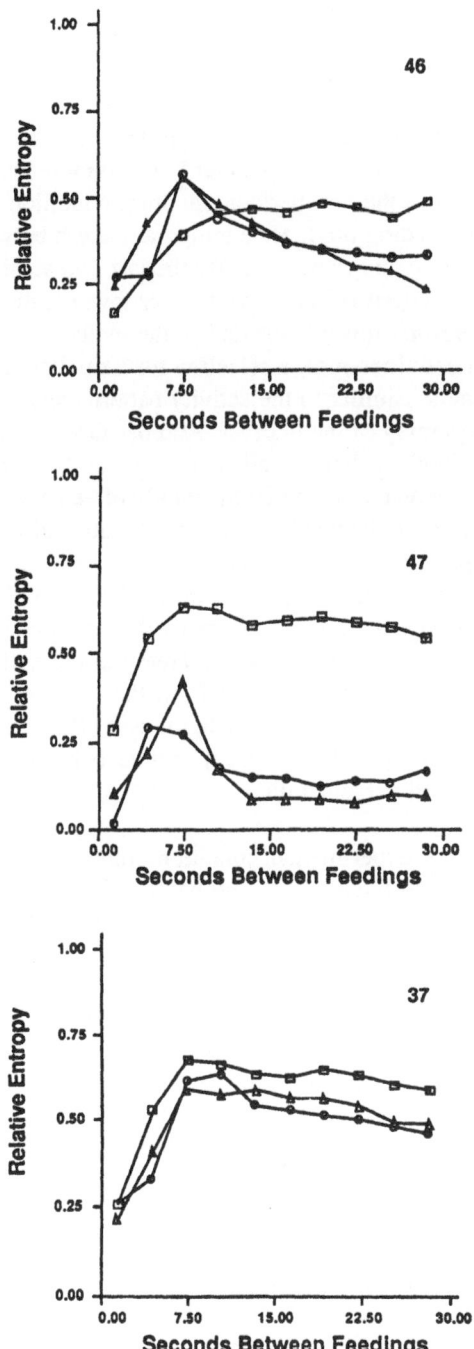

FIG. 12.9. Brillouin entropy patterns at each point in a fixed feeding interval. Shown are the first feeding session and the last two (14th and 15th).

feedings are shown. Relative entropy is plotted on the ordinate at every 3 seconds before feeding for the various conditions.[1] In this case each point in the figure represents an index of the distribution of all 17 behaviors. Unlike activity, relative entropy decreases from the first periodic feedings condition to the last. The first session of periodic feeding produces a temporal pattern between feedings that is negatively accelerating and monotone. By the last two sessions, particularly for W47, the temporal pattern is bitonic and lower towards the end of the interval. This decrease of entropy towards the end of the interval reflects a sharpening of the distribution of activities, i.e., a selection for a smaller set of behaviors at the expense of the others. Similar to the activity patterns there is a concentration of diversity at the beginning of the interval. And because activity measures a small subset of these behaviors that are all under temporal control, it may be more informative and constructive to model the family of behaviors that occur moment to moment in order to understand the temporal control of a molar measure such as general activity.

A surprising aspect of this temporal control of diversity is that noncontingent presentation of food sharpens behavior simply through continued exposure. Note that this selection effect is not due to an increase in the null category at the end of the interval because it was excluded from the moment by moment entropy calculations (including it did not have a significant effect on the patterns; except for W47, whose feeding pattern raised); active behaviors were emerging as dominant prior to food presentation.

Summary.   All three experimental conditions reinforce the fact that diversity seems highest during unlikely or low probability feeding situations. The single feeding a day experiments showed the greatest transitory change in diversity after feeding, although the habituation to periodic reinforcement Boltzmann curves showed that the stationary behavior distribution sharpens as more reinforcement is present. Finally, the temporal patterns during feeding suggests that diversity concentrates during post feeding periods, where reinforcement is least likely. Thus, these more abstract measures of behavior variability confirm Staddon and Simmelhag's observations and at least at a behavioral level suggest complex, "elicitive" (behavior variational principles) and "directive" (principles of selection) properties of reinforcers (cf. Killeen, Hanson, & Osborne, 1978; Staddon & Simmelhag, 1971).

In the next part of the chapter we open up the discussion of principles of behavioral variation and search principles that arise in connectionist models. In particular, we attempt to use the notion that the learning rule must somehow incorporate intrinsic variability of the response hypothesis. This in turn will allow the connectionist model to explore various kinds of search strategies during the

---

[1]Similar in spirit to V-C dimension. See Baum & Haussler (1989). We consider the dichotomization capacity using linear *fan-in* in the next section.

learning process. But first we introduce some of the basic concepts behind connectionism.

CONNECTIONISM

As pointed out in the introduction, connectionism finds many of its roots in animal learning theory, however, it is a blend or synthesis of many other traditions as well (cognitive psychology, computer science, physics, etc.). This apparent eclectic nature and its recent appearance make precise consensual definition difficult. Nonetheless, there are three aspects that recur through most connectionist models: (a) simple homogeneous processing elements, (b) local, incremental, learning rules and (c) dynamic representational properties (e.g., "hidden units"). We claim these three properties are critical for the definition of connectionism and provide some sort of coherence in this rapidly developing field. These three properties taken together contrast connectionism from other fields and especially from artificial intelligence area. The key difference has to do with a *reduction* of the degrees of freedom available in the modeling language. After all, connectionist models employ a programming language of a very special sort; one that forces the modeler to use very simple representational elements and to design adaptive procedures for the emergence of complex behavioral and cognitive phenomena. This underscores a central focus of most connectionist models that concerns the nature of the interaction between representation and learning. This sort of interaction has been in the past ignored by animal learning theory and side-stepped by later theories of memory organization and problem solving from cognitive psychology. Thus, one important, novel aspect of connectionism is its preoccupation with simple representational aspects of computation and how they may interact with equally simple modification or learning rules.

REPRESENTATION AND LEARNING

Although representation elements in connectionist models are typically modeled by simple threshold units, it is possible to consider many more complex variations of the same idea. The dot product or weighted linear combination of input variables produces one of the simplest sort of *fan-in* functions that might be used. There are several consequences of this fan-in function for representation. First, hidden unit integration is a linear projection of the input variables, which can allow for simple interpretation of hidden units as in principle components of the input variance. Second, linear fan-in leads to simple derivatives and consequently, simple learning rules. And finally, linear fan-in produces simple hyper-plane partitions in the feature space (cf. Hanson & Burr, 1990; see Figure 12.10). Nonetheless, there is every reason to believe that the actual biophysics of neural

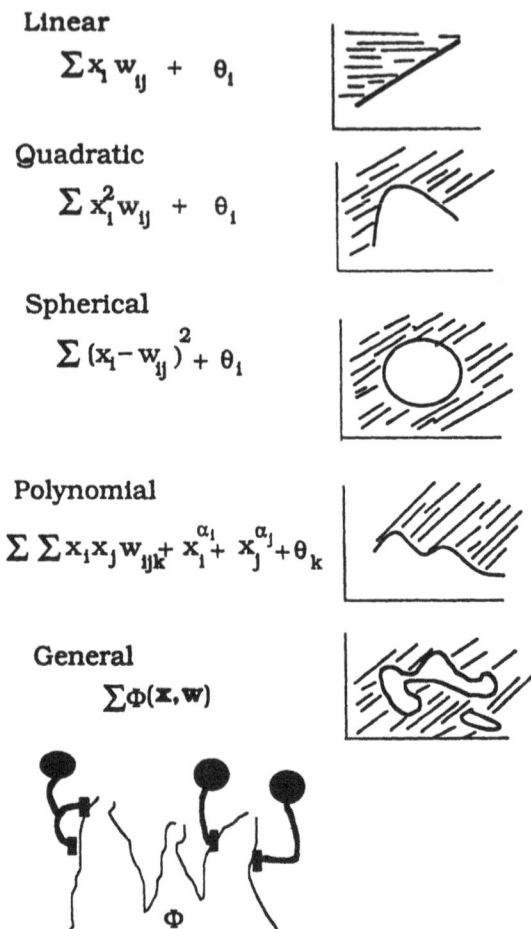

Linear

$$\Sigma x_i w_{ij} + \theta_i$$

Quadratic

$$\Sigma x_i^2 w_{ij} + \theta_i$$

Spherical

$$\Sigma (x_i - w_{ij})^2 + \theta_i$$

Polynomial

$$\Sigma \Sigma x_i x_j w_{ijk} + x_i^{\alpha_i} + x_j^{\alpha_j} + \theta_k$$

General

$$\Sigma \Phi(\mathbf{x}, \mathbf{w})$$

$$\Phi$$

FIG. 12.10.   Different possibilities for synaptic-dendritic interactions. Shown are fan-in function and effect on feature space in terms of partition boundary shape.

integration is probably *not* simple linear (cf. Shepherd & Brayton, 1987). In Figure 12.10 we show the range of possibilities and their effects on the elemental partition in feature space, for example, simple quadratics provide for curved areas, whereas spherical units allow for volumes to be bounded by a single unit. The last two cases indicate that very general fan-in functions are possible; the polynomials can allow characterization of a general *capacity* of the networks[1], whereas the last case simply reinforces our ignorance concerning the exact nature of the fan-in function of actual neural systems—but that very complex synaptic-dendritic interactions are likely.

Dichotomization Capability: Network Capacity

Using the linear *fan-in* we can characterize the degrees of freedom inherent in a network of units with threshold output. For example, with linear units, consider four points, well distributed in a two-dimensional feature space. As shown in Figure 12.11, there are exactly 14 linearly separable dichotomies that can be formed with the four target points. However, there are actually 16 (2^4) possible dichotomies of four points in two dimensions consequently, the number of possible dichotomies or arbitrary categories that are linearly implementable can be thought of as a capacity of the linear network in k dimensions with n examples. The general category capacity measure can be written as:

$$C(n,k) = 2\sum_{j=0}^{k} \frac{(n-1)!}{(n-1-j)!j!}, n > k + 1 \tag{17}$$

Note the dramatic growth in C as a function of k, the number of feature dimensions, for example, for 25 stimuli in a five dimensional feature space there are 100,670 linear dichotomies. Undetermination in these sorts of linear networks is the rule not the exception. This makes the search process and the nature of constraints on the search process critical.

Search

Search problems that involve high dimensionality, a-priori constraints and nonlinearities are hard. Unfortunately, learning problems in biological systems involve just these sorts of properties. Worse, one can characterize the sort of problem that organisms probably encounter in the real world as one that does not

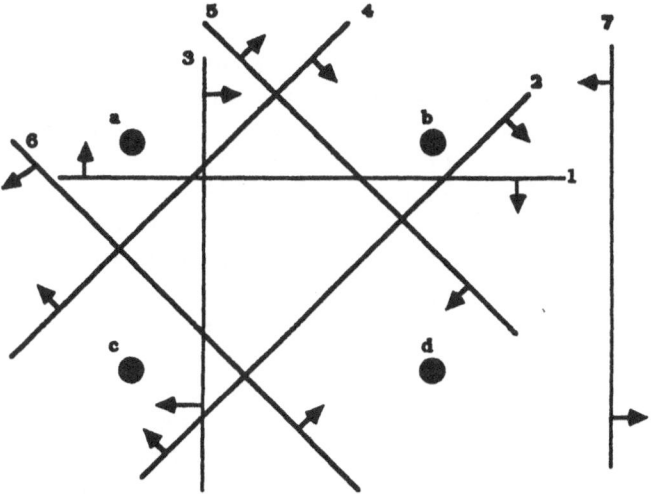

FIG. 12.11. Number of different ways four well distributed points can be dichotomized by a linear partition (14).

admit solutions that involve, simple averaging, optimality, linear approximation or complete knowledge of data or nature of the problem being solved. We would contend there are three basic properties of real learning that result in an ill-defined set problems and solutions:

- Data are continuously available but incomplete; the learner must constantly update parameter estimates with stingy bits of data that may represent a very small sample from the possible population.
- Conditional distributions of response categories with respect to given features are unknown and must be estimated from possibly unrepresentative samples.
- Local (in time) information may be misleading, wrong, or nonstationary, consequently there is a poor tradeoff between the present use of data and waiting for more and possibly flawed data—consequently updates must be small and revocable.

These sorts of properties actually represent only one aspect of the learning problem faced by real organisms in real environments. Nonetheless, they begin to underscore why "weak" methods—methods that assume little about the environment that they are operating in—are so critical.

LEARNING RULES

In connectionist models there are generally thought to be two classes of learning rules: supervised and unsupervised. Supervised models are actually "open loop" procedures and include classical conditioning procedures. For example, every input has a fixed a-priori target or teacher value. Similarly every CS is associated with a UCS. Backpropagation, which is based on the delta-rule or Widrow-Hoff (1960) rule, can be shown to be formally identical with the Rescorla-Wagner compound stimulus rule (see Sutton & Barto, 1981).

Unsupervised or self-supervised rules are learning rules based on internal or prespecified criterion that preserve some invariant or measured quantity about the input features. For example, a simple auto-associator is a self-supervised model that has the property that requires the output states to be identical to the input states. Still another possibility might be to find the linear combination of input features that are associated with the greatest input feature variance, then find the linear combination associated with the next greatest input feature variance and so on. This sort of criterion is identical to principle components analysis, which itself could be considered to be an unsupervised learning method. Behaviorally, it is not clear what procedure would be an example of a connectionist unsupervised learning system. Possibly any sort of pre-exposure, sensitization procedure or sensory pre-conditioning in animal learning procedures are actually analogous to unsupervised learning. Whereas in human experimental procedures, probably sorting tasks without feedback are most similar to the unsupervised task given the computation model.

Procedures that might be intermediate in the feedback information provided the learning system are what are typically called "reinforcement" learning. Work done in the engineering field on reinforcement learning was originally related to proposals by Skinner (1938), Hull (1943), and actual formal models later proposed by Bush and Mosteller (1955) and Estes (1959). The essence of such formalizations was to define a probability distribution over a finite set of possible responses. Then a stochastic response generation procedure could be defined. Given a response, the environment or an abstract trainer would provide signals of reward or punishment, these may or may not be hedonistic or secondary (like payoff values as in the case of Game Theory), based on a-priori contingencies or specific feedback function (Baum, 1969). If a positive reinforcement is given the probability for the response it follows is increased whereas if punishment (many recent connectionist modelers seem to confuse this with negative reinforcement where response probability is increased by a *removal* of the hedonistically negative stimulus) results, the probability of the response it follows decreases. These reward and punishment functions were typically stated as one parameter linear difference equations, thus leading to simple exponential growth and decay functions. Hull proposed such "Habit strength" functions in the form of what appeared to be a deductive account although the number of functions and parameters that he proposed seemed to quickly outweigh the available data. Skinner himself had a quasi-formal model that recently Killeen (1988) has explicated. Skinner later eschewed and criticized such modelling not to mention many other approaches in psychology (e.g., all of cognitive psychology).

We explore the connectionist versions of classical conditioning and operant conditioning (a.k.a. "reinforcement learning") later and discuss variations based on local noise production in the network itself.

Classical Conditioning

The simplest version of classical conditioning models in connectionist models is the Hebb rule, in which local correlations between units cause a strengthening of their connections. This positive feedback leads to exponential growth that is usually checked by assuming some sort saturation property. As Tesauro (1990) has pointed out Hebb models cannot provide differences in performance between forward and backward pairing conditions because there is no temporal information required for strengthening of the connections between units. However, versions of the delta rule, in which the target is interpreted as an "expected" firing rate, in which a temporal difference is formed between it and the actual postsynaptic firing rate can provide asymmetric outcomes.

The delta rule or LMS rule and its generalization are easy to state. Given an output $(\hat{y}_i)$ and a target (y_i) an error can be computed that can be accumulated over a sample of input-output tokens.

$$E = \frac{1}{2}\sum_s \sum_i (\hat{y}_{is} - y_{is})^2 \tag{18}$$

A gradient can be found for the error as function of synaptic weights between units, assuming we start with a suitable fan-in function—say dot product $(x_i = \sum_j w_{ij} y_j)$—and a reasonable fan-out function—smooth, differentiable and monotone. We require that this gradient be decreasing.

$$\frac{\partial E}{\partial w_{ij}} = 0 \tag{19}$$

The weight update in the network is proportional to this gradient.

$$w_{ij}(n + 1) = \alpha\left(\frac{\partial E}{\partial w_{ij}}\right) + w_{ij}(n) \tag{20}$$

For the back-propagation algorithm, the weight updates for output distal parts of the network are based on the output gradient that is recursively passed back through the network by weighting it with a connection strength most recently responsible for the error. Thus errors are precisely and deterministically sent back through the network affecting connections as a function of their predictability to the output. Back-propagation is a generalization of the delta rule and is also an algorithm in a class of multivariate, nonlinear regression methods using dynamic feature selection and extraction.

Successful versions of these models typically use time derivative information for prediction of the UCS. For example, Sutton and Barto's (1990) recent version of their temporal differences model provides an elegant nexus between the past modeling attempts in animal learning theory and more recent connectionist models. Sutton's model also provides a large coverage over various classical conditioning models, including ISI effects, conditioned inhibition, primacy effects, and second order conditioning.

All such models however, are predicated on the assumption that the weight update is fixed, deterministic and globally homogeneous throughout the network. This seems at best biologically unlikely, and we can *weaken* these assumptions and ask what the consequences are for the delta rule in terms of learning. Consequently, suppose that weight updates are diverse, random, and globally heterogeneous throughout a network. This set of assumptions gives rise to a family of learning rules that are stochastic and potentially more efficient in their search properties than the standard delta rule.

Stochastic Delta Rule

Actual mammalian neural systems involve noise. Transmission of excitation through neural networks in living systems is essentially stochastic in nature. The activation function, typically a smooth monotonically increasing function used in neural network/connectionist models must reflect an integration over some time interval of what neurons actually produce: discrete randomly distributed pulses of finite duration. In fact, the typical activation function used in connectionist models must be assumed to be an average over many such intervals, since any particular neuronal pulse train appears quite random (in fact, Burns, 1968; Poisson, distributed; Tomko & Crapper, 1974).

It is possible that noise introduced into the system at some point could be useful in improving both the starting point of the system and the stability of convergence to minima that represent solutions to the given problem. However, noise must be judiciously introduced and subsequently removed from the system as the solution is approached. Simulated annealing (Kirkpatrick, Gelatt, & Veechi, 1983) is an attractive method for finding global minima, however, the method can approach such global minima quite slowly (Ackley, Hinton, & Sejnowski, 1985). This is partly due to the fact that simulated annealing methods as implemented in Boltzmann machines, for example, introduce noise throughout the network independently of the learning process.

This suggests that a particular neural signal in time may be modeled by a *distribution* of synaptic values rather then a single value. Further this sort of representation provides a natural way to affect the synaptic efficacy in time. In order to introduce noise adaptively, we require that the synaptic modification be a function of a random increment or decrement proportional in size to the present error signal. Consequently, the weight delta or gradient itself becomes a random variable based on prediction performance. Thus, the noise that seems ubiquitous and apparently useless throughout the nervous system can be turned to at least three advantages in that it provides the system with mechanisms for (a) entertaining multiple response hypotheses given a single input (b) maintaining a coarse prediction history that is local, recent, and cheap, thus providing punctate credit assignment opportunities and finally, (c) revoking parameterizations that are easy to reach, locally stable, but distant from a solution.

Although it is possible to implement the present principle a number of different ways, we chose to consider a connection strength to be represented as a distribution of weights with a finite mean and variance (see Figure 12.12).

A forward activation or recognition pass consists of randomly sampling a weight from the existing distribution calculating the dot product and producing an output for that pass.

$$x_i = \sum_j w_{ij}^* y_j \tag{21}$$

where the sample is found from,

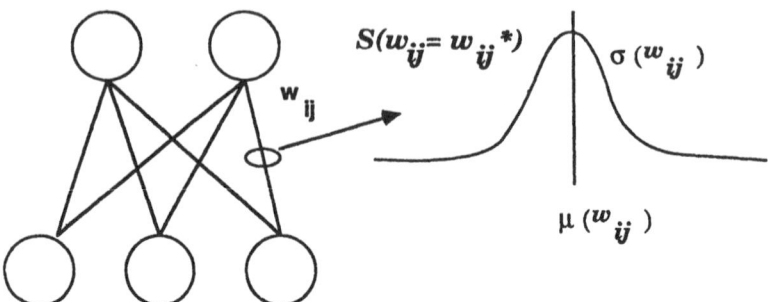

FIG. 12.12. Weights as they are used in the Stochastic Delta rule and stochastic reinforcement systems. Each weight is associated with a distribution of values and a mean and standard deviation.

$$S(w_{ij} = w_{ij}^*) = \mu_{w_{ij}} + \sigma_{w_{ij}}\phi(w_{ij}; 0,1) \tag{22}$$

Consequently $S(w_{ij} = w_{ij}^*)$ is a random variable constructed from a finite mean $\mu_{w_{ij}}$ and standard deviation $\sigma_{w_{ij}}$ based on a normal random variate (ϕ) with mean zero and standard deviation one. Forward recognition passes are therefore one to many mappings, each sampling producing a different weight depending on the mean and standard deviation of the particular connection while the system remains stochastic.

In the present implementation there are actually three separate equations for learning. The mean of the weight distribution is modified as a function of the usual gradient based upon the error, however, note that the random sample joint is retained for this gradient calculation and is used to update the mean of the distribution for that synapse.

$$\mu_{w_{ij}}(n + 1) = \alpha \left(\frac{\partial E}{\partial w_{ij}^*}\right) + \mu_{w_{ij}}(n) \tag{23}$$

Similarly the standard deviation of the weight distribution is modified as a function of the gradient, however, the sign of the gradient is ignored and the update can only increase the variance if an error results. Thus errors immediately increase the variance of the synapse, to which they may be attributed.

$$\sigma_{w_{ij}}(n + 1) = \beta \left|\frac{\partial E}{\partial w_{ij}^*}\right| + \Theta_{w_{ij}}(n) \tag{24}$$

A third and final learning rule determines the decay of the variance of synapses in the network,

$$\sigma_{w_{ij}}(n + 1) = \xi\sigma_{w_{ij}}(n), \xi < 1. \tag{25}$$

As the system evolves for ξ less than one, the last equation of this set guarantees that the variances of all synapses approach zero and that the system itself becomes deterministic prior to solution. For small ξ the system evolves very rapidly to deterministic, whereas larger ξs allow the system to revisit chaotic states as needed during convergence. A simpler implementation of this algorithm involves just the gradient itself as a random variable (hence, the name "stochastic delta rule"), however this approach confounds the growth in variance of the weight distribution with the decay and makes parametric studies more complicated to implement.

There are several properties of this algorithm that we have touched on implicitly in earlier discussion that are worth making explicit: (a) adaptive noise injection early in learning is tantamount to starting in parallel many different networks and selecting one that reduces error the quickest, thus primarily dissociating initial random starting point from eventual search path; (b) noise injections are punctate and accumulate to parts of the network primarily responsible for poor predictions, this allows the network to maintain a cheap prediction history concerning the consequences of various response hypotheses; (c) given enough time in a *ravine* or *mesa*, a single synapse can accumulate enough noise to perpetuate variance increase throughout the entire network thus inducing a "chaotic episode"; (d) on average the network will follow the gradient as in a "drunkard's walk" allowing synapses of the network with greater certainty (lower variance) to follow the gradient exactly whereas other synapses of less certainty to explore nearby and during chaotic episodes distant parts of weight space; (e) finally, punctate synaptic noise injection introduces noise indirectly into the unit activation, output predictions and thus, errors, and on average implements a local, learning dependent, simulated annealing process.

As a simple example of how this method can speed learning a number of runs were made with the parity predicate. Learning curves are shown in Figure 12.13 from four different cases of the parity predicate with back-propagation (solid lines) paired on random starting value with stochastic delta rule (dashed line). Note that in the xor (2-bit parity case) only 3 of the 10 runs of back-propagation converged with the 500 sweeps, whereas all but one of the stochastic delta rule runs converged within 100 sweeps. In the 3-bit parity case 3 of the 6 runs converged within 5,000 sweeps, however all 6 runs of the stochastic delta rule converged within 1,000 sweeps. In the 4-bit parity case, neither back-propagation run converged in 10,000 sweeps whereas noise is actually regenerated during late phases of convergence in one of the stochastic delta rule runs soon followed by attaining a global minimum. Finally, shown in the 6-bit parity run with a lowered learning rate are two back-propagation runs that do not converge in 2,000 sweeps. Although the stochastic delta rule runs both converge within 1,000 sweeps, one producing a single chaotic episode near sweep 995, which apparently "shakes" it loose and convergence soon follows on sweep 998.

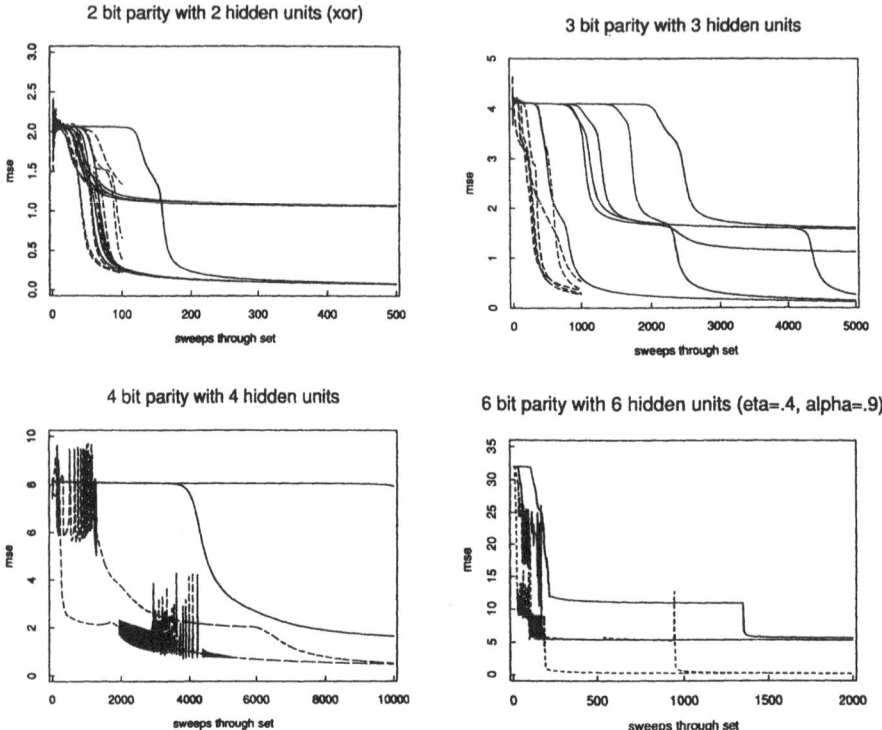

FIG. 12.13.   Various runs of the stochastic delta rule algorithm (dashed lines) compared to back propagation (solid lines) with different sizes of the parity problem. See text for more details.

Operant Conditioning

As stated earlier, formal models of the reinforcement process took the form of linear difference equations, so for example, Estes stimulus sampling model appeared as

$$p_{n+1} = \theta p_n + (1 - \theta) \qquad (26)$$

In Estes' model θ represented the "sampling rate" of stimulus elements in the environment, it was meant to be a theoretical construct that could be estimated from the data. Other models offered other similar parameters. What was missing from this sort of model as well as in the classical conditioning models such as the Rescorla-Wagner account is some sort of connection between the input states of the system to internal states of affairs, internal representation, or cognition more generally. Furthermore, architectures inasmuch as you could call them that, were one dimensional and consequently what they could learn was much too coarse about the stimulus environment. In fact, in all such models the *trial* level was the only independent variable. Later models allowed more complexity in that "states" of the system such as "knowledge" or conditioning states, or stimulus array elements could be expressed in a multiple state markov chain. Notwithstanding, complex analyses of the responses or their probabilities the "input representation" remained impoverished, and dissociated with possible theoretical characterizations of the stimulus environment.

Sutton and Barto (see Barto, 1985, pp. 238–239) realized that the sort of reinforcement models used in engineering contexts that were also close relatives of the psychological learning models, possessed this stimulus representation problem. They characterized it as a deficiency in "stimulus discrimination." In other words, given a stimulus as a set of features, how could a reinforcement system learn a discrimination? They argued that the probability of the response should be a function of other parameters that directly reference the input features. One such function is a linear discriminate, or a single neural unit. Thus, updates to weights, update a *representation* of the stimuli in the environment by positioning a hyperplane in the feature space separating one set of stimuli from another.

Sutton and Barto essentially based their model on Thorndike's "Law of Effect," in which response probabilities could be modified by both "satisfyers" and "annoyers." Consequently, they implemented their model with rules that determined both increments and decrements to weights based on successful responses or errorful responses. They termed their learning rule associative reward-penalty (A_{r-p}). For positive reinforcement $= +1$ and punishment $= -1$:

$$\Delta w_{ij} = p(S^r y_i - E y_i \,|\, y_i] \, x_i)$$ (27)

if $S^r = +1$ else scale update with λ.

Essentially, this rule compares the predicted response $S^r y_i$ to the typical or expected response given the correct response. Williams (1988) provides a more general characterization of these type of reinforcement models and discusses several sorts of logical variations. However, all such models are predicated on a "strengthing" metaphor whereas more contemporary reinforcement theory has been developing hypotheses concerning behavioral variation and its relation to the selection of behavior as discussed earlier in this chapter. For example, one simple hypothesis (traceable to Skinner) is that reinforcement acts to weaken the competition from other responses consequently the target response appears to increase in probability merely because all other responses have decreased in probability. These sorts of competition theories have appeared implicitly in recent behavioral regulatory (e.g., optimal) or equilibrium models (Hanson & Timberlake, 1983; Rachlin & Burkhard, 1978; Staddon, 1979; Staddon & Zhang, chap. 11 in this volume).

Of course it is possible to incorporate such competition mechanisms in connectionist models such that increases in one weight automatically renormalize other weights (Grossberg, 1987; Rumelhart & Zipser, 1985) producing competition within the network representation. In operant conditioning, such mechanisms must be incorporated stochastically, such that the response generation process also reflects the selection of features that are predictive of reinforcement. The following section discusses one such possibility that is a close relative to the Stochastic Delta Rule discussed before.

Stochastic A_{r-p}

Response competition can result because response probabilities themselves trade off as in a linear constraint: $\sum_i p_i = 1$. Or it is possible to embed the competition in the variability of the response generation process. For example, assume that weights are represented by a distribution of values (like SDR) so that both a mean and variance can be used to characterize the strength of output of any unit. The reinforcement process will strengthen the weight as in the A_{r-p} rule above and it will reduce the variability of the weights responsible for the response leading to reinforcement and increase the variability of weights responsible for responses that do not lead to reinforcement. If $S^r = +1$, then

$$\Delta\sigma(w_{ij}) = \begin{cases} -\beta\,|\,\Delta reinf\,|\text{ if } S^r = +1 \\ \beta\,|\,\Delta reinf\,|\text{ if } S^r = -1 \end{cases} \tag{28}$$

Where *reinf* represents the standard A_{r-p} reinforcement signal from Equation 27.

A simple categorization problem is shown in Figure 12.14, which involves three features: color, shape, and orientation (and an arbitrary number feature; after Dennis, Hampton, & Lea, 1971). Category membership is determined by an M out of N rule. For example in this problem category A is determined by any two of the features white, square or symmetric around a 90 *degree* axis. This is a linearly separable problem so a single stochastic A_{r-p} unit is sufficient for

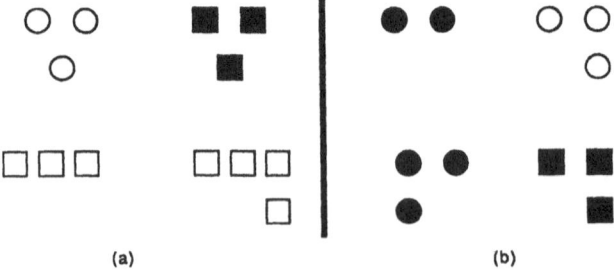

(a)   (b)

FIG. 12.14. Polymorphy problem. Categories are formed by an M out of N rule, in which features of shape, color and orientation are used to construct tokens.

learning. The average (10 runs) learning curves for the probability of reinforcement obtained on each trial is shown in Figure 12.15. Notice that the stochastic A_{r-p} is faster in this learning problem than the Sutton and Barto A_{r-p}. We hypothesize that this improvement is due to efficiency with which the stochastic version explores solutions per reinforcement.

CONCLUSIONS

One apparent feature of biological systems is the diversity of their actions. However, it is not clear whether the inherent stochastic nature of the nervous system also necessarily makes it a useful property. One thesis of the present chapter is that informational states, those concerning "success" or "failure" or are motivated in some such way lead to a transient goal dependent increase in the diversity of behavior. Increases in randomness or variability are not necessarily associated with useful functions. Simple random search is known to be poor in even very simple optimization problems and simulated annealing although globally convergent can be tortuously low.

In spite of these failures of noise introduction in search problems, there is reason to believe that the diversity of behavior is related to various biological activities including foraging, predation, sexual selection, and perhaps even gen-

Polymorphy Problem (Arp, solid; SArp, dashed)

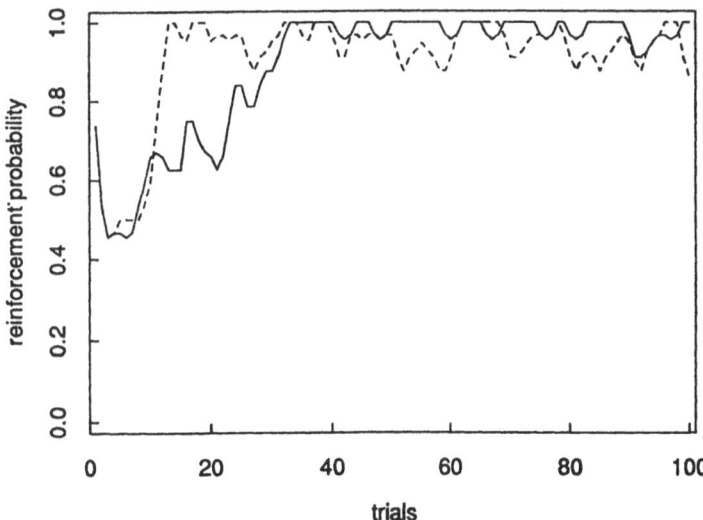

FIG. 12.15. Runs comparing a Stochastic A_{r-p} unit (dashed line) and Sutton and Barto's A_{r-p} (solid line) on the polymorphy problem.

eral problem solving. Indigenous noise as presented here may be useful in hypothesis generation as the network is learning. It may help reduce the likelihood of paths the network learning takes that may be slow, unstable, and locally convergent.

Thus, diversity in behavior leads to investigating *purposeful* introduction of noise into connectionist learning systems. One heuristic for noise introduction as suggested by the stochastic connectionist systems discussed in this chapter has to do with how credit assignment and noise are related. In the systems discussed, note that the noise or diversity was introduced under conditions where improvement was possible and solution related constraints were available. Thus, the noise acted as variation on tendencies in the network that were already selected for. This same sort mechanism may provide insights into problem solving and learning as well as the nature of the exploitation and generation of learning strategies.

ACKNOWLEDGMENTS

Part of the chapter was based on a dissertation submitted to Arizona State University. I thank Carl Olson for conversations concerning possible neural transmission mechanisms, Mark Gluck for conversations about delta rule, and the Bellcore connectionist group for support and interesting discussions.

REFERENCES

Ackley, D., Hinton, G., & Sejnowski, T. (1985). A learning algorithm for Boltzmann machines, *Cognitive Science, 9,* 147–169.

Barto, A. G. (1985). Learning by statistical cooperation of self-interested neuron-like computing elements. *Human Neurobiology, 4,* 229–256.

Baum, E., & Wilczek, F. (1988). *Supervised learning of probability distributions by neural networks.* Neural Information Processing Systems, American Institute of Physics.

Baum, E., & Haussler, D. (1989). What size net gives valid generalization?, In D. Touretzsky (Ed.), *Advances in Neural Information Processing* (Vol. 1).

Baum, W. M. (1969). Optimization and the matching law as accounts of instrumental behavior. *Journal of the Experimental Analysis of Behavior, 36,* 387–403.

Bolles, R. C. (1967). *Theory of motivation.* New York: Harper & Row.

Breland, K., & Breland, M. (1961). The misbehavior of organisms. *American Psychologist, 16,* 681–684.

Brillouin, L. (1962). *Science and information theory.* (2nd Ed.). New York: Academic Press.

Burns, B. D. (1968). *The uncertain nervous system.* London: Edward Arnold Ltd.

Bush, R. R., & Mosteller, F. (1955). *Stochastic models for learning.* New York: Wiley.

Dennis, I., Hampton, J. A., & Lea, S. E. G. (1971). A new problem in concept formation. *Nature, 243,* 101–102.

Elman, J. L. (April, 1988). *Finding structure in time,* CRL Technical Report 8801.

Estes, W. K., (1959). The statistical approach to learning theory. In Koch, S. (Ed.), *Psychology: a study of a science* (Vol. 2.) New York: Mcgraw-Hill.

Falk, J. L. (1969). The motivational properties of schedule-induced polydipsia. *Journal of the Experimental Analysis of Behavior, 9,* 19–25.

Fast, J. D. (1962). *Entropy.* Eindhorne, Holland: Philips Technical Library.

Feller, W. (1968). *An introduction to probability theory and its applications.* New York: Wiley.

Gluck, M. A., & Bower, G. H. (1988). From conditioning to category learning: An adaptive network model. *Journal of Experimental Psychology: General, 117,* 3, 225–244.

Gluck, M. A., & Thompson, R. F. (1987). Modeling the neural substrates of associative learning and memory: a computational approach, *Psychological Review, 94,* 176–192.

Gibbon, J. (1977). Scalar expectancy theory and Weber's law in animal learning. *Psychological Review, 84,* 279–325.

Grossberg, S. (1987). Competitive Learning: From interactive activation to adaptive resonance. *Cognitive Science, 11,* 23–64.

Hanson, S. J. (1980). *Studies of diversity in the pigeon.* Unpublished doctoral dissertation, Arizona State University.

Hanson, S. J., & Burr, D. J. (1988). Minkowski Back-propagation: Learning in connectionist models with non-euclidian error signals, Neural Information Processing Systems, American Institute of Physics.

Hanson, S. J., & Burr, D. J. (1990). What connectionist models learn: Learning and representation in connectionist networks. *Behavioral and Brain Sciences, 3.*

Hanson, S. J., & Kegl, J. (1987). Parsnip: A connectionist model that learns natural language grammar from exposure to natural language sentences. In *Proceedings of ninth annual meeting of the Cognitive Science Society.* Hillsdale, NJ: Lawrence Erlbaum Associates.

Hanson, S. J., & Timberlake, W. (1983). Regulation During Challenge: a general model of learned performance under schedule constraint. *Psychological Review, 90,* 262–282.

Hearst, E., & Jenkins, H. M. (1974). *Sign tracking: The stimulus reinforcer relation and directed action.* Austin, TX: Monograph of the Psychonomic Society.

Hinson, J. M., & Staddon, J. E. R. (1979). Behavioral competition: A mechanism for schedule interactions. *Science, 202,* 432–433.

Hull, C. L. (1943). *The principles of behavior.* New York: Appleton-Century-Crofts.

Kehoe, E. J. (1989). A layered network model of associative learning: Learning to learn and configuration. *Psychological Review.*

Killeen, P. R. (1978). Superstition: A matter of bias, not detectability. *Science, 199,* 88–90.

Killeen, P. R. (1979). Arousal: Its genesis, modulation and extinction. In P. Harzem & M. D. Zeiler (Eds.), *Reinforcement and the temporal organization of behavior.* New York: Wiley.

Killeen, P. R., Hanson, S. J., & Osborne, S. R. (1978). Arousal: Its genesis and manifestation as response rate. *Psychological Review, 85,* 571–581.

Killeen, P. R. (1988). The reflex reserve. *Journal of the Experimental Analysis of Behavior, 50,* 319–350.

Kirkpatrick, S., Gelatt, C. D., & Veechi, M. (1983). Optimization by simulated annealing, *Science, 220,* 671–680.

Krebs, C. J. (1978). Ecology: *The experimental analysis of distribution and abundance.* New York: Harper & Row.

Pielou, E. C. (1977). *Mathematical ecology.* New York: Wiley.

Premack, D. (1965). Reinforcement theory. In D. Levine (Ed.), Nebraska symposium on motivation. Lincoln: University of Nebraska Press.

Rachlin H. C., & Burkhard, B. (1978). The temporal triangle: Response substitution in instrumental conditioning. *Psychological Review, 85,* 22–47.

Rumelhart D. E., Hinton G. E., & Williams R. (1986). Learning internal representations by error propagation. *Nature, 323,* 1986.

Rumelhart, D. E., & McClelland, J. J. (eds.). (1986). *Parallel distributed processing: Explorations in the microstructure of cognition. Vol. 1: Foundations.* Cambridge, MA: Bradford Books/MIT Press.

Rumelhart, D. E., & Zipser, D. (1985). Feature discovery by competitive learning. *Cognitive Science*, *9*, 75–112.

Shettleworth, S. J. (1975). Reinforcement and the organization of behavior in golden hamster: Hunger, environment, and food reinforcement. *Journal of Experimental Psychology: Animal Behavior Processes*, *104*, 56–87.

Shepherd, G. M., & Brayton, R. K. (1987). Logic operations are properties of computer-simulated interactions between excitable dendritic spines, *Neuroscience*, *21*, 151, 166.

Skinner, B. F. (1938). *The behavior of organisms*. New York: Appleton-Century.

Skinner, B. F. (1948). "Superstition" in the pigeon. *Journal of Experimental Psychology*, *38*, 168–172.

Staddon, J. E. R. (1979). Operant behavior as adaptation to constraint. *Journal of Experimental Psychology: General*, *108*, 48–67.

Staddon, J. E. R. (1977). Schedule-induced behavior. In W. K. Honig & J. E. R. Staddon (Eds.), *Handbook of operant behavior*. Englewood Cliffs, NJ: Prentice-Hall.

Staddon, J. E. R., & Simmelhag, V. L. (1971). The "superstition" experiment: A re-examination of its implications for the principles of adaptive behavior. *Psychological Review*, *78*, 3–43.

Stevens, S. S. (1951). Mathematics, measurement and psychophysics. In S. S. Stevens (Ed.), *Handbook of experimental psychology*. New York: Wiley.

Stevens, S. S. (1959). Measurement, psychophysics, and utility. In C. W. Churchman & P. Ratoosh (Eds.), *Measurement and definitions and theories*. New York: Wiley.

Sutton, R. S., & Barto, A. G. (1981). Toward a modern theory of adaptive networks: Expectation and prediction. *Psychological Review*, *88*, 135–171.

Sutton, R. S., & Barto, A. G. (1990). Time-Derivative models of Pavlovian reinforcement. In J. W. Moore & M. Gabriel (Eds.), *Learning and computational neuroscience*.

Tesauro, G. (1990). Neural models of classical conditioning: A theoretical viewpoint. In S. J. Hanson & C. R. Olson (Eds.), *Connectionist modeling and brain function: The developing interface*. Cambridge, MA: MIT Press/Bradford.

Timberlake, W. D. (1969). Continuous coding of general activity in the rat during repeated exposure to a constant environment and to stimulus change. Unpublished doctoral dissertation, University of Michigan.

Tinbergen, N. (1960). *A study of instinct*. London: Oxford University Press.

Tomko, G. J., & Crapper, D. R. (1974). Neural variability: Non-stationary response to identical visual stimuli, *Brain Research*, *79*, p. 405–418.

Widrow, B. E., & Hoff, M. E. (1960). *Adaptive switching circuits*. Institute of Radio Engineers, Western Electronic Show and Convention, Convention Record, Part 4, 96–104.

Williams, D. R., & Williams, H. (1969). Auto-maintenance in the pigeon: Sustained pecking despite contingent nonreinforcement. *Journal of the Experimental Analysis of Behavior*, *12*, 511–520.

Williams, R. J. (1988). *Toward a theory of reinforcement-learning connectionist systems*. Technical Report NU-CCS-88-3.

Zipf, G. K. (1949). *Human behavior and the principle of least effort*. New York: Hafner Publishing Co.

Author Index

Subject Index